**is**

Now you will find Saunders College Publishing's distinguished innovation, leadership, and support under a different name ... a new brand that continues our unsurpassed quality, service, and commitment to education.

We are combining the strengths of our college imprints into one worldwide brand: Harcourt

Our mission is to make learning accessible to anyone, anywhere, anytime—reinforcing our commitment to lifelong learning.

We are now Harcourt College Publishers. Ask for us by name.

One Company
"Where Learning Comes to Life."

www.harcourtcollege.com
www.harcourt.com

A First Course in Algebra
An Interactive Approach

Alison Warr

Catherine Curtis

Penny Slingerland

Mt. Hood Community College

Harcourt College Publishers

FORT WORTH · PHILADELPHIA · SAN DIEGO · NEW YORK · ORLANDO
AUSTIN · SAN ANTONIO · TORONTO · MONTREAL · LONDON · SYDNEY · TOKYO

Publisher: Emily Barrosse
Executive Editor: Angus McDonald
Marketing Strategist: Julia Downs-Conover
Developmental Editor: James LaPointe
Project Management: TSI Graphics
Production Manager: Alicia Jackson
Cover Credit: Mt. Hood, Oregon (© William McKinney/FPG International LLC)

A First Course In Algebra: An Interactive Approach 1/e, Alison Warr, Catherine Curtis, and Penny Slingerland

ISBN 0-03-024592-3
Library of Congress Catalog Card Number: 131480

Copyright © 2001 by Alison Warr, Catherine Curtis, and Penny Slingerland

All rights reserved. No part of this publication may be reproduced or transmitted in any form or by any means, electronic or mechanical, including photocopy, recording, or any information storage and retrieval system, without permission in writing from the publisher.

Requests for permission to make copies of any part of the work should be mailed to the following address: Permissions Department, Harcourt, Inc., 6277 Sea Harbor Drive, Orlando, Florida 32887-6777.

Address for domestic orders:
Saunders College Publishing, 6277 Sea Harbor Drive, Orlando, FL 32887-6777
1-800-782-4479
email: collegesales@harcourt.com

Address for international orders:
International Customer Service, Harcourt Brace & Company
6277 Sea Harbor Drive, Orlando, FL 32887-6777
(407) 345-3800
Fax (407) 345-4060
email: hbintl@harcourtbrace.com

Address for editorial correspondence:
Saunders College Publishing, Public Ledger Building, Suite 1250,
150 S. Independence Mall West, Philadelphia, PA 19106-3412

Web Site Address
http://www.harcourtcollege.com

This material is based upon work funded in part by the National Science Foundation under Grant No. DUE-9454627.

Any opinions expressed are those of the authors and not necessarily those of the National Science Foundation.

Some material in this work previously appeared in *Interactive Mathematics II* copyright © 1998 by Alison Warr, Catherine Curtis, and Penny Slingerland. All rights reserved.

Printed in the United States of America

0123456789   048   10 987654321

Contents

Preface   ix

Chapter 0   AN INTRODUCTION TO LEARNING MATHEMATICS, PROBLEM SOLVING, AND TEAMWORK   1

Chapter 1   NUMERICAL CONCEPTS   13

- 1.1   The Language of Mathematics and Term-by-Term Evaluation   14
- 1.2   Estimation   24
- 1.3   Approximate Numbers and Error   31
- 1.4   Unit Conversion   50
- Chapter 1 Summary   61

Chapter 2   AN INTRODUCTION TO ALGEBRA—NUMERICALLY, GRAPHICALLY, AND ALGEBRAICALLY   63

- 2.1   Using Block Patterns to Introduce Variables and Graphs   65
- 2.2   The Rectangular Coordinate System   78
- 2.3   Graphing Equations   87
- 2.4   Modeling with Tables, Graphs, and Equations   92
- 2.5   Evaluating Expressions with Units   107
- 2.6   Formula Evaluation   115
- Chapter 2 Summary   126

Chapter 3   WORKING WITH ALGEBRAIC EXPRESSIONS AND EQUATIONS   127

- 3.1   Simplification of Algebraic Expressions   128
- 3.2   The Distributive Property   138
- 3.3   Solving Linear Equations   145
- 3.4   Solving Linear Literal Equations   157
- 3.5   Solving Linear Literals Using Common Factoring   165
- Chapter 3 Summary   172

CHAPTER 1–3 REVIEW   174

Chapter 4 — LINEAR EQUATIONS IN TWO VARIABLES 185

- 4.1 Linear Relationships: Numerically and Graphically 186
- 4.2 The Graphs of Linear Equations 200
- Chapter 4 Summary 219

Chapter 5 — POSITIVE INTEGER EXPONENTS 221

- 5.1 Language and Evaluation 222
- 5.2 Properties and Simplification 227
- Chapter 5 Summary 239

Chapter 6 — WORKING WITH ALGEBRA 241

- 6.1 Multiplication of Multiterm Expressions 242
- 6.2 Division 252
- 6.3 Simple Quadratics 258
- 6.4 Rational Equations 266
- Chapter 6 Summary 277

CHAPTER 4–6 REVIEW 278

Chapter 7 — MORE LINEAR EQUATIONS 285

- 7.1 Solving Linear Equations Graphically 286
- 7.2 Writing Equations of Lines 290
- 7.3 Systems of Linear Equations 303
- 7.4 Solving Linear Inequalities in One Variable 316
- Chapter 7 Summary 324

Chapter 8 — ZERO AND NEGATIVE INTEGER EXPONENTS 327

- 8.1 Language and Evaluation 328
- 8.2 Properties and Simplification 335
- Chapter 8 Summary 344

Chapter 9 — GEOMETRY 345

- 9.1 Angles and Triangles 346
- 9.2 Polygons and Circles 358
- 9.3 Similar Triangles 369
- 9.4 Scale Drawings 380
- 9.5 Geometric Solids and Surface Area 390
- 9.6 Volumes of Prisms and Cylinders 401
- Chapter 9 Summary 414

Chapter 10   STATISTICS AND PROBABILITY   417

- 10.1   Graphical Displays of Data   418
- 10.2   Descriptive Statistics   427
- 10.3   Probability   448
-        Chapter 10 Summary   467

CHAPTER 7–10   REVIEW   469

Appendices

- APPENDIX A   THE REAL NUMBER SYSTEM   A-1
- APPENDIX B   ARITHMETIC WITH SIGNED NUMBERS   A-6
- APPENDIX C   READING AND MEASURING WITH RULERS   A-15
- APPENDIX D   PERIMETER AND AREA CONCEPTS   A-23

- GEOMETRY REFERENCE   A-30
- CONVERSION TABLES   A-34
- SELECTED ANSWERS   ANS-1
- INDEX   I-1
- INDEX OF STRATEGIES   I-5

Preface

Overview

This overview will acquaint you with the features of this book and concepts covered as well as our philosophy in teaching mathematics. Following that, we will give you a view of the classroom that works with this philosophy and this text. Included are discussions of how the class is organized, what goes on in the classroom, the instructor's role, and the students' role.

Keep in mind that what we are describing is the structure that has worked for us. Some of the elements are crucial in making this text work with students. Each of us, however, adapts the situation to our own comfort level and experience.

Assumptions

We are making the following assumptions of the students who are utilizing this text.

- Students are able to calculate and do applications with signed numbers.
- Students are able to do applications involving basic perimeter and area concepts.
- Students are able to measure with both English and metric rulers.

A review of these topics can be found in the appendices.

Features of the Text

Activity Sets. Nearly every section of the text begins with an Activity Set. These activities are designed to guide students to discover, at an introductory level, the major concepts of the section. While some of the activities can be done by students at home, they are most effective when done in class in teams. The team often ensures success; what one student cannot figure out, a team most often can.

Discussions. In this portion of the text, the concepts of the section are discussed, when possible, in the context of applications. It is crucial that the students read the discussion part of the textbook. Looking briefly at the examples will not be sufficient for understanding and success! This textbook should be read by students with paper, pencil, and calculator in hand.

Definitions. As the concepts are presented, new vocabulary is introduced and important ideas are highlighted.

Examples. As mentioned above, the concepts of the section are discussed, when possible, in the context of applications. This means that some examples in the textbook are longer and more involved than in a traditional textbook. As students read the text, it is very important that they work through each and every example.

Strategy Boxes. Throughout the text, strategy boxes are included in appropriate places. These boxes are intended to provide problem-solving strategies for skill-oriented tasks. There is an index of these strategies in the back of the book.

Problem Sets. The problem sets include skill problems, conceptual questions, applications, and open-ended questions. In these problem sets, we are asking students to communicate in writing their thought processes, conjectures, comparisons, and mathematical arguments.

Cumulative Reviews. There are three cumulative reviews in the book. These sections allow students to test their understanding of the concepts learned in the previous chapters. Students must realize that although a concept may have been tested when a chapter was completed, the skills learned are needed throughout the course.

Topics

The topics in this textbook will look somewhat different from a more traditional beginning algebra text. They are in line with the National Council of Teachers of Mathematics Standards and the American Mathematics Association of Two Year Colleges Crossroads document. Topics such as estimation, approximate numbers, and formula evaluations are in response to making the mathematics more applicable and the transfer of mathematics to other fields more assured.

Your choices of topics will be driven by your own curriculum. We hope that you will consider including some of the less familiar topics. As topics are added to the curriculum, other topics need to be omitted or de-emphasized. We are constantly asking ourselves, "What do students need to know and when do they need to know it?" Some traditional topics have been moved to other places in our curriculum based on this question. One example is the moving of polynomial long division to precalculus when students are studying rational functions and need to use long division to rewrite some functions. Difficult, but rewarding, discussions are needed to make these hard choices. We are in continual dialogue with our colleagues, in and out of the mathematics division, as to the use of topics down the road and how our decisions may affect students in their future classes. We feel we are making responsible choices but welcome continued dialogue on this issue.

While we would urge their inclusion in your course, there are some sections that could be omitted without harm to future sections.

- Section 1.2, Estimation
- Section 1.3, Approximate Numbers and Error
  These sections can be skipped. We would, however, urge at least a passing discussion of Section 1.3. Throughout the textbook, answers are rounded reasonably, either based upon the particular application or on the guidelines introduced in Section 1.3.
- Section 1.4, Unit Conversions
  While unit conversion may not be part of your curriculum, we would encourage you to do this section because unit conversion plays a role throughout the textbook, particularly in Chapter 9, Geometry.
- Section 2.5, Evaluating Expressions with Units
- Section 2.6, Formula Evaluation
  Although we find them very valuable, these two sections could be skipped.
- Chapter 9, Geometry
- Chapter 10, Statistics and Probability
  Chapters 9 and 10 can stand by themselves, They can be included anytime after Chapter 6 or skipped without affecting the other material.

Course Content.   At Mt. Hood Community College, we use Chapters 1 through 5 in a 3-credit, ten-week course that meets five hours per week. Chapters 6, 8, 9, and 10 are then covered in second 3-credit, ten-week course that also meets five hours per week. Chapter 7 is covered later in our curriculum. These two courses have replaced our elementary algebra sequence and are required for both transfer and professional technical students.

Pacing of the Course.   One section of this textbook does not correspond to one class period. It is organized around concepts rather than time.

Desired Results.   We have tried to focus on a set of desired results in writing this textbook and in planning what goes on in our classroom. To achieve these results, pedagogical changes have taken place in the classroom. The desired results and pedagogical changes are listed next. Some of these items are discussed in more detail later in the Preface.

Desired Student Results

- **Patterning**
  Mathematics is the study of patterns. Students' patterning skills are developed, giving them an additional tool to investigate and solve problems.

- **Multiple Models**
  Students are introduced to algebra with verbal descriptions, numerical models, graphical models, and equations. This approach emphasizes the connections between the models from the beginning.

- **Verification**
  Students are required to verify simplification results and solutions to equations and formulas. Students must always check the reasonableness of their results and accuracy of their models. These processes empower students in their learning of mathematics.

- **Communication**
  Students are expected to communicate their ideas, processes, and understanding orally and in writing.

- **Units**
  Unit conversion is incorporated throughout the curriculum, enabling students to analyze and solve problems in context. Emphasis is placed on the correct use of units in all situations.

- **Problem Solving**
  Applications are fully integrated. Students are expected to clearly communicate their mathematical processes, results, and verifications.

- **Interpretation**
  Mathematics is an abstract system used to represent situations. Whether the information is presented graphically, numerically, or algebraically, it is only useful if it can be correctly interpreted in the particular scenario. Therefore, students are continually asked to interpret information represented mathematically.

Pedagogical Changes

- **Visual Learning**
  Visual learning is incorporated throughout the curriculum to address the needs of visual learners.

- **Hands-on Materials**
  Hands-on materials are utilized to make abstract mathematics more concrete.

- **Collaborative Learning**
  Students benefit from communicating mathematics, sharing different approaches, and being more actively involved. They attain a deeper understanding by explaining concepts to others. Teamwork creates a more positive learning environment in class; students enjoy attending.

- **Integrated Review**
  Once a concept has been introduced, related problems are sprinkled throughout the course.

Technology

We believe technology should impact curriculum and pedagogy at all levels of mathematics. Our textbook integrates appropriate calculator technology without trying to teach specific keystrokes for specific calculators. Our goal is to teach mathematics; the technology is taught as it applies to the mathematics. This means that we do not spend time up front on the calculator, but rather, we teach various features as the need arises.

It is assumed throughout the textbook that students are using at least a scientific calculator. This is crucial for verification and checking as well as for routine calculations. While a scientific calculator is adequate for most of the text, we would recommend a graphing calculator for Chapter 4, and it is required for Chapter 7. We have gone to the graphing technology at this level for several reasons. It is far easier for students to learn correct calculator entry of numerical expressions when they are able to see what they have entered. While some of the scientific calculators display one line, the graphing calculator allows viewing of several lines and even multiple expressions. Editing for errors is easy rather than frustrating. Another reason for our choice of a graphing calculator at this level is the table feature. This is a powerful tool. The third reason for this choice of technology at this level is the graphing feature. Although it is used extensively only in Chapters 4 and 7, it is useful in introducing students to the topics of graphing relationships and scaling of graphs.

Student Involvement

This course requires students to be much more actively involved in their learning than they may have been before. One of the problems that sometimes surfaces is the complaint that they do not have enough time to give the two hours per credit per week that is needed for outside of class work. We all are encountering more students who are trying to work, raise a family, and go to school. Students need to be realistic about how much they can take on at one time. One exercise that we have found helpful is to do the following reality check with the class early in the term. It serves to bring the discussion out in the open and is a real eye-opener for many students. We try to get through to them that taking the course once with enough time devoted is much less time consuming in the long run, less frustrating, and cheaper than taking the course three times!

> *Reality Check*
> 1. Number of hours of work per week _____
> 2. Number of hours of sleep per night times 7 days per week _____
> 3. Number of hours spent eating and bathing per day times 7 day per week _____
> 4. Number of credit hours this term times three _____
> 5. Number of hours commuting each week _____
> 6. **Total of 1–5** _____
> 7. Total number of hours in a week _____
> 8. Number of hours remaining in a week for family, household chores, entertainment, exercise, etc. _____

A Typical Day in the Classroom

Student teams are essential to the success of the program. To see our philosophy of teams and how they are used, read Chapter 0. We try to set up the class teams as soon in the term as possible.

Beginning of Class. We feel it is essential at the beginning of each class period to have time set aside for students, in their teams, to compare their homework. We circulate during that time to keep teams focused and to find which questions need to be addressed for the entire class. Many minor questions will be handled by students in their teams. One of the reasons that this is a crucial part of class time is that very few answers are included in the back of the textbook. This is a chance for students to compare and see where they disagree. We want students to become confident problem solvers, able to verify in many ways that they are correct. We give them many methods for verifying results throughout the term.

While this process sometimes takes a lot of time, it is well worth it. Student understanding is assured, and the instructor can quickly assess the progress of the class. Also, time used this way cuts down on the time spent answering homework questions in front of the whole class on problems that only a handful of students may have.

Active Student Involvement. What happens next will depend on where the class ended the previous class period.

If a new section is being introduced that includes an activity set, students would now begin that in their teams. These activity sets are designed to guide students to discover, at an introductory level, major concepts of the section. While some of the activities can be done by students at home, most are more effective when done in class in teams. The team often ensures success; what one student cannot figure out, a team most often can.

During an activity, students are expected to do several things.

- They are to be actively involved with their teammates following the directions in the activities.
- They are then expected to make their best effort in discussing their observations of patterns and in making conjectures. This takes some training for many of them, who are used to being told to "work alone and keep your eyes on your own paper."
- They are expected to rearrange their chairs to easily work with one another. Unless you are fortunate enough to have tables for students to work around, chairs should be formed into a circle to allow everyone to hear and to participate. Some will resist this—but you should insist.
- They are expected to take notes on their efforts that allow them to contribute to the class follow-up of the activity. They will not always have the precise mathematical language. Let them find ways to say it in their own words and supply the correct new vocabulary during the follow-up discussion sessions.

While students are working on the activities, the instructor is busy circulating through the room watching and listening to what the teams are doing. Different actions by the instructor may be appropriate.

- The instructor can watch for students who are off topic and give them a gentle nudge to get back on task.
- If a team is going way off the intended course, the instructor can ask questions to get them focused and back on track.
- If a team has a major misconception, it may be advisable to correct them.
- The instructor needs to practice reading the students to see how they are feeling and what their individual level of tolerance for exploration is. When some students direct a question to the instructor, it is appropriate to put the question back in the lap of the team for them to grapple further with. With other students, that action may cause them to put up a wall and

tune out. An answer or partial answer with a focused question may be more appropriate. There is no easy way to know the correct intervention. Time and practice will make it easier!

- If it becomes clear that several teams are all having the same question or point of confusion, it may be appropriate for the instructor to pull the class back together for a couple of minutes to clarify the point or the directions.

When the activities are taken seriously, they give the student a sense of ownership of the concepts. They will have experienced the mathematics, not just memorized someone else's findings. Even when students do not reach all of the conclusions intended in the activities, attempting the activities gives students a feel for the questions that the section hopes to answer. Knowing what questions to anticipate gives the student some insight and confidence as they begin reading the section.

The activities are not designed to build particular skills. They are designed to get students involved in doing mathematics and creating their own understanding. Students should not be held responsible for particular results from the activities but for their genuine effort in doing what the activities asks them to do. A great deal of learning occurs even when the final result is not what the instructor might have had in mind. The effort by the team to describe their findings and to make generalizations, forces the students to use the language of mathematics in communicating with each other and in writing down their conclusions. The activities lay the groundwork for the discussions that follow.

Follow-up Discussion.

When most teams have had time to finish the activity, the instructor needs to pull the class back together for an all-class follow-up of the activity. This follow-up discussion after an activity is the key to the success of the activity.

During the follow-up, the instructor directs the discussion making sure to include several elements.

- The instructor solicits input from the students based on their observations. This is a good time to point out the variety of responses from student teams. If a team is reluctant to share something you know they observed, they can sometimes be coaxed or you can share their observations with the class.

  At times two teams will come up with seemingly different but equivalent observations. This is an ideal time for the instructor to show mathematically that they are equivalent. It also reinforces the fact that more than one right answer is possible. There are times when a student or a team will be adamant that one of their incorrect observations is correct. This is a wonderful time for a well-selected counter-example. If you are unable to come up with one on the spot do not hesitate to promise to bring one in the next day—and then don't forget. The instructor is put on the spot much more often in this type of classroom than with a lecture followed by a predictable discussion. There is a definite loss of control of just where a discussion may lead.

- Next the instructor should elaborate on the student input to make sure all the points of the objectives are covered.
- This is the time that the instructor should offer the mathematical justification for the observations and conjectures. Students must realize that two to four observations of a pattern do not form enough evidence to formulate a rule in mathematics.
- During the discussion, the instructor should provide students with correct mathematical language.

The instructor should be prepared to conclude the discussion with a problem or example for the students to do that reinforces the concepts covered in the activity.

During class time, there is a constant movement back and forth between students working on problems or activities and instructor-led discussions or explanations. This is true whether or not there is an activity.

Whatever the class has worked on during the period, the instructor should plan time to pull things together before students leave for the day. The class should leave at a common point of understanding.

Homework Assignments. At the end of the class period, students are given an assignment to be done for the next class meeting. This will usually include both a reading assignment and appropriate problems from the problem set. What can be assigned may change based on how far the class got during a certain day. Instructor flexibility is the key!

Reading the Text. Students should be expected to read carefully the discussion and examples in the textbook. In this portion of the textbook, the concepts of the section are discussed, when possible, in the context of applications. This means that some examples in the textbook tend to be longer and more involved than in a traditional textbook. It is crucial that the students read the discussion part of the textbook. Looking briefly at the examples will not be sufficient for understanding and success! The textbook was written to be read by students with paper, pencil, and calculator in hand. They should read in detail, answering any questions that are asked, working through the examples as they read them, and making notes in the margins. And most important, they should write down any questions that they have and be sure that they get their questions answered. As the concepts are presented, new vocabulary is introduced and important ideas are highlighted. Each presentation concludes with a summary of the key points of the section.

Problem Sets. The problem sets generally have a few straightforward skill problems. Because we want the problem sets to give students a chance to apply the problem solving tools that they are developing in the course, the problems often do not just mimic the examples in the discussion. The problem sets include conceptual questions, applications, and also open-ended questions. In these problem sets, we are asking students to communicate in writing their thought processes, conjectures, comparisons, and mathematical arguments. We want our students to see mathematics differently. In the presentation of the material and in the problem sets, problems are presented numerically, graphically, and algebraically with an emphasis on the connection between the three models.

Selected Answers. Many of the answers included in the back of the textbook are those which cannot be checked or verified by some other method. This textbook models checking and verifying all results using one or more of the three models (numeric, graphic, or algebraic).

Most answers for problems that can be checked or verified are not included in the back of the book for three main reasons.

1. Many students gain a false sense of confidence by working backwards from the answer. Because the exams they will be taking and life in general does not start with the answer, the skills that they are developing of solving a problem by working backwards will not transfer to real life situations.
2. Checking and verifying problems are important skills to develop and use. If answers are included for problems that the students can verify themselves, students will rely on the answers in the back rather than learning and practicing these important skills.
3. When students are able to check or verify their results themselves they gain confidence in their mathematical ability.

Supplements

Instructors who adopt this text may receive, free of charge, the following items.

An Instructors' Resource Manual is available to all instructors. This manual provides complete solutions to all the exercises in the problem sets. Additionally, the manual contains additional material from the authors on how to effectively teach from this text.

The Printed Test Bank contains prepared tests for each chapter of the text. Final exams are also included. For each section of the text, sample problems are provided that can be used to give students additional practice.

The Computerized Test Bank includes all the test bank questions and allows instructors to prepare quizzes and examinations quickly and easily. Instructors may also add questions or modify existing ones. A gradebook feature is available for recording and tracking student's grades. Instructors have the opportunity to post and administer a test over a network or on the Web. A user-friendly printing capability accommodates all printing platforms.

Students using *A First Course in Algebra: An Interactive Approach* may purchase the Students Solution Manual. This manual contains complete solutions to all problem numbers or letters in blue.

A Web site (URL to come later) has specifically been created for the first edition of *A First Course in Algebra: An Interactive Approach.* This Web site offers additional resources to instructors and students in conjunction with the adoption of the text.

Acknowledgements

We would like to express our appreciation to the Mt. Hood Community College Mathematics Department who agreed to use these materials before the project was complete. Their support and feedback has been invaluable.

Thanks to Steve Bernard, Wini Benvenuti, Bill Covell, Gary Grimes, Harold Hauser, Teresa Kuntz, Pamela Matthews, Mike McAfee, Kory Merkel, Maria Miles, Paul Porch, Gina Shankland, Frank Weeks, Sara Williams, and Steve Yramategui.

We are also grateful to all of our part-time colleagues who have used preliminary editions of this text and offered comments. Additionally, we wish to thank the faculty from other departments who shared their expertise in helping us to write realistic applications from other disciplines. Others who have offered moral support and advice are Betty Brace, Brenda Button, Lynn Darroch, Andres Durstenfeld, Gretchen Schuette, Mike Shaughnessy, Bert Waits, and Bob Wesley.

Thanks to the hundreds of students who have used these materials in their beginning stages. Their constructive input has led to many improvements.

In addition, there are many mathematics instructors across the United States who have reviewed previous versions of these materials. We thank them for their comments and suggestions.

   Chris Allgyer, Mountain Empire Community College
   Richard Broomfield, Eastern Connecticut State University
   O. Robert Brown, Montgomery College-Takoma Park
   Jim Chesla, Grand Rapids Community College
   Mishka Chudilowsky, DeAnza Community College
   Linda Green, Santa Fe Community College
   Laurel Grindy, Columbia College
   Bernadette Kocyba, J. Sargeant Community College
   Shannon Lienhart, Palomar College
   Charles Miller, Albuquerque State University
   Jim Oman, University of Wisconsin Oshkosh
   Kathy Perino, Foothill College
   Sharon Ross, Georgia Perimeter College
   Bernadette Sandruck, Howard County Community College

Karen Tabor, College of the Desert
Steve Winters, University of Wisconsin Oshkosh
Alice Wong, Miami-Dade Community College
Bill Wunderlich, Cincinnati State Technical and Community College

We would like to thank the National Science Foundation for their support that made significant work on this project possible.

We also wish to acknowledge our editors and their assistants at Saunders College Publishing. In particular, we want to thank our current editors, Jim LaPointe and Angus MacDonald, for their support, humor, and understanding through the final stages.

A special thank-you to everyone at TSI Graphics, and especially to Donna Cullen-Dolce for her hard work, concern, and understanding.

We are grateful to Lauri Semarne who examined (and corrected where necessary) the activity sets, examples, and problem sets.

We know we may have inadvertently missed others who have had an impact. Please know that we appreciate all the help we have received.

Finally, we wish to express our heartfelt appreciation to Jack, Mike, Erik, and Beth for supporting, enduring, and encouraging us through the development of this project.

Alison Warr
Cathy Curtis
Penny Slingerland

April 2000

Chapter Zero

An Introduction to Learning Mathematics, Problem Solving, and Teamwork

The goal of this text may be different from that of others you have used in the past. Don't be surprised if you find the presentation of the material and the overall content unique as well. The purpose of this chapter is to communicate the goals of the text and to give you some insight into our thoughts about learning mathematics and the reason we approached the writing of this textbook as we did.

Learning Mathematics

One of the major goals of this textbook is to help you become a better problem solver. This means not only solving equations and typical mathematics problems but also being able to look at unfamiliar problems, choosing strategies, and applying mathematics to solve these problems.

A second goal of this textbook is to present mathematical ideas in meaningful ways. Mathematics is a useful tool in solving problems in many different fields, and we want you to see this value in mathematics.

Many of you may have learned mathematics in a lecture format. From our experience we have found several drawbacks to a mathematics course taught primarily from a lecture. Many students become bored by mathematics taught in this format and consider the subject matter as a collection of abstract, disjointed facts that they are unable to use in real-life situations. They therefore feel that mathematics is an unimportant hoop for them to jump through to receive a degree.

A more important issue is the quality of learning and understanding that students achieve through this approach. Many students who take mathematics in a lecture format comment that "It looks so easy when you do it, but I can't." In a lecture, an instructor typically shows the students many examples and works each of them out correctly. Of course it looks simple! The instructor knows exactly how to do the mathematics. It is much like watching carpenters work. They know exactly how to perform the skills and have hours of experience. It is an entirely different situation when an inexperienced person tries to perform the job. You do not know where you are going to have difficulty until you try. Mathematics is similar. Until you dig into the material, you may not know what you can do or where you have questions.

For this reason, this textbook expects you to actively participate in learning mathematics. In doing so you will not only learn the skills necessary to perform specific tasks, but also obtain an understanding of the concepts behind the mathematics and know when and where to apply the concepts and skills.

Learning anything new takes effort. Learning mathematics is no exception. In using this textbook you will be asked to work. You will be involved in doing mathematics both in and outside the classroom. You will be asked to discuss mathematics with your classmates and your instructor. To be successful in this course, you cannot be a passive recipient of mathematical facts. We are asking you to do more than put in time doing assignments. We are asking you to *think!* You will be asked to think about the mathematics that you are doing, as well as what is going on in your mind as you do mathematics. You will need to decide when to struggle with a mathematics problem and when to ask for help. In essence, we are asking you to take responsibility for your own learning.

In return, we promise that this textbook will provide you with the opportunity to learn mathematics in a meaningful way. If you use it properly and your course is consistent with the principles on which the textbook is written, then you will begin to claim these mathematical ideas as your own. Mathematics will no longer be a mystery or a mere collection of facts. It will be a collection of principles and concepts that you can understand. You will find yourself less frequently staring at a collection of symbols not knowing how to proceed. More often you will see many possibilities in choosing where to start a problem. Finally, we believe that you will begin to see the use, value, power, and beauty of mathematics.

Discovery Learning

In this textbook, we adopted a discovery approach to learning. Most sections begin with an Activity Set. These activities were written with the assumption that you will be working on them with other members of your class. Each activity has been written to provide you with the opportunity to "dig into" the mathematics. Although we do not expect that you will learn all the mathematics through these activities, we do expect you to be actively involved with others in your class and to make your best effort in discussing your observations of patterns and in making conjectures.

Following a set of activities, your instructor will typically lead a discussion of the observations and conjectures made by the class. This discussion may include vocabulary, generalizations to formulas, or applications of the ideas learned in the activities. Even when you do not reach all of the conclusions intended in the activities, attempting the activities will make this discussion more meaningful and useful.

How to Use This Text

> We have tried to produce a textbook that
> - demands greater understanding and less routine manipulation
> - covers less material in greater depth
> - presents concepts numerically, graphically, and algebraically
> - develops concepts through commonsense investigations rather than abstract definitions
> - incorporates the use of technology and expects technology to be available at all times
> - is written to be read
> - is written for students to discover concepts, not as a reference for those who know the concepts

We wrote this textbook with you, the student, in mind. We believe that the combination of doing discovery activities, participating in class discussions, and actively reading the textbook will lead to your success in learning mathematics. Most sections of this textbook begin with activities, are followed by a text discussion, and end with a problem set. Some of the discussions read like a standard textbook, but there are some important differences. Many of the examples emphasize mathematics in context. We de-emphasize the rote learning of mathematics. Questions are raised and left to you, the reader, to answer. In addition, appendices in the back of the textbook include review material.

To successfully use this book, you must read it with a pencil and calculator in hand. The activities at the beginning of each section *should* be done first; however, you are not penalized for looking ahead. Once you have done the activities, you should participate in any class discussion that may follow, read the textbook in detail, answer any questions that are asked, work through the examples as you read them, and make notes in the margins. Most importantly, write down any questions that you have, and be sure that you get your questions answered.

Mathematics was created to solve real-world problems like balancing a checkbook or determining the load that a roof can bear. Because mathematics has many different applications, it is not possible to learn specific methods or templates to solve every type of application. Instead we need to learn problem-solving skills that will allow us to tackle all sorts of problems. For this reason, you will not find worked examples for every type of problem you will encounter in the problem sets. In addition, for a given problem, many different solution processes are often correct. We encourage you to discuss your solutions with your classmates. As you see different ways of approaching and solving problems, you will learn new strategies that can be used on future applications.

What Is a Good Problem Solver?

One of the goals of this course is to make you a better problem solver, but what is a good problem solver? The following is a list of suggestions for becoming a better problem solver written by students who successfully completed a course in problem solving.

> *How to Become a Better Problem Solver*
> - Accept the challenge of solving a problem.
> - Take time to explore, reflect, think, . . .
> - Look at the problem in a variety of ways.
> - If appropriate, try the problem using simple numbers or break the problem into smaller steps.
> - Develop good problem-solving helper skills. Don't give solutions; instead provide meaningful hints.
> - Write up your solutions neatly and clearly enough so that you could understand your solution if you reread it in ten years.
> - Help others by giving hints. You will find that you develop new insights.
> - Don't hesitate to take a break. Many problems require an incubation period. But remember to return to try again!
> - Be persistent. Don't give up!
> - Don't just sit there, do something!

When solving a problem in mathematics (or elsewhere), you may begin by asking "How do I get started?" There is not one answer to this question but several possible ways to get started. The following is a list of strategies for trying to solve a problem. These are just the beginning. As you solve more and more problems, you will be adding to this list.

Write Down Everything You Know
Write down what you know about the problem and the question you are trying to answer. This can help organize the information.

Draw a Picture
Can you draw a picture of the problem situation? Often you may find that in the process of drawing and labeling a picture you gain some insight into relationships that can lead to the solution of the problem, or at least to the next step in the process.

Make a Table of Values
A table of values can give you some concrete information from which to discover patterns. In the textbook, you will see how to use this powerful tool.

Guess
If you don't know where to begin, go ahead and guess. Many problems have been solved by someone saying "What if . . . ?" Once you make a guess, your next step is to determine whether or not your guess is correct. This can be done by substituting your guess into the problem situation. If you guessed correctly, great! If you did not, you may find the insight that you need to solve the problem, or you may be able to eliminate several possibilities.

Make a Physical Model
Many of us need to physically see the situation to understand what is happening. This may mean we need to represent the problem using concrete objects. For example, in Problem 2,

"Moving Dots" in the Problem Set at the end of this chapter, you may want to represent the shaded dots with pennies and the unshaded dots with dimes. Then you can physically try solving the problem using the coins. Other times you may want to build a model. For example, if the problem is about a swimming pool, it may be helpful to construct a model of the swimming pool out of cardboard.

Throughout this course, you will encounter many mathematical problems. Some you will be able to solve with what you already know or what you are learning. Others you will need to develop more ideas and tools to solve. The time to conclude that you are unable to solve a problem is *after you have given it a royal try — not before*.

As you struggle with various problems throughout the course, the most important thing for you to remember is not to think of these problems in terms of success and failure. When you are exploring new ideas and the results that you get do not completely solve the problem, think of this as an opportunity to learn and improve your understanding. Often a strategy that does not lead to the solution will lead to new strategies.

Why Work in Teams?

In this class, you may be expected to work in a team. Students often ask why they are being asked to do this. There are many reasons for working in a team in this class. Working in a team gives you a safe environment to try out new ideas and problem-solving strategies. By working with your team, you will learn to communicate more clearly your ideas and your understanding of concepts. Your facility with the language of mathematics will improve as you get more practice using it. You will also learn to listen carefully to other people's ideas. You will be challenged to explain concepts so others can understand them. The give and take of ideas will improve your own reasoning and critical thinking abilities.

In addition, many recent school-to-work studies call on schools to teach students more of the skills that business and industry find their employees need. The general ability to work with others and communicate clearly, orally and in writing, are always on the lists of desired skills. Other expectations are that workers know how to develop new skills, can manage themselves, and are responsible for their actions. Working in teams in class gives you one more place to develop and refine these skills. There is more to school than learning specific academic skills.

Production Team Versus Learning Team

Although many of the skills learned by working in teams are appropriate for both school and work settings, many important differences exist. In a workplace team, the team may have been formed by pulling together people with different skills and expertise to work together to produce a product. In this situation, the team may do some initial planning together. They may then go off to produce their individual part of the project before the team comes back together to integrate the separate pieces into the finished product. Each member of the team relies heavily on the differing expertise of the team members. Although the team members count on each other for support and trouble shooting, they may not be working closely together every day. This type of team could be called a **production team.**

In the classroom, everyone on the team is expected to learn the same concepts. The product they are producing is mutual understanding of the course material — not a consumer product for others. It is important that all members of the team gain a full understanding of all concepts. The main point of the team is to help in the learning process. We refer to this type of team as a

learning team. We need to keep in mind that in this class we are forming learning teams and not production teams.

Team Selection

The manner in which teams are selected will vary from class to class and instructor to instructor. Many instructors look for a balance of gender, previous experience, age, and motivation. Other considerations that may be taken into account are times you are available to study outside of class or what kind of graphing calculator you have. The object is to get a mix that allows all of you to have a good learning experience.

What Makes a Team Work?

For a team to be effective, it is important that each member knows what is expected of him or her. Often ground rules for team behavior are established to help with this process. A sample set of ground rules is included in the following box.

> *Ground Rules for Teams*
> - Attend class regularly, and come prepared.
> - Stay focused on the team task.
> - Work cooperatively with other team members.
> - Reach a team decision for each problem.
> - Make sure each person on the team understands the solution before the team moves on.
> - Listen carefully to others, and try to build on their ideas.
> - Share the leadership of the team.
> - Make sure that everyone participates and no one dominates.
> - Take turns recording team results.

It is important that team members agree on the ground rules for the class. It is the responsibility of team members to monitor each other in following the agreed-on rules. Although the instructor can help in this process, it is not possible for the instructor to be in all teams at all times.

Team Building

Being an effective team member takes some practice. The problems at the end of this section are designed for you to practice team-building skills in a nonthreatening environment. As you work on these activities, keep in mind the ground rules. Remember that the goal is to learn to work together and communicate clearly so that you can use these skills to help each other effectively learn mathematics.

Many of these activities may not seem mathematical to you. They are all in some way related to thinking and visualizing skills that are important in mathematics. Sometimes you will want to work together on a problem from the start—brainstorming approaches and strategies. At other times, you may want to spend some time working individually before you share your ideas. Be aware of the needs of your team members as well as your own preferences. Some of the activities that follow will be easier with some concrete props — coins, or pieces of paper, for instance. Don't hesitate to use any tools that make the problem solving or explaining easier.

Last Words of Wisdom

Don't get discouraged if you don't always feel successful. Frustration is a natural part of any learning experience. Jump in, get involved, and enjoy yourself! Solving problems is a positive experience.

Problem Set O

1. **Coins**   If you have $11.52 in your coin jar and there is an equal number of pennies, nickels, and dimes, how many of each coin do you have?

2. **Moving Dots**   Start with the arrangement of dots shown here.

   Try to produce this arrangement, given the following three restrictions.

   - The solid dots can move only to the right, and the striped dots can move only to the left.
   - A dot can move into an adjacent space.
   - A dot can jump over *one* other dot — either solid or striped — into an empty space.

   Once you can do the problem, work together on a way to clearly communicate your solution in writing. Assume that the solution is being written for someone who is familiar with the rules of the problem but has not necessarily solved it.

3. **Paths**
   a. How many paths are there from points *A* to *C* that are 8 units long?
   b. How many of the paths pass through the point *B*?

   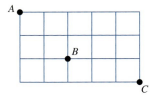

4. **Last Digit**   What is the last digit of 4^{80}?

5. **How Many Marbles?**   Carla placed five containers on a table at the front of the classroom. She said, "There are a total of 50 marbles in these containers. In the first and second containers there are a total of 20 marbles. The second and third containers hold a total of 23 marbles. In the third and fourth containers there are a total of 21 marbles. In the fourth and fifth containers there are a total of 15 marbles." How many marbles are in each container?

6. Following Directions
   a. Print the words DON'T JUST SIT THERE without the apostrophe and without space between the words.
   b. If a letter occurs more than once, reading from the left, delete all but the first one.
   c. In your result from part b, exchange the first and the seventh letters.
   d. In your result from part c, change the last letter to the letter that follows it in the alphabet.
   e. In your result from part d, change the T to an E.
   f. In your result from part e, exchange the third and the seventh letters.
   g. In your result from part f, change the letter J to match the first and last letter.
   h. In your result from part g, change the third vowel from the left to an I.
   i. In your result from part h, change the last vowel to the first vowel in the alphabet.
   j. In your result from part i, change the letter O to the next vowel in the alphabet.
   k. In your result from part j, delete the second letter I, from the left.
   l. In your result from part k, change the letter H to the letters CL.
   m. In your result from part l, change the letter D to the letter that precedes it in the alphabet.
   n. In your result from part m, if there is a vowel on both sides of a consonant, double that consonant.
   o. In your result from part n, double the last letter.

   After following these directions, what will you have?

7. **Occupations** Alice, Brenda, Carl, and Daniel are all sitting around a square table. Their occupations are artist, biologist, chemist, and dentist, but the first letter of their first names and their occupations do not match.

   • Alice and Brenda are female. Carl and Daniel are male.
   • Brenda is sitting across the table from the dentist.
   • The biologist is sitting across from Carl.
   • The dentist and the artist are Brenda's mother and father (but not necessarily in that order).

   Match each person with his or her occupation.

8. **Connections** Connect each pair of dots (*A* to *A*, *B* to *B*, and so on). You may not retrace or cross any drawn line. All lines must be contained within the boundary of the drawing.

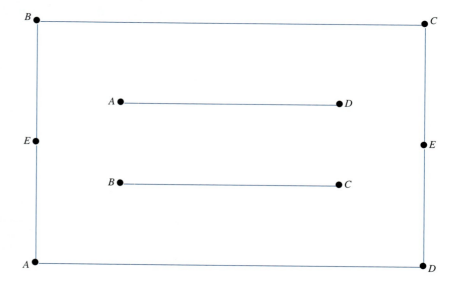

9. **Toothpick Problems** In these problems follow rules i and ii.
   i. No toothpicks are to be broken.
   ii. All of the toothpicks will be part of the figure described.

   a. Use 12 toothpicks to make the pattern shown here. Move three toothpicks to form a pattern of three squares.

   b. Use 16 toothpicks to make the following pattern. Change the position of three toothpicks to form a pattern of only four squares.

   c. Use 18 toothpicks to make the following pattern. Remove four toothpicks so that five triangles remain.

   d. Using 35 toothpicks, form the spiral pattern shown here. Move four of the toothpicks to form exactly three squares.

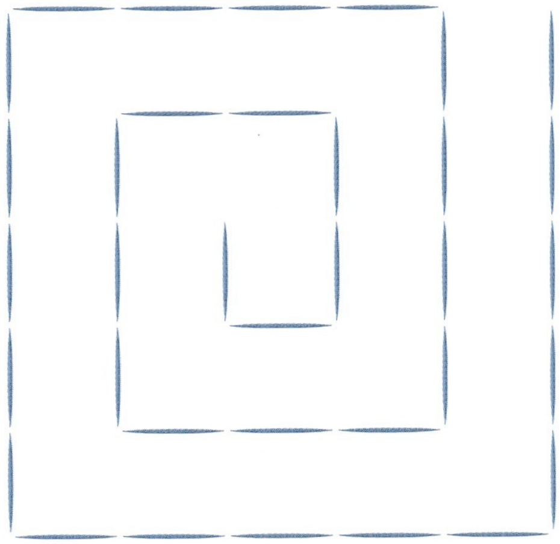

10. Pages To number the pages of a book, a printer used 3385 digits starting on page 1. How many pages were in the book?

11. More Toothpick Problems In these problems follow rules i and ii.

   i. No toothpicks are to be broken.

   ii. All of the toothpicks will be part of the figure described.

   a. Seven identical triangles occur in the pattern shown. Move six toothpicks to form six identical rhombuses.

   b. Sixteen toothpicks make the following pattern of five identical squares. By moving only two toothpicks, make an array of four identical squares.

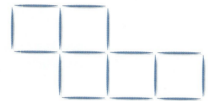

   c. Thirteen toothpicks are arranged to make the following six identical rectangular enclosures.

   • Remove one of the toothpicks. Arrange the remaining 12 into six enclosures of the same size and shape.

   • Remove another toothpick. Arrange the remaining 11 into six enclosures of the same size and shape.

12. **Balance the Scale** Assume that the first three scales shown here balance perfectly. How many of the same shaped solid objects (all balls, all rectangular blocks, or all cones) will it take to balance the fourth scale?

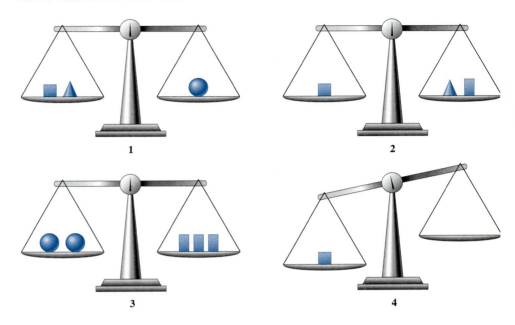

Chapter One

Numerical Concepts

A good understanding of numbers is necessary for the study of algebra and geometry. The goal of this chapter is to build a strong numerical understanding of mathematics. We will learn how to simplify and estimate expressions using terms. The ideas of precision, accuracy, and the correct rounding of numbers will be discussed. We will learn how to analyze units and convert between different units of measurements. The geometrical concepts of measurement, perimeter, and area will be included throughout the chapter. For additional background on geometry concepts, consult the appendices.

1.1 The Language of Mathematics and Term-by-Term Evaluation

Discussion 1.1

As we begin to learn the language of mathematics, we must become familiar with its symbols, vocabulary, and rules for writing and reading the language. The six basic operations used in working with numbers are summarized in the following table along with the terminology and notation associated with them.

Operation	Symbols Used	Parts	Result
Addition	$2 + 3$	Term + Term	Sum
Subtraction	$5 - 2$	Term − Term	Difference
Multiplication	$2*3,\ 2\cdot 3,\ 2(3),\ 2\times 3$	Factor * Factor	Product
Division	$\dfrac{6}{2},\ 6 \div 2$	$\dfrac{\text{Dividend}}{\text{Divisor}}$	Quotient
Exponentiation	2^3	$\text{Base}^{\text{exponent}}$	Power
Root extraction	$\sqrt[3]{8},\ \sqrt{9}$	$\sqrt[\text{index}]{\text{radicand}}$	Root

The table includes alternative notations for multiplication and division. The notations $2 * 3$ and $2(3)$ are the most commonly used in this textbook. This is because the symbol $\cdot$ can easily be confused with a decimal point and the symbol $\times$ with the letter x.

Notice that the operations come in pairs. Addition and subtraction are called inverse operations of each other. Multiplication and division are also inverses of each other. By inverse we mean that what one operation does, the other one undoes. For example, if we start with the number 5 and we add 3 to it and then subtract 3 from it, we will be back to the starting number of 5.

Term-by-Term Evaluation

The following additional definitions are needed as we begin to evaluate expressions.

> **Definition**
> The symbols [ ], ( ), — (fraction bar), and $\sqrt{}$ (radical sign) are **grouping symbols** and are used to denote the result of the operation inside the grouping symbols.

> **Definition**
> A **numerical expression** is any single number or any group of numbers combined with mathematical operations.

The following all represent numerical expressions.

$$4, \quad 5(8+3), \quad \text{and} \quad \frac{12}{2-5}$$

$$5 + 3 * 8^2, \quad 10 - 5(8+3), \quad 3 * 8^2 + 4 \quad \text{and} \quad 7 + \frac{12}{2-5}$$

> **Definition**
> **Terms** of an expression are the parts of sums or differences.

Strategy — *Identifying the Terms of an Expression*

Because terms are the parts of a sum or a difference, *addition and subtraction* symbols *outside of grouping symbols* separate terms. Therefore, to identify the terms of an expression, find all of the addition and subtraction symbols that are not in grouping symbols, and the collections of symbols on either side of these will be the terms.

In the numerical expression $5 + 3 * 8^2$, there is one addition symbol.

$$\underline{5} + \underline{3 * 8^2}$$

This addition symbol separates the terms of the expression. Therefore, 5 is a term, and $3 * 8^2$ is a term.

In the numerical expression $10 - 5(8 + 3)$, one subtraction symbol is *not in grouping symbols*.

$$\underline{10} - \underline{5(8 + 3)}$$

This subtraction symbol separates the terms of the expression. Therefore, 10 is a term, and $5(8 + 3)$ is a term.

In the numerical expression $7 + \dfrac{12}{2 - 5}$, one addition symbol is *not part of a grouping*. The subtraction symbol in the denominator does not separate terms because it is in the grouping implied by the fraction bar.

$$\underline{7} + \underline{\dfrac{12}{2 - 5}}$$

Therefore, 7 is a term, and $\dfrac{12}{2 - 5}$ is a term.

If there are no addition or subtraction symbols that separate the terms, then the expression is a single term. For example, 4, $5(8 + 3)$, and $\dfrac{12}{2 - 5}$ are all single-term expressions.

> **Definition**
> To **evaluate** a numerical expression means to find its value.

We will first look at evaluating single-term expressions.

Example 1

Evaluate each of the following expressions.

a. $3 * 8^2$
b. $(3 * 8)^2$
c. $3 * 4 * 5$
d. $5\sqrt{9}$

Solution

a. The expression $3 * 8^2$ means 3 times the square of 8. The power is done before the multiplication because the base of the exponent 2 is only the 8.

$$3 * 8^2$$
$$= 3 * 64$$
$$= 192$$

b. The expression $(3 * 8)^2$ means the square of the product of 3 and 8. Therefore, in $(3 * 8)^2$ the multiplication is done first.

$$(3 * 8)^2$$
$$= 24^2$$
$$= 576$$

c. The expression $5\sqrt{9}$ means 5 times the square root of 9.

$$5\sqrt{9}$$
$$= 5 * 3$$
$$= 15$$

d. The expression $3 * 4 * 5$ is the product of the three factors 3, 4, and 5. Multiplication and division within a term are done in order from left to right.

$$3 * 4 * 5$$
$$= 12 * 5$$
$$= 60$$

When an expression has more than one term, we need a strategy for evaluating the entire expression.

Strategy

Evaluating an Expression Term by Term

To evaluate an expression term by term, first find the value of each term and then add or subtract terms from left to right.

Within a given term, find the value of all expressions inside grouping symbols first. Next, evaluate any exponents. And finally, perform any multiplication or division in order from left to right.

Earlier, we underlined the terms of the expressions in the example. We make use of that underlining now as we evaluate the expressions.

1.1 The Language of Mathematics and Term-by-Term Evaluation

Example 2

Evaluate each of the following expressions.

a. $5 + 3 * 8^2$ b. $10 - 5(8 + 3)$ c. $7 + \dfrac{12}{2-5}$

Solution

a.
$$5 + 3 * 8^2$$
$$= 5 + 3 * 64$$
$$= 5 + 192$$
$$= 197$$

b.
$$10 - 5(8 + 3)$$
$$= 10 - 5(11)$$
$$= 10 - 55$$
$$= {}^-45$$

c.
$$7 + \dfrac{12}{2-5}$$
$$= 7 + \dfrac{12}{{}^-3}$$
$$= 7 + {}^-4$$
$$= 3$$

We will find that identifying the terms in a numerical expression aids in the evaluation process. Let's look at more numerical expressions.

Example 3

For each of the following numerical expressions, identify the terms and evaluate the expressions.

a. $5 - 3[1 + (5 - 3)^2]$
b. $2 - 4(5 - 3)^3 + \sqrt{4 + 5}$

Solution

a. The numerical expression $5 - 3[1 + (5 - 3)^2]$ has one subtraction symbol that is not in grouping symbols.

$$5 - 3[1 + (5 - 3)^2]$$

This subtraction symbol separates the terms of the expression. Therefore, 5 is a term, and $3[1 + (5 - 3)^2]$ is a term.

The second term of this expression is somewhat complex. In this case we will look at the expression inside the parentheses and determine the terms of it. These are called **subterms.**

The expression inside the parentheses, $1 + (5 - 3)^2$, has two terms; 1 is a term and $(5 - 3)^2$ is a term. Therefore, these are subterms of the original expression. We will underline these to help us evaluate the expression. When evaluating a given term, evaluate the subterms first.

$5 - 3[1 + (5 - 3)^2]$	Underline the terms and subterms.
$= 5 - 3[1 + (2)^2]$	Because the first term is completely simplified, we begin by simplifying the second term. We start by simplifying the subterms inside the square brackets.
$= 5 - 3[1 + 4]$	Now that the subterms are completely simplified, we can add the 1 and 4.
$= 5 - 3[5]$	Finish simplifying the second term by multiplying the 3 and 5.
$= 5 - 15$	
$= {}^-10$	

b. The numerical expression $2 - 4(5-3)^3 + \sqrt{4+5}$ has three terms;

$$2 - 4(5-3)^3 + \sqrt{4+5}$$

2 is a term, $4(5-3)^3$ is a term, and $\sqrt{4+5}$ is a term.

Now, we will evaluate the expression term by term.

$\underline{2} - \underline{4(5-3)^3} + \underline{\sqrt{4+5}}$ Underline the terms.

$= \underline{2} - \underline{4(2)^3} + \underline{\sqrt{9}}$

$= \underline{2} - \underline{4*8} + \underline{3}$ Cube the 2 in the second term.

$= \underline{2} - \underline{32} + \underline{3}$

$= {}^-30 + 3$ Add and subtract terms from left to right. Two minus 32 is $^-30$.

$= {}^-27$

NOTE: In this example it is important that after finding the value of each term, the addition and subtraction is done in order from *left to right*. If we were to add the 32 and 3 first, we would get $2 - 35 = {}^-33$, not the correct answer of $^-27$.

Understanding how expressions are evaluated and being able to identify terms enables us to read mathematical expressions correctly.

Translating Between English Phrases and Numerical Expressions

One way to get more familiar with the language of mathematics is to practice translating between English phrases and numerical expressions.

Example 4

Translate each English phrase into a numerical expression.

a. six times the sum of negative two and nine
b. the reciprocal of the square of five
c. the product of the sum of sixteen and two and the quotient of six and two

Solution a. In the phrase "six times the sum of negative two and nine," we call *the sum of negative two and nine* a **subphrase**. One way to identify the subphrase is to think about what we expect to follow the words "six times." We would normally expect to hear *six times some number*. This can be visualized as

$6 * \bigcirc$

The subphrase, *the sum of negative two and nine,* $(^-2 + 9)$ is what goes in the space.

The phrase "six times the sum of negative two and nine" translates to the numerical expression

$6 * (^-2 + 9)$.

b. Recall the reciprocal of a number is the quotient of one and the number. For example, the reciprocal of 7 is $\frac{1}{7}$. When we see the word "reciprocal," we find ourselves asking "the reciprocal of what?" This can be visualized as

The subphrase, *the square of five*, translates as 5^2. This is what goes in the space. Therefore, the "reciprocal of the square of five" translates to the numerical expression $\frac{1}{5^2}$.

c. As you translate more complex phrases, look first at the general form of the expression. Then identify the subphrases. For example, to translate "the product of the sum of sixteen and two and the quotient of six and two" first notice this expression is a product that can be visualized as

◯ * ◯

Underline the subphrases to help see how to write the expression "the product of *the sum of sixteen and two* and *the quotient of six and two*." The underlined phrases go into the spaces.

$(16+2) * \left(\frac{6}{2}\right)$

Therefore, the phrase "the product of *the sum of sixteen and two* and *the quotient of six and two*" translates to the numerical expression

$$(16+2) * \left(\frac{6}{2}\right)$$

It is always safe to put the subphrases in parentheses to indicate that they must be done first when we do the calculation.

Reading a numerical expression is really translating from the language of mathematical symbols to English. When we translate in this direction, we discover that, whereas mathematics is a precise language, English often has many ways to say the same thing. For example, $3 - 7$ could be translated as seven less than three, the difference of three and seven, or three minus seven. Understanding what operation is to be done first is very important in reading numerical expressions correctly. Identifying terms and subterms can be helpful. In the expression $14 - 2 * 5$, the term $2 * 5$ is calculated before the subtraction is done. We need to keep this in mind when reading the expression.

The expression $14 - 2 * 5$ is often read as fourteen minus two times five. This reading could lead some people to translate the expression as $14 - 2 * 5$, and others to interpret it as $(14 - 2) * 5$. Because people may come to different conclusions, it is better to avoid this reading. One correct way to read the expression is *fourteen minus the product of two and five*. Another correct way to read $14 - 2 * 5$ is the *difference of fourteen and the product of two and five*.

The following table summarizes some of the correct and ambiguous readings of the expression.

	$14 - 2 * 5$
Correct	• fourteen minus the product of two and five • the difference of fourteen and the product of two and five
Ambiguous	• fourteen minus two times five

Notice that in each correct reading, the term 2 * 5 is grouped by using the word "product."

Expressions that include exponents require special care in reading. Recall the expression $3 * 8^2$ that we encountered in Example 1. The meaning of these mathematical symbols tells us that the eight must be squared before we multiply by three.

If we say three times eight squared, it is unclear whether we mean $3 * 8^2$ or $(3 * 8)^2$. If we read $3 * 8^2$ as three times the square of eight, there is no possible confusion about what we mean. The expression $(3 * 8)^2$ would be read as the square of the product of three and eight. This makes it very clear that the product must be done first.

The following table summarizes correct and ambiguous readings for the expression $3 * 8^2$.

$3 * 8^2$	
Correct	three times the square of eight
Ambiguous	three times eight squared

In general, reading the exponent just prior to its base will make the meaning clear.

Confusion also commonly arises in distinguishing between a negative number and taking the opposite. Consider the expressions $^-2^4$ and $(^-2)^4$. We must understand the difference in the meaning of these two expressions to evaluate them correctly and to read them correctly.

The first expression $^-2^4$ is the opposite of the fourth power of two. In other words, we first find the fourth power of two, which is 16, and then take the opposite of it. Our result is then $^-2^4 = ^-16$.

$^-2^4$	
Correct	the opposite of the fourth power of two
Ambiguous	negative two to the fourth power

The second expression $(^-2)^4$ is the fourth power of negative two. We are raising negative two to the fourth power and we find $(^-2)^4 = 16$.

$(^-2)^4$	
Correct	the fourth power of negative two
Ambiguous	negative two to the fourth power

Learning to read expressions correctly now will make many concepts easier to understand as we move through the term.

In this section, we have introduced mathematics as a language. We have looked at the names of operations, the symbols associated with them, and the names given their results and their parts. We have examined how to evaluate expressions term by term and how to correctly read numerical expressions. Understanding the language of mathematics can clarify many concepts that might otherwise be confusing.

Problem Set 1.1

1. For each numerical expression do the following.

  i. Underline the terms.
  ii. Without using your calculator, evaluate the expression showing all steps.
  iii. Evaluate the expression *in one step* on your calculator. Compare your results with those from part ii. If your results are different, find where you have made an error.

  a. $5 + \frac{1}{2} * 6$

  b. $16 - 14 \div {}^-2$

  c. $^-14 + 2(2 - 9)$

  d. $(5 * 2)^3 + 6$

  e. $^-3^2 - 12 \div 3$

  f. $500 - 100 * 3^2$

  g. $27 - 3(4^2 - 20)$

  h. $23 - 2\sqrt{25}$

  i. $10 + \frac{2}{3}\sqrt{36} - 2^4$

  j. $4800 \div 40 - 0.5\sqrt{6400}$

  k. $\frac{-12}{3} * (2 + {}^-5)$

  l. $\left(\frac{4 - 12}{4}\right)^3$

  m. $30 * 10^3 - \frac{4000 - 40{,}000}{3}$

2. For each numerical expression do the following.

  i. Underline the terms.
  ii. Without using your calculator, evaluate the expression showing all steps.
  iii. Evaluate the expression *in one step* on your calculator. Compare your results with those from part ii. If your results are different, find where you have made an error.

  a. $24 + 4 * 5$

  b. $(24 - 8) * 3$

  c. $\frac{1}{2} * 4^3 + 6$

  d. $\frac{12 + 8}{5 - 9}$

  e. $\sqrt{9 + 16}$

  f. $^-80 * 10 - 100 * {}^-5$

  g. $\sqrt{9} + \sqrt{16}$

  h. $(^-4)^2 - 7 + 12$

  i. $\frac{480 - 80 * 5}{\sqrt{25(4)}}$

  j. $(2 - 6)^2$

  k. $\sqrt{5 * 9 - 3^2}$

  l. $^-9^2 - 4(2 - 5)^2$

  m. $3\sqrt{36} - \frac{12}{2 * 3}$

3. Identify the terms in the expression $^-73.56 * 13.08 - 78.96 \div 2.45$.

4. Identify the subterms in the numerator of the expression.

$$\frac{3 * 57 + 7 * 88 + 4 * 62 + 4 * 98}{18}$$

5. For each numerical expression below do the following.
   i. Underline the terms and subterms.
   ii. Without using your calculator, evaluate the expression showing all steps.
   iii. Evaluate the expression *in one step* on your calculator. Compare your results with those from part ii. If your results are different, find where you have made an error.

   a. $\dfrac{24 + 5(^-2)}{5 + 2}$ b. $\dfrac{40 * 10 + {}^-5 * 60}{^-5}$ c. $3 + \dfrac{1}{2}[5 + 3(5 - 2)]$

6. Translate each phrase into a numerical expression. Do not find the value.
   a. the sum of fifteen and four
   b. the product of fifteen and twelve
   c. the quotient of twenty and negative four
   d. the difference of thirty-five and fourteen
   e. twice twenty-five
   f. the reciprocal of fourteen
   g. the opposite of negative five

7. For each pair of phrases, do the following.
   i. Translate each phrase into a numerical expression.
   ii. Explain how the two expressions are different.
   iii. Evaluate both expressions.

   a. two times the square of three
      the square of the product of two and three
   b. the square root of the product of four and nine
      four times the square root of nine
   c. the square of negative three
      the opposite of the square of three
   d. the reciprocal of the sum of three and two
      the sum of three and the reciprocal of two

8. Translate each phrase into a numerical expression. Do not find the value. You may find it helpful to underline subphrases.
   a. twenty minus the sum of five and four
   b. twelve times the sum of twenty-one and fifteen
   c. twenty-five plus the product of nine and negative five
   d. the square root of the sum of seven and nine
   e. the square of the sum of negative seven and nine
   f. the quotient of thirty-six and the sum of five and four
   g. twenty-four divided by the product of four and two
   h. five plus the square of negative three
   i. the cube of the product of five and seven
   j. negative five plus the reciprocal of ten
   k. the reciprocal of the sum of negative five and ten
   l. the product of the sum of eight and sixteen and the difference of eleven and three

9. Translate each numerical expression into an English phrase.
   a. $5^2 + 3^2$ b. $(5 + 3)^2$ c. $(3 * 5)^2$ d. $3 * 5^2$ e. $\sqrt{25 - 16}$ f. $\sqrt{25} - \sqrt{16}$

10. **Rental Income** Sara manages an office building. There are three offices on the first floor with areas (measured in square feet; ft^2) of 400 ft^2, 320 ft^2, and 850 ft^2. The two offices on the second floor are each 875 ft^2. The monthly rent for a first floor office is $1.20 per square foot, and the monthly rent for a second floor office is $0.92 per square foot. Write one numerical expression to determine Sara's monthly rental income when all of the offices are rented. Include units in your expression. Evaluate your expression.

11. **Which Doesn't Belong** Each of the following sets contains five items. One of the five does not belong. Cross out the one that doesn't fit, and describe what the other four have in common.

a.	$6 \div 2 - 8$	$3(5) + 9$	$8 + 4$	$8 + 9 + 9$	$\frac{6}{3} + 3$
b.	$\frac{2}{4+9}$	$2 * 4 + 9$	$2\sqrt{4+9}$	$2 \div (4+9)$	$2(4+9)$
c.	term	factor	dividend	divisor	quotient
d.	πr^2	LW	$2\pi r$	$\frac{h(b_1 + b_2)}{2}$	$\frac{1}{2} bh$
e.	meter	inch	foot	acre	mile
f.	$\frac{8}{4}$	$4\overline{)6}$	$4 * 6$	$\frac{3}{5}$	$5 \div 1$
g.	circle	sphere	cylinder	cone	box
h.	$2 * 4$ in. $+ 9$ in.	$3 * 6$ cm	3 m $+ 12$ m $+ 4$ m	4 ft $* 8$ ft	4 cm $+ 3$ cm $+ 1$ cm $+ 6$ cm

1.2 Estimation

Discussion 1.2

It is often appropriate to estimate a numerical value for a problem. At times, an estimate is all that is necessary. For example, suppose you are shopping for groceries and want to be sure you have enough cash with you when it is time to check out. It would not be necessary to use the exact costs of each item you decide to buy. By rounding each price *up* to the nearest dollar amount you could estimate the total cost of the groceries. Because you are always rounding up, you are certain to have enough money to pay the cashier.

In other cases, exact measurements are not available or are very difficult to obtain. Suppose you needed to resurface a gravel driveway. Using your truck's odometer (not a very accurate measuring tool) you find the driveway to be about three quarters of a mile long. The driveway is just a little wider than your small pickup, which you estimate to be about 6 feet wide. By multiplying the length of the driveway, the width of the driveway, and the depth of the gravel, you could determine the amount of gravel to purchase. Even though you may use a calculator to determine the amount of gravel, the measurements are estimated.

In many practical situations, such as the driveway example, an exact value is not important and an estimate will do. At other times we want an exact value and will be using a calculator or computer to do the calculations. We would then like a method of verifying the calculator results without having to do the entire problem by hand.

The following strategy can be used to estimate a numerical expression. When we are verifying a calculator entry, this process must be done without entering the expression in the calculator. Although the estimation must be done without using a calculator, it is usually helpful to write out the steps.

Strategy

Estimating a Numerical Expression

1. Round all numbers so that the arithmetic can be done easily without a calculator.
   a. When adding or subtracting, round numbers to the same place value.
   b. When multiplying, round numbers to the leading digit. When we have a factor between -1 and 1, we need to round the number to a reasonable *fraction* or *single-digit decimal*. Otherwise, the estimate may be extremely far from the actual value of the expression.
   c. When dividing, round the numbers so that the division is easy to perform.

2. Perform the calculations in each step mentally, continuing to round as needed so the mental arithmetic is easy.

Example 1

Estimate the following expressions.

a. $45{,}671 + 7180$ b. $45{,}671 * 7180$ c. $\dfrac{45{,}671}{7180}$

Solution

a. $45{,}671 + 7180$
  $\approx 46{,}000 + 7000$ Round each term to the nearest thousand, and add mentally.
  $= 53{,}000$

b. $45{,}671 * 7180$
  $\approx 50{,}000 * 7000$ Round each factor to the leading digit, and multiply mentally.
  $= 350{,}000{,}000$

c. $\dfrac{45{,}671}{7180}$
  $\approx \dfrac{42{,}000}{7000}$ Round to numbers that are easy to divide mentally.
  $= 6$

NOTE: Notice the use of the symbols $\approx$ and $=$. The symbol $\approx$ is used when the expression is approximately equal to the previous step. The symbol $=$ is used when the expression is exactly equal to the previous step.

In the three expressions in the previous example, the first number 45,671 is rounded differently each time. In the sum $45{,}671 + 7180$, each number is rounded to the same place value. Because the leading digit in the first number is in the ten-thousands place and the leading digit in the second number is in the thousands place, both numbers are rounded to the nearest thousand. Therefore, we rounded 45,671 to 46,000 in the sum.

In the product $45{,}671 * 7180$, both numbers are rounded to the leading digit to make the product easy to compute mentally. Therefore, 45,671 is rounded to 50,000 in the product.

In the quotient, 45,671 was rounded to a number that was easily divisible by 7000. Note that there is not a "best" way to round for everyone. In part a of Example 1, we could have estimated the sum by rounding each number to the nearest ten thousand, in which case we would obtain the following estimate:

$45{,}671 + 7180$
$\approx 50{,}000 + 10{,}000$
$= 60{,}000$

The estimates would then be quite different. So how much can we tell from an estimate if there is more than one way to round our work? In this case we would be safe in saying that the exact sum is in the ten thousands and somewhere between 50,000 and 60,000.

Example 2 For the numerical expression at the end of these instructions, do the following.

a. Estimate the value of the expression. (It may be helpful to underline the terms in the expression.)
b. Evaluate the expression in one step using a calculator. Round results to the nearest whole number.
c. Compare your estimate with your calculated value.

$$0.63 * 19^3 - \frac{4607 - 8210}{2}$$

Solution a. First round each number so that the arithmetic can be done easily.

$$0.63 * 19^3 - \frac{4607 - 8210}{2}$$

$$\approx \frac{1}{2} * 20^3 - \frac{4600 - 8200}{2}$$
Because 0.63 is a factor between $^-1$ and 1, we need to round 0.63 to a reasonable fraction or a single-digit decimal. In this case we will round to the fraction $\frac{1}{2}$.

$$= \frac{1}{2} * 8000 - \frac{^-3600}{2}$$

$$= 4000 - {^-1800}$$

$$= 4000 + 1800$$

$$= 5800$$

Let's see how our estimate would change if we round 0.63 to 0.6.

$$0.63 * 19^3 - \frac{4607 - 8210}{2}$$

$$\approx 0.6 * 20^3 - \frac{4600 - 8200}{2}$$
Because 0.63 is a factor between $^-1$ and 1, we need to round 0.63 to a reasonable fraction or a single-digit decimal. In this case we will round to 0.6.

$$= 0.6 * 8000 - \frac{^-3600}{2}$$

$$= 4800 - {^-1800}$$
Because $6 * 8 = 48$, $0.6 * 8000 = 4800$.

$$= 4800 + 1800$$

$$= 6600$$

Depending on how we round our fractional factor, we obtain an estimate of 5800 or 6600.

b. For this part, we need to enter the expression in one step on our calculator. The second term of the original expression is a quotient written in fraction form as

$$\frac{4607 - 8210}{2}$$

Because many calculators will not accept entries of this form, we must rewrite this quotient in a horizontal format using the division operator. Because the fraction bar of the quotient groups the numerator and denominator, we will need to include parentheses to accomplish this same grouping. To enter the expression on our calculator we will type something similar to

$$0.63 * 19^3 - (4607 - 8210) \div 2$$

In doing so we will obtain ≈ 6123.

c. Our estimate of the expression is 5800 or 6600, depending on how we round our fractional factor, and our calculated value is about 6123. These values are close, which indicates that our calculator entry is probably correct.

Why is it important to round fractional factors to fractions or decimals rather than to whole numbers? Consider the first term in Example 2, $0.63 * 19^3$.

If we round 0.63 to the reasonable fraction $\frac{1}{2}$, we obtain

$$0.63 * 19^3$$
$$\approx \frac{1}{2} * 20^3$$
$$= \frac{1}{2} * 8000$$
$$= 4000$$

If we round 0.63 to a single-digit decimal, we obtain

$$0.63 * 19^3$$
$$\approx 0.6 * 20^3$$
$$= 0.6 * 8000$$
$$= 4800$$

If we round 0.63 to a whole number, we obtain

$$0.63 * 19^3$$
$$\approx 1 * 20^3$$
$$= 1 * 8000$$
$$= 8000$$

The exact value of $0.63 * 19^3$ is 4321.17. We can see that by rounding the fractional factor to a whole number our estimate is almost twice the exact value. Rounding fractional factors to fractions or decimals will therefore result in more reasonable estimates.

Example 3

A team of students are evaluating a numerical expression. None of the results agree! Checking their calculator entries they found that all three had entered the expression differently, but they could not decide on whose entry was correct. Estimate the expression to help decide which student's answer is correct.

Student 1 $\quad \dfrac{4.99(41.3 + 16.7)}{2.21 * 3.60} \approx 471.5$

Student 2 $\quad \dfrac{4.99(41.3 + 16.7)}{2.21 * 3.60} \approx 233.3$

Student 3 $\quad \dfrac{4.99(41.3 + 16.7)}{2.21 * 3.60} \approx 36.4$

Solution

$$\dfrac{4.99(41.3 + 16.7)}{2.21 * 3.60}$$
$$\approx \dfrac{5(40 + 20)}{2 * 4}$$
$$= \dfrac{5(60)}{8}$$
$$= \dfrac{300}{8}$$
$$\approx \dfrac{320}{8}$$
$$= 40$$

Note: The 300 was "rounded" to 320 so that the division can be done easily because $32 \div 8 = 4$.

Given the calculated results for the three students, our estimate of 40 allows us to see that the third student's result of 36.4 is the only reasonable choice.

In the second to last step in Example 3, we could have "rounded" the 8 to 10 to make the division easy to do mentally. This would have resulted in an estimate of

$$\dfrac{300}{8}$$
$$\approx \dfrac{300}{10}$$
$$= 30$$

Both 30 and 40 are good estimates of the expression and allow us to easily see that the correct calculated value is about 36.4.

In this section we saw that estimates can be used in practical situations where an exact value is not necessary. An estimate can also be used to verify exact calculator calculations. To estimate the value of an expression we first round the numbers so that we can do the arithmetic mentally. This means that the estimated values of an expression may differ depending on how the expression is rounded. The key points when estimating are to round the numbers so that the *arithmetic is easy to do mentally* and when rounding a factor that is between $^-1$ and 1, round to a reasonable fraction or single-digit decimal rather than to the nearest whole number.

Problem Set 1.2

1. Identify the terms in the expression.

   $$63.8 + 12.4\sqrt{22.3 + 28.1}$$

2. Identify the terms in the expression.

   $$2.3 + 3.4(4.7 + 6.7) + 1.99 * 5.03^2$$

3. Identify the subterms in the numerator of the expression.

   $$\frac{3.7 + 9.1 * 2}{\sqrt{\frac{84}{11.2}}}$$

4. For the expressions a through j do the following.

    i. Estimate the expression without using a calculator. Show your steps. (It may be helpful to underline the terms in the expression.)

    ii. Evaluate the expression in one step using a calculator. Round the calculator results to the nearest hundredth.

    iii. Check your estimate and calculator results. Are they close? If not, try to find any mistakes. A mistake can occur either in estimating or in calculator entry.

    a. $124.62 + 317.20 - 487.15 + 159.98$

    b. $\dfrac{2015.6}{680.4}$

    c. $\dfrac{3*57 + 7*88 + 4*62 + 4*98}{18}$

    d. $^-73.56 * 13.08 - 78.96 \div 2.45$

    e. $34 * 9.6^3 - \dfrac{4607 - 38{,}210}{3.2}$

    f. $63.8 + 12.4\sqrt{22.3 + 28.1}$

    g. $0.22\sqrt{83.4} - \dfrac{12.2}{2.7 * 4.1}$

    h. $2.3 + 3.4(4.7 + 6.7) + 1.99 * 5.03^2$

    i. $\dfrac{3.7 + 9.1 * 2}{\sqrt{\dfrac{84}{11.2}}}$

    j. $\dfrac{\sqrt{48.5 + 72.6}}{\dfrac{10.2}{68.4}}$

5. **a.** Underline the subterms in the numerator of the expression.

$$\frac{18.7 * 4.3 - 2.9\sqrt{65.4}}{0.47 * 23.4}$$

b. A team of students evaluated the expression

$$\frac{18.7 * 4.3 - 2.9\sqrt{65.4}}{0.47 * 23.4}$$

on their calculators. Their results did not agree, so they evaluated the expression again. They obtained a total of six different answers. Use your estimation skills to determine which of the following six answers is reasonable.

i. $\dfrac{18.7 * 4.3 - 2.9\sqrt{65.4}}{0.47 * 23.4} \stackrel{?}{\approx} 2835.76$ iv. $\dfrac{18.7 * 4.3 - 2.9\sqrt{65.4}}{0.47 * 23.4} \stackrel{?}{\approx} 31{,}207.90$

ii. $\dfrac{18.7 * 4.3 - 2.9\sqrt{65.4}}{0.47 * 23.4} \stackrel{?}{\approx} 78.28$ v. $\dfrac{18.7 * 4.3 - 2.9\sqrt{65.4}}{0.47 * 23.4} \stackrel{?}{\approx} -85.07$

iii. $\dfrac{18.7 * 4.3 - 2.9\sqrt{65.4}}{0.47 * 23.4} \stackrel{?}{\approx} 5.18$ vi. $\dfrac{18.7 * 4.3 - 2.9\sqrt{65.4}}{0.47 * 23.4} \stackrel{?}{\approx} -1087.22$

6. Give an example in which an estimate would be useful and exact calculation may not be needed (other than as a verification of a calculation).

7. Without using a standard measuring tool, estimate the area of the floor of your classroom and the volume of your classroom. Compare your team's results with those of another team. If your results are different, try to explain why.

8. In this problem you need to make several estimations. Be sure to explain how you made your estimates and any assumptions you made in order to answer the questions. No standard measuring tools should be used on this problem.

   a. Estimate the amount of blackboard area in your mathematics classroom.

   b. Suppose you have a sheet of tin foil with the same area as your result from part a. How many Hershey's chocolate kisses would you be able to cover?

   c. Estimate the volume of chocolate needed to make the number of Hershey's chocolate kisses from part b.

9. **a.** Using your ruler, draw a rectangle with an area of 1 in^2. *Note:* A square is a special type of rectangle.

   b. Using your ruler, draw a rectangle with an area of 3 in^2.

   c. Using your ruler, draw a rectangle with an area of (3 in.)^2.

10. The area of the following rectangle is 6 square centimeters (cm^2).
    a. Use the area of the rectangle to *estimate* the area of the circle.
    b. Use the area of the rectangle to *estimate* the area of the triangle.
    c. Measure the circle, and determine its area to the nearest square centimeter. How does your calculated value compare with your estimate?
    d. Measure the triangle, and determine its area to the nearest square centimeter. How does your calculated value compare with your estimate?

6 cm^2

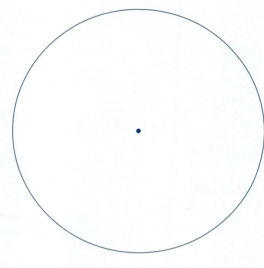

1.3 Approximate Numbers and Error

Activity Set 1.3

1. The figure shows four magnified metric rulers that can be used to measure pencils A, B, C, and D. Use the figure to determine the measurements of pencils A, B, C, and D to the nearest millimeter (mm).

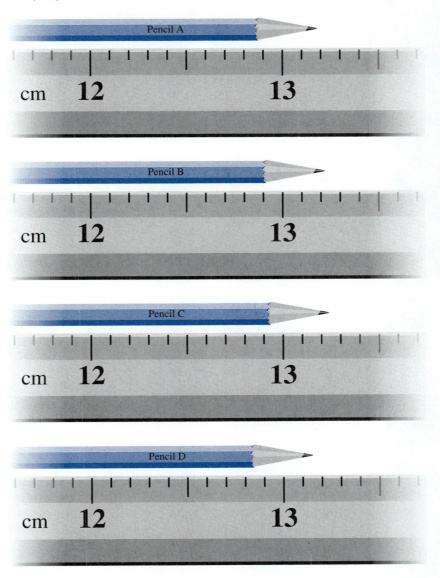

2. a. Are pencils A, B, C, and D exactly the same length?
   b. Did all of the pencils have the same measurement to the nearest millimeter?
   c. Explain any discrepancies between your answers to parts a and b.

3. a. What is the shortest an object can be and still measure 132 mm when measured to the nearest millimeter?
   b. What is the longest an object can be and still measure 132 mm when measured to the nearest millimeter?
   c. If you know that the measurement of an object to the nearest millimeter is 132 mm, then how much longer than 132 mm *can* the object be? How much shorter than 132 mm *can* the object be?

4. **a.** What is the shortest an object can be and still measure 13.2 cm when measured to the nearest tenth of a centimeter?

   b. What is the longest an object can be and still measure 13.2 cm when measured to the nearest tenth of a centimeter?

   c. If you know that the measurement of an object to the nearest tenth of a centimeter is 13.2 cm, then how much longer than 13.2 cm *can* the object be? How much shorter than 13.2 cm *can* the object be?

5. Determine the perimeter and area of the following triangle by making the appropriate measurements. Measure all lengths to the nearest *tenth of a centimeter*.

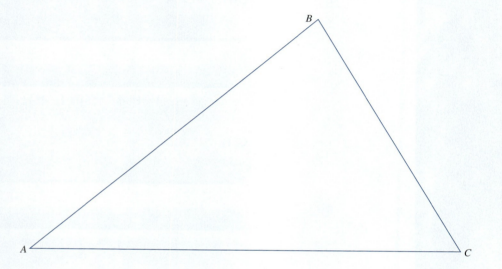

6. Make the appropriate measurements of the following circle to determine its circumference and area. Make all measurements to the nearest *millimeter*.

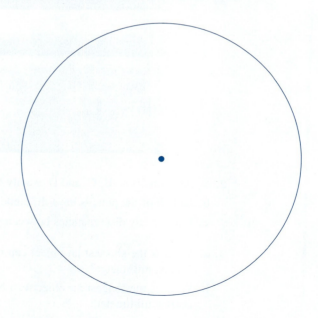

7. Make the appropriate measurements of the following figure to determine its perimeter and area. Make all measurements to the nearest tenth of a centimeter.

Discussion 1.3

Approximate Numbers

We usually think of numbers as exact quantities. When we read the number 13.2, we assume this means exactly thirteen and two tenths. However, in the activities, we saw that all of the pencils were of different lengths, but we called them all 13.2 cm when measuring to the nearest tenth of a centimeter. In this context the number 13.2 is clearly not exact because it can represent any length from 13.15 to 13.25 cm.

Activity 5 asked you to measure the sides of the triangle *ABC*. You probably noticed that your measurements did not match all of the other students' measurements. Your results for the perimeter and area and those of other students would therefore most likely differ. The measurement of side *AC* may have ranged from 11.3 to 11.5 cm. We certainly cannot call these results exact because they do not agree. In this section we will discuss numbers in context. We will need to decide from the context whether a number is exact or not. We begin our discussion with the following definition.

Definition
Exact numbers are defined as those numbers with no uncertainty.

Exact numbers occur in *counting* and in *definitions*. We know, for example, that a car has exactly four tires because we can count them. By definition, a minute has exactly 60 seconds (sec), and an inch has exactly 2.54 cm. Your class may include exactly 35 students; we know this because the number can be counted. If I report that the population of Gresham is about 72,200, this number is an estimate and, therefore, not exact. **Approximate numbers** occur whenever a number represents a *measured* or an *estimated* quantity. When solving problems involving approximate numbers we must *round our results reasonably*. Before we begin this discussion it is helpful to introduce some vocabulary.

Precision

We have seen that precision can be determined by your measuring tool. If you are recording measurements using a tool that measures to a tenth of a milliliter (mL), your list may look something like the following:

34.5 mL, 20.4 mL, 12.0 mL, and 0.7 mL

Notice that the third number is written as 12.0 mL; the zero in the tenths place is not dropped. By recording the number as 12.0 mL, you are communicating the precision of the measurement.

Significant Digits

In our discussion of approximate numbers we need to know both the precision of a number and how many significant digits are in a number. Significant digits are those digits that represent measured or counted amounts. The digits 1, 2, . . . , 9 are always significant, and the digit zero is sometimes significant. We will look at several examples to see when zeros are and are not significant.

- *All zeros lying between significant digits are considered significant.* For example, the number 103 has three significant digits because the number 0 in the tens place represents exactly 0 tens.

  8003 has four significant digits.

  540.008 has six significant digits.

- *Trailing zeros in numbers that do not contain a decimal are not significant unless specifically indicated.* If I claim that my tax refund this year will be about $500, the zeros are holding place values. I am not claiming that my tax refund will be exactly $500, but that it will be close to $500. The zeros in the tens and units place are not known exactly; they are merely place holders. Because the zeros in the number 500 are not measured quantities, they are not significant.

  500 has one significant digit.

  37,000 has two significant digits.

  10,500 has three significant digits.

- *All trailing zeros to the right of the decimal point are significant.* The numbers 0.5 inch, 0.50 inch, and 0.500 inch all represent one-half of an inch. We cannot drop the trailing zeros, however, because they tell us how precise the measurements are. The first measurement, 0.5 inch, is only precise to the nearest tenth of an inch; however, the number 0.500 is precise to a thousandth of an inch. When we write 0.500 inch we know that 5 tenths, 0 hundredths, and 0 thousandths of an inch were measured. Therefore, 0.500 inch has three significant digits. The number 0.5 in. has only one significant digit.

  0.5 has one significant digit.

  0.50 has two significant digits.

  0.500 has three significant digits.

  39.00 has four significant digits.

  60400.00 has seven significant digits.

- *Leading zeros to the right of the decimal point are not significant.* Suppose you measure a very small quantity in a laboratory as 25 milligrams (mg). Clearly, 25 mg has two significant digits. We could convert 25 mg to 0.025 grams (g) or to 0.000025 kilograms (kg). The quantities 0.025 g and 0.000025 kg should also have two significant digits, the same as our original measurement. In the numbers 0.025 g and 0.000025 kg, the leading zeros to the right of the decimal point are not significant. The zeros are merely holding place values relative to the units we select for the number and are not considered significant.

    0.057 has two significant digits.

    0.00057 has two significant digits.

    0.00903 has three significant digits.

    0.04070 has four significant digits.

Note that 0.04070 is numerically equal to 0.0407. This means that we can write 0.04070 without the last zero; therefore, if we include the last zero when we write 0.04070, it *must* be significant.

NOTE: 0.04070 can also be written as .04070 and still is considered to have four significant digits. The zero to the left of the decimal point is to help you see the decimal point.

Suppose that in measuring a large sheet of aluminum, we determine the dimensions to be 304 cm by 500 cm. If our measuring device is precise to the nearest centimeter, the number 500 should be read as having three significant digits rather than one. We can indicate that the trailing zeros in 500 are significant in three ways.

1. Because the zero in the ones place is significant, we can place a decimal point after this zero. Therefore, 500. is considered to have three significant digits.

    500. has three significant digits.

2. We can place a bar over the last significant digit. So, $50\bar{0}$ is considered to have three significant digits.

    $50\bar{0}$ has three significant digits.

3. We can also use **scientific notation.** A number written in scientific notation is written as a product of a number between 1 and 10 and a power of 10. For example, if we want to have three significant digits, 500 could be written as $5.00 * 10^2$. In this notation, the trailing zeros are significant, and therefore $5.00 * 10^2$ has three significant digits.

    $5.00 * 10^2$ has three significant digits.

Example 1

Determine the number of significant digits in the following numbers.

a. 209 cm   c. 0.008 km   e. 3070. m

b. 17.00 ft   d. 0.03010 m

Solution

a. The digits 2 and 9 are significant. Any zero between these numbers is significant; therefore, 209 cm has three significant digits.

b. The digits 1 and 7 are significant. Trailing zeros in numbers that contain a decimal are significant. Therefore, 17.00 ft has four significant digits.

c. The digit 8 is significant. The leading zeros are not considered significant because they are merely place holders; therefore 0.00<u>8</u> km has one significant digit.

d. The digits 3, 1, and the 0 between are significant. The 0 to the right of 1 denotes the precision of the measurement and is significant. The zeros to the left of 3 are place holders and are not significant; therefore, 0.0<u>3010</u> m has four significant digits.

e. Notice the decimal point at the end of the number. This indicates that the 0 in the units place is significant. All of the digits between 3 and the 0 in the units place are significant; therefore, <u>3070</u>. m has four significant digits.

Definition

A number written in **scientific notation** is written as the product of two factors. The first factor is a number greater than or equal to one and less than ten. The second factor is a power of ten.

Writing a Number in Scientific Notation

The decimal point in the first factor always occurs immediately to the right of the leftmost nonzero digit. (This is the first significant digit when we read digits from left to right.) All of the digits of the first factor of a number written in scientific notation are significant. This means that if a number has four significant digits, the first factor of this number in scientific notation will have exactly four digits.

The first factor of a number written in scientific notation is a number greater than or equal to one and less than ten. The second factor is a power of ten that makes the number written in scientific notation equal to the number when it is written in ordinary notation.

Example 2

Write the following numbers in scientific notation.

a. 85,000   b. 85,0$\overline{0}$0   c. 0.00004730

Solution

a. To write 85,000 in scientific notation, we must first determine the number of significant digits. Because 85,000 has only two significant digits, we write only two digits in the first factor. The decimal point in the first factor is always placed to the right of the first significant digit, in this case to the right of the digit 8. The first factor is written as 8.5, and we then need to move the decimal point four places to the right to obtain 85,000. Therefore, we multiply 8.5 by 10^4.

$$85000 = 8.5 * 10^4$$
<u>4 places</u>

Therefore, 85,000 written in scientific notation is $8.5 * 10^4$.

b. For 85,0$\bar{0}$0 to be written in scientific notation, the first factor must have four significant digits. Remember, a bar over a zero denotes that zero as a significant digit. The first factor is written as 8.500, and we then move the decimal point four places to the right to obtain 85,000. Therefore, we multiply 8.500 by 10^4.

$$850\underbrace{00}_{4 \text{ places}} = 8.500 * 10^4$$

Therefore, 85,0$\bar{0}$0 written in scientific notation is $8.500 * 10^4$.

c. To write 0.00004730 in scientific notation, we first determine that the number has four significant digits (0.0000$\underline{4730}$). The decimal point is placed to the right of the first significant digit, so the first factor is written as 4.730. To obtain 0.00004730, we then move the decimal point five places to the left. We therefore multiply 4.730 by 10^{-5}.

$$0.0\underbrace{00004}_{5 \text{ places}}730 = 4.730 * 10^{-5}$$

Therefore, 0.00004730 written in scientific notation is $4.730 * 10^{-5}$.

Notice that when a number greater than ten is written in scientific notation, the power of ten is positive and when the number is less than one, the power of ten is negative.

> **Definition**
>
> The **precision** of a measurement, provided the measuring tool is partitioned in powers of ten, is the place value of the rightmost significant digit.

If a measurement is given as a fraction, it is necessary to know the precision of the measuring tool. We cannot determine the level of precision from how the number is written. For example, a measurement given as $3\frac{1}{2}$ inches does not tell us that the precision of the measurement is to the nearest half of an inch. A measuring tool precise to the nearest eighth or sixteenth of an inch or better could have been used, and we would still record the measurement as $3\frac{1}{2}$ inches.

Example 3

Assuming that the following numbers are written to indicate precision, what is the precision of each number?

    a. 29 cm   b. 36,000 gal   c. 14.00 in.   d. 0.007 L

Solution

a. Because the rightmost significant digit is in the units place, 29 cm is precise to the nearest centimeter.

b. Because the rightmost significant digit is in the thousands place, 36,000 gallons (gal) is precise to the nearest thousand gallons.

c. Because the rightmost significant digit is in the hundredths place, 14.00 in. is precise to the nearest hundredth of an inch.

d. Because the rightmost significant digit is in the thousandth place, 0.007 liter (L) is precise to the nearest thousandth of a liter.

Example 4

a. Round 89.99719 to the nearest ten.
b. Round 89.99719 to the nearest whole number.
c. Round 89.99719 to the nearest tenth.

Solution

a. The number 89.99719 falls between 80 and 90. (These are the closest values measured to the tens place.) It is closer to 90, so

$$89.99719 \approx 90 \text{ rounded to the nearest ten}$$

b. The number 89.99719 falls between 89 and 90. (These are the closest values measured to the units place.) It is closer to 90., so

$$89.99719 \approx 90. \text{ rounded to the nearest whole number}$$

Notice that we must include the decimal point in 90. to indicate that this measurement is rounded to the nearest whole number, rather than the nearest ten.

c. The number 89.99719 falls between 89.9 and 90.0. (These are the closest values measured to the tenths place.) It is closer to 90.0, so

$$89.99719 \approx 90.0 \text{ rounded to the nearest tenth}$$

The zero to the right of the decimal is significant because it informs us that the number is rounded to the tenths place.

Error in Measurement

Earlier we said that approximate numbers occur when a number represents a measured quantity. If we are computing values based on these approximate numbers, it is important to round our results reasonably. We have seen that the precision of a number written in decimal form can be determined by how it is written. So how should we round our results so that the final answer is reasonable? To help answer this question we will look at some examples. First, suppose that you were asked to measure the metal rod shown here using the measuring tool provided. (The ruler is magnified.)

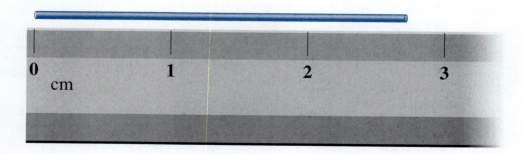

We could say that the rod measures 2.7 cm, 2.8 cm, or possibly 2.75 cm. We are guessing the length of the rod. The only thing we can say for certain (using the ruler shown) is that the rod is between 2 and 3 cm and it is closer to 3 cm. We can therefore say that the rod is 3 cm, measured to the nearest centimeter. The measuring tool allows us to measure with certainty to the nearest centimeter. Let's look at the same rod measured with the next ruler shown. This ruler divides each centimeter into ten equal parts. With this ruler, we can measure with certainty to the nearest tenth of a centimeter.

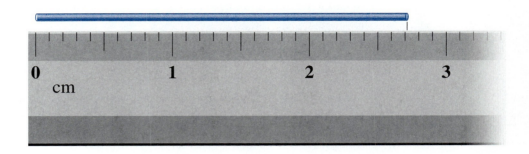

The rod measures between 2.7 cm and 2.8 cm, and it is closer to 2.7 cm. We would therefore say that the rod measures 2.7 cm to the nearest tenth of a centimeter.

Note that the measuring tool determines the way we report the length of the rod, that is, to the nearest centimeter or tenth of a centimeter. We call this the precision of the measuring tool.

As you saw in the activities, when the measurement of an object is recorded as 13.2 cm, the object might have been slightly smaller or slightly larger than 13.2 cm. However, you could be sure that the object's actual length would fall between 13.15 cm and 13.25 cm. Another way to say this is that the actual length of the object can differ from 13.2 cm by at most 0.05 cm or that the error in the measurement is at most ±0.05 cm. In general, the actual length of an object can differ from the approximate value by at most *half of the precision of the measurement*. This means that the error of any measurement is within ±*half of the precision of the measurement*.

Example 5

Given the following approximate measurements, what is the smallest value possible for the actual measurement? What is the largest?

a. 18 cm   b. 36.2 mm   c. 14.0 sec   d. 0.005 mL

Solution

a. Because 18 cm is measured to the nearest centimeter, the error is within ±0.5 cm.

   18 cm − 0.5 cm = 17.5 cm
   18 cm + 0.5 cm = 18.5 cm

   The smallest possible value for the actual length is 17.5 cm, and the largest possible value for the actual length is 18.5 cm.

b. Because 36.2 mm is measured to the nearest tenth of a millimeter, the error is within ±0.05 mm.

   36.2 mm − 0.05 mm = 36.15 mm
   36.2 mm + 0.05 mm = 36.25 mm

   The smallest possible value for the actual length is 36.15 mm, and the largest possible value for the actual length is 36.25 mm.

c. Because 14.0 sec is measured to the nearest tenth of a second, the error is within ±0.05 sec.

   14.0 sec − 0.05 sec = 13.95 sec
   14.0 sec + 0.05 sec = 14.05 sec

   The smallest possible value for the actual time is 13.95 sec, and the largest possible value for the actual time is 14.05 sec.

d. Because 0.005 mL is measured to the nearest thousandth of a milliliter, the error would be within ±0.0005 mL.

$$0.005 \text{ mL} - 0.0005 \text{ mL} = 0.0045 \text{ mL}$$
$$0.005 \text{ mL} + 0.0005 \text{ mL} = 0.0055 \text{ mL}$$

The smallest possible value for the actual volume is 0.0045 mL, and the largest possible value for the actual volume is 0.0055 mL.

Example 6

We encounter the following measurements every day. How precise are they, and what error would we expect for each of the measurements?

  a. outside temperature
  b. a person's body temperature
  c. number of textbooks in a student's backpack
  d. the length and width of a room

Solution

a. Outside temperature is usually measured to the nearest degree, so we would say the temperature is precise to the nearest degree (°). The error is ±0.5°.

b. A person's body temperature is usually measured to the nearest tenth of a degree. The error is ±0.05°.

c. The number of textbooks in a student's backpack will result in a whole number. But is it reasonable to say that the precision is a whole number of textbooks and the error is ±0.5 textbook? Of course not! The number of textbooks is countable and therefore is an exact number. We do not say that an exact number has a degree of precision or error.

d. It is reasonable to assume that the length and width of a room are measured to the nearest inch, and therefore the error is ±0.5 in.

When working with approximate numbers in computations, we need a set of guidelines to help us determine where to round the results of our computations. Can we just round our results anywhere, or is there a reasonable place to round to? Does it make sense not to round at all when we calculate using approximate numbers? To answer these questions, we revisit Activity 5 using results from one class.

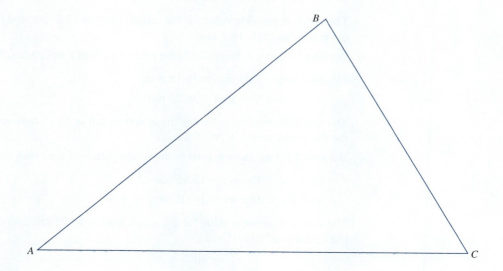

1.3   Approximate Numbers and Error

MEASUREMENTS RECORDED			
Side AB (cm)	Side BC (cm)	Side AC (base) (cm)	Height (cm)
9.9	7.4	11.3	6.5
10.0	7.4	11.4	6.4
10.0	7.5	11.4	6.5
9.9	7.3	11.5	6.4
9.8	7.3	11.3	6.3
9.9	7.4	11.4	6.3

You can see from the table that the students do not all agree on their measurements. Because the numbers in the table were measured, the numbers and any computations using these numbers are approximate. How much variability in measurements and the resulting computations of perimeter and area should we expect? The basic rule for approximate numbers is that some disagreement can occur in the last decimal place only. The numbers in the table are measured to the nearest tenth of a centimeter. Looking at the table, we see that the digits in the tenths place vary, but most of the digits in the units place agree. You may notice, however, that because side AB is very close to 10.0 cm, it varies between 9.8 and 10.0 cm and therefore does not agree in the units place. Notice that 10 cm is recorded as 10.0 cm, so that it is clear that this measurement is also measured to the nearest *tenth* of a centimeter.

Notice that in side BC

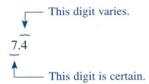

> *Approximate Numbers Basic Rule*
> Some disagreement can occur in the last digit of any approximate number. All other digits are reliable.

The following table looks at perimeter computations for triangle ABC.

Perimeter (cm)
28.6
28.8
28.9
28.7
28.4
28. 7

To decide where to round the perimeter, we begin reading the numbers from the left. Where do the numbers agree? At what place value do we find disagreement in the numbers?

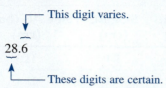

We can see that all of the perimeter results match in the tens and units place. In the tenths place, the numbers differ. Because the basic rule for approximate numbers allows them to disagree in the last digit, it is reasonable to keep all of the perimeter results to a tenth of a centimeter. You can see that we wrote the perimeter results to the same precision as the original measurements. In general, when we are adding (or subtracting) measurements, we round our results to the same place value as the least precise measurement. In determining the perimeter of this triangle we need to add the measures of the sides; therefore, rounding to the same precision as the measurements makes sense.

> **Adding and Subtracting Guideline for Rounding Approximate Numbers**
>
> When adding or subtracting approximate numbers, round the results to the same place value as the least precise measurement.

Next, we look at area computations for triangle *ABC*.

Area (cm^2)
36.725
36.48
37.05
36.8
35.595
35.91

The results for area computations are not consistent concerning which decimal place should be considered the last. Depending on the students' measurements, the results ended in tenths, hundredths, or thousandths. Beginning from the left, the numbers all agree in the tens place but differ in the units place. Using the basic rule for approximate numbers, we round our results to the units place.

Original Area Results (cm^2)	Rounded to Nearest Square Centimeter (cm^2)
36.725	≈ 37
36.48	≈ 36
37.05	≈ 37
36.8	≈ 37
35.595	≈ 36
35.91	≈ 36

It may seem odd that we did not round our results to a tenth of a square centimeter because the original measurements were precise to a tenth of a centimeter. If we draw a square that measures 1 cm² (as in the following figure), it seems reasonable that our area computations can only be accurate to 1 cm², not to 0.1 cm².

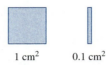

1 cm²   0.1 cm²

We saw that the precision of the original measurements was not helpful in determining where to round our area results. You may have noticed, however, that the number of significant digits for each of the rounded areas is the same as the number of significant digits in most of the original measurements. In general, when we are multiplying (or dividing) measurements, we round our results to the same number of significant digits as the measurement with the least number of significant digits. In determining the area of this triangle, we needed to multiply the base and height; therefore, rounding to the same number of significant digits as the measurement with the least number of significant digits works.

> *Multiplying and Dividing Guideline for Rounding Approximate Numbers*
>
> When multiplying or dividing approximate numbers, round results to the same number of significant digits as the measurement with the least number of significant digits.

NOTE: In the formula for the area of a triangle ($A = \frac{1}{2}bh$), $\frac{1}{2}$ is an exact number, not a measurement. You should *not* round the area of the triangle to one significant digit because of this factor.

In this class, you will be expected to round results of application problems reasonably. We see from the previous example that looking at the place value and the number of significant digits in a number can help us decide what is reasonable. If only one operation occurs in an expression, selecting the appropriate guideline (place value or number of significant digits) is easy. When expressions become more complex, it is sometimes more difficult to pick which guideline to follow. In geometrical applications, when we are computing a perimeter, we follow the first guideline for adding approximate numbers summarized in the following box. If we are computing area or volume, we follow the second guideline for multiplying approximate numbers. For other applications, if both guidelines 1 and 2 might apply, perform the calculation in one step using a calculator and follow guideline 2 when rounding your result. For expressions that include powers or roots, follow the guideline for multiplication and division. In all cases, ask yourself if the rounded result is reasonable given the context of the problem.

> *Summary of Guidelines for Rounding Approximate Numbers*
>
> 1. When adding or subtracting approximate numbers, round results to the same place value as the least precise measurement.
> 2. When multiplying or dividing approximate numbers, round results to the same number of significant digits as the measurement with the least number of significant digits.

In the next example, we will look at a more precise method to determine where to round calculations using approximate numbers. Because this is a different method for determining how to write a result, we will not apply the guidelines. We will want to compare the results from this new method to the results we would get if we were using the guidelines.

Example 7

The radius of the circular plate shown here is measured to the nearest thousandth of an inch.

a. Using the measurement given, calculate the area of the plate.
b. Using the smallest possible value for the actual radius, calculate the minimum value for the area of the plate.
c. Using the largest possible value for the actual radius, calculate the maximum value for the area of the plate.
d. Based on your results for b and c, round your answer to part a reasonably.

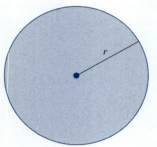

Radius = 0.912 in.

Solution

a. Area = πr^2

$A \approx \pi * (0.912 \text{ in.})^2$

$A \approx 2.61300084 \text{ in.}^2$

The area of the plate appears to be approximately 2.61300084 square inches; however, this is clearly not rounded reasonably.

b. The radius is measured to the nearest thousandth of an inch, which means that the error is at most ±0.0005 in. The minimum value for the radius could be 0.9115 in. Using this value we can calculate the minimum value for the area.

$A \approx \pi * (0.9115 \text{ in.})^2$

$A \approx 2.610136493 \text{ in.}^2$

The minimum value for the area is approximately 2.610136493 square inches.

c. We found that the error is at most ±0.0005 in., so the maximum value for the radius could be 0.9125 in. Using this value we can calculate the maximum value for the area.

$A \approx \pi * (0.9125 \text{ in.})^2$

$A \approx 2.615866758 \text{ in.}^2$

The maximum value for the area is approximately 2.615866758 square inches.

d. We can see that some difference exists among the results of parts a, b, and c. If we compare the numbers by reading from the left, we see that they agree in the units place, the tenths place, and the hundredths place. The numbers differ in the thousandths place. Therefore we round all of the results to the thousandths place.

From part a $A \approx 2.61300084$ in.$^2 \approx 2.613$ in.^{2}
From part b $A \approx 2.610136493$ in.$^2 \approx 2.610$ in.^{2}
From part c $A \approx 2.615866758$ in.$^2 \approx 2.616$ in.^{2}

You may notice that this is one more significant digit than 0.912 in., our original measurement. Therefore, if we had followed the guidelines for rounding approximate numbers for multiplication, we would have come to a slightly different conclusion. This is why we do not refer to the guidelines as *rules*. If answers are reasonable, they will be acceptable. The area could reasonably be rounded to the thousandths place using error analysis or the hundredths place using the guidelines. Either answer is acceptable. Because error analysis is time-consuming and difficult, we will rely on the guidelines and common sense to round reasonably.

Example 8

Martin has just completed 25 laps of the college's 50-m pool in 30 min. Determine how far Martin swam.

Solution   First we will determine how far Martin swam using unit conversions.

$$25 \text{ laps} * \frac{50 \text{ m}}{1 \text{ lap}} = 1250 \text{ m}$$

The result of our computation is 1250 m, but should we round the result because this is an application involving measured quantities? If we follow the guidelines, we would round to just one significant digit because, as it is written, 50 m has only one significant digit. But is this reasonable? We need to ask ourselves whether a 50-m pool used in swimming competitions would be measured precisely to the nearest 10 m? Clearly the pool is measured more precisely than this. We will assume that the pool is measured to the nearest centimeter or hundredth of a meter. This means that the length of a 50-m pool would be written as 50.00 m if we are indicating the precision and number of significant digits. Because 25 laps is an exact number and 50.00 m has four significant digits, we would say that Martin swam 1250. m.

In most technical and scientific fields, the precision of numbers is extremely important, and they are reported so as to make the precision and number of significant digits obvious. As we saw in the previous example, many of our everyday applications are not as clear about the degree of precision on measured quantities. In all cases, we must decide what is reasonable when rounding results. If you are unsure of the degree of precision in an application, write down the assumptions you make when rounding your results.

Example 9

Each of the following numbers appears to have just one significant digit. Rewrite the following measured quantities with an appropriate number of significant digits.

a. 2 L Coke   b. a living room is 20 ft long

Solution   a. We assume that a 2-L bottle of Coke is measured to a hundredth of a liter, so we should write this quantity as 2.00 L.

b. A living room is probably measured to the nearest inch. An inch is approximately equal to a tenth of a foot. We should write this number as 20.0 ft.

Numbers in applications of mathematics can be **exact** or **approximate.** Any number that represents a measured quantity is approximate. The **precision** of an approximate number is determined by the tool used to make the measurement. It is important to understand that error exists when we work with approximate numbers. It is possible to determine the minimum and maximum value for any quantity calculated using approximate numbers. However, this method is time-consuming, and we will not use this method on every problem in this course. Many disciplines have created guidelines for rounding approximate numbers, but these guidelines can change from discipline to discipline. You have encountered one such set of guidelines in this section. When problems are presented in context, you should always make an attempt to round final results reasonably, but this can often be done by using common sense!

Problem Set 1.3

1. Determine whether the numbers given are exact or approximate. Explain your response.
   a. The Santos own 1.54 acres of land.
   b. There are 26 letters in the alphabet.
   c. There are 16 oz of Coke in a bottle.
   d. There are 47 keys on a HP-38G calculator.
   e. The Busicks' deck is 16 ft wide.

2. Given the following approximate measurements, what is the smallest value possible for the actual measurement? What is the largest?
   a. 14 m   b. 3.02 cm   c. 6 miles   d. 12.0 mm

3. List at least five different everyday measuring tools and what they typically measure. What is the degree of precision of the tools on your list?

4. If a measurement is recorded as $3\frac{1}{2}$ in. to the nearest half of an inch, what is the smallest value possible for the actual measurement? What is the largest?

5. If a measurement is recorded as $5\frac{3}{8}$ in. to the nearest eighth of an inch, what is the smallest value possible for the actual measurement? What is the largest?

6. Determine the number of significant digits in the following numbers.
   a. 0.00801   b. 78,000   c. 503,000,000   d. 0.00005600

7. Determine the number of significant digits in the following numbers.
   a. 4509   b. 4500   c. 0.073   d. 0.010010

8. Write the following numbers in ordinary notation.
   a. $1.713 * 10^5$ b. $7.800 * 10^2$ c. $1.041 * 10^{-6}$ d. $3.8000 * 10^{-3}$

9. Which of the following numbers are written in scientific notation?
   a. $0.895 * 10^3$ b. $4.06 * 10^{-5}$ c. $53 * 10^{12}$ d. $3 * 10^{-6}$

10. Write the following numbers in scientific notation.
    a. 0.00801   b. 78,000   c. 503,000,000   d. 0.00005600

11. Write the following numbers in scientific notation.
    a. 505,000   b. 0.0040   c. 4500.   d. 0.000566

12. Which of the following numbers have the same value?
    a. $5 * 10^3$ b. $5 * 10^2$ c. $50 * 10^2$ d. $0.5 * 10^2$ e. $0.5 * 10^4$

13. Which of the following numbers have the same value?
    a. $0.27 * 10^{-4}$
    b. $2.7 * 10^{-5}$
    c. $2.7 * 10^{-3}$
    d. $27 * 10^{-6}$
    e. $0.027 * 10^{-3}$
    f. 0.000027

14. List the following numbers in order from smallest to largest.
    a. $2.19 * 10^3$
    b. $1.08 * 10^4$
    c. $7.98 * 10^2$
    d. $5.46 * 10^3$
    e. $9.90 * 10^{-4}$

15. List the following numbers in order from smallest to largest.
    a. $4.1 * 10^6$
    b. $2.005 * 10^2$
    c. $2.7 * 10^2$
    d. $5.6 * 10^{-4}$
    e. $4.15 * 10^{-5}$
    f. $1.2 * 10^{-5}$

16. List the following numbers in order from smallest to largest.
    a. $3.54 * 10^3$
    b. $25.7 * 10^2$
    c. $0.123 * 10^4$
    d. $0.056 * 10^{-2}$
    e. $0.760 * 10^{-4}$

17. a. Round 56,799.495 to the nearest hundred.
    b. Round 56,799.495 to the nearest whole number.
    c. Round 56,799.495 to the nearest tenth.
    d. Round 56,799.495 to the nearest hundredth.

18. **How Small Is the Rectangular Room?** A rectangular room measures 128.3 ft by 73.5 ft. Using the smallest possible value for the actual length and width, calculate the minimum values for the perimeter and area of this room.

19. **How Large Is the Sheet of Metal?** The lengths of the sides of a triangular sheet of metal are 7.315 m, 5.825 m, and 1.857 m. When the base is 7.315 m, the altitude (height) of this triangle is 0.986 m. Using the largest possible values for each for each of the measurements, calculate the maximum values for the perimeter and area.

20. a. **Tim's Measurements** In reading the following blueprint, Tim measured the piece as shown in the figure. His measurements were to the nearest thousandth of an inch. To determine the overall length of the piece he calculated the sum of those measurements. Using the minimum value and maximum value for each of Tim's measurements, determine the smallest value and the largest value for the overall length using Tim's data.
    b. Charn decided to measure the overall length and measured this to be 7.032 in. What is the minimum value and maximum value for the overall length using Charn's data?
    c. In general, would one measurement or several smaller measurements provide a more accurate result? Explain.

Tim's measurements.

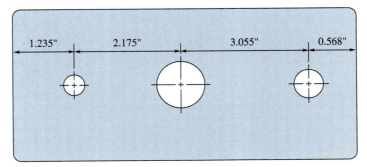

21. **Tool and Die Machine** In machine tool making, a part that must be made very precisely costs more to make than a piece that is made less precisely. Jela's Tool and Die Machine Shop has orders to produce rods with the diameters listed in the table. Which of the rods would cost the most to produce, if any? Explain.

Rod	Diameter (in.)
1	0.1
2	0.10
3	0.100
4	0.1000

22. In this problem you are given several geometrical shapes and asked to compute the perimeter or area.

    a. Compute the perimeter or circumference of the figures pictured here. Round results according to the guidelines.

    b. Compute the area of the figures pictured here. Round results according to the guidelines.

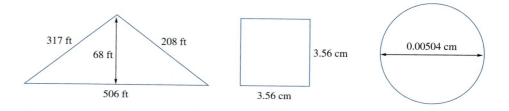

Summary of Guidelines for Rounding Approximate Numbers

1. When adding or subtracting approximate numbers, round results to the same place value as the least precise measurement.
2. When multiplying or dividing approximate numbers, round results to the same number of significant digits as the measurement with the least number of significant digits.

23. Compute the perimeter and area of the figure formed from a rectangle and a semicircle. Round your results according to the guidelines.

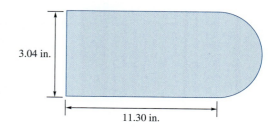

24. Compute the volume of a rectangular solid whose dimensions are 34 in. by 56 in. by 12 in. Round results according to the guidelines.

25. a. Make appropriate measurements of the following figure to be able to determine its perimeter and area. Sketch the figure, and label it with your measurements. Include all units on the sketch.

  b. Using your measurements from part a, calculate the perimeter and area of the figure. Round your results reasonably.

26. a. Write a single expression, including units, for the area of the following figure.

  b. Evaluate the expression in one step using a calculator. Round your results reasonably.

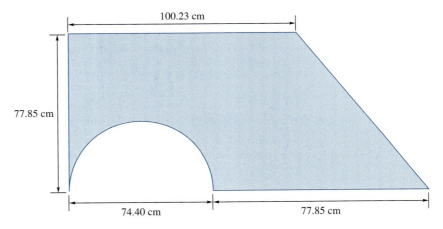

27. a. Estimate the following expression without using a calculator. Show all of your steps.

  b. Evaluate the expression in one step using a calculator. Round your result to three significant digits.

$$\frac{24.8 + 0.261 * 500.47}{\sqrt{0.261 * 389}}$$

28. a. Estimate the following expression without using a calculator. Include the units in all of your work. Show all of your steps.

  b. Evaluate the expression in one step using a calculator. Assuming this expression came from a geometry application, round the result reasonably.

$$(56.75 \text{ cm})(42.80 \text{ cm}) + \frac{(79.56 \text{ cm})^2}{2} - 907.50 \text{ cm}^2$$

Chapter 1 Numerical Concepts

1.4 Unit Conversion

Activity Set 1.4

We can measure dimensions such as length, area, weight, time, and speed. In measuring these dimensions we use units. For example, we can measure length in inches or meters, and we can measure weight in pounds or kilograms. It is important to record both the numerical value and its units of measurement. In addition, the units in a problem situation can be useful in solving the problem. In this section, we will be using unit analysis and unit fractions to convert between different units of measurement.

Before we continue, we need to discuss the different uses of the word "unit." A **unit** is any fixed quantity or amount used as a standard. For example, when we measure someone and record his or her height as 6 ft, the unit for this measurement is feet because the standard measure for this was 1 foot. Similarly, gallons, miles, kilograms, and miles per hour are all examples of units of measurement. The word "unit" is also used to represent one whole. We will see in the next few definitions both uses of the word "unit."

> **Definition**
>
> **Unit analysis** or **dimensional analysis** involves analyzing the units (feet, liters, grams, and so on) in a problem situation so that the dimensions match that of the desired answer.

> **Definition**
>
> **Unit fractions** are fractions in which the numerator and denominator are equivalent measurements and have units associated with them. Unit fractions are therefore always equal to 1.

For example, because 12 inches is equal to 1 foot, both of the fractions $\frac{12 \text{ in.}}{1 \text{ ft}}$ and $\frac{1 \text{ ft}}{12 \text{ in.}}$ are unit fractions. When multiplying factors, the units cancel just as numerical factors cancel. When we ask you to convert between different units of measurement, you need to use unit analysis together with unit fractions to show your work. We will discuss this process in the remainder of this section.

Suppose a car is traveling at an average speed of 45 miles per hour. How far will the car travel in 3 hours? We know that to determine the distance we calculate the product of 45 and 3, but how do the units work? What do units of miles per hour look like? To travel at a rate of 45 miles per hour means that in 1 hour the car travels a distance of 45 miles. This rate is the ratio of 45 miles to 1 hour, which we write as the fraction $\frac{45 \text{ miles}}{1 \text{ hour}}$. So the units on 45 are $\frac{\text{miles}}{\text{hour}}$. In general, the word "per" translates to division. Now let's include the units on our product of 45 and 3.

$$\frac{45 \text{ miles}}{\text{hour}} * 3 \text{ hours}$$
$$= \frac{45 \text{ miles}}{\cancel{\text{hour}}} * \frac{3 \cancel{\text{ hours}}}{1}$$
$$= 135 \text{ miles}$$

The hours in the denominator of the first factor cancel the hours in the numerator of the second factor, which produces miles as the resulting unit.

Consider the unit fractions $\frac{12 \text{ in.}}{1 \text{ ft}}$ and $\frac{1 \text{ yd}}{3 \text{ ft}}$. When we multiply these two fractions we obtain

$$\frac{12 \text{ in.}}{1 \text{ ft}} * \frac{1 \text{ yd}}{3 \text{ ft}} = \frac{12 \text{ in. yd}}{3 \text{ ft ft}}$$

which is not terribly useful! If we multiply $\frac{12 \text{ in.}}{1 \text{ ft}}$ and $\frac{3 \text{ ft}}{1 \text{ yd}}$, however, we obtain

$$\frac{12 \text{ in.}}{1 \text{ \color{gray}ft}} * \frac{3 \text{ \color{gray}ft}}{1 \text{ yd}} = \frac{36 \text{ in.}}{1 \text{ yd}}$$

Notice that the ft in the denominator of the first fraction cancel the ft in the numerator of the second fraction. The result is $\frac{36 \text{ in.}}{1 \text{ yd}}$, which we know to be a unit fraction because there are 36 inches in 1 yard.

1. Follow the directions to convert 3.5 yd to centimeters.
   a. On a card provided, write down the given information.
   b. From the conversion table in the back of the book, write down all of the conversion facts that you think might be helpful in this problem.
   c. For each conversion from part b, write the conversion as a unit fraction on a card. On the back of the card, write down the reciprocal of the unit fraction. We will call these cards unit fraction cards.
   d. Using the cards that you made from parts a and c, create a product whose units will cancel appropriately and result in the desired units.
   e. Write down and evaluate the product.

2. Following the directions in Activity 1, convert 2 miles to yards.

3. Following the directions in Activity 1, determine the number of pounds in 1 metric ton. (*Note:* A metric ton is different from a ton.)

4. In a Martian mathematics book, we found the following conversions.

   2.7 pip = 1 wor   3 nat = 2 wor   5 nat = 3 muz

   a. Make unit fraction cards for these conversions.
   b. Use your unit fraction cards to determine how many pip are in 10 muz.

5. Suppose we want to know how many square miles are in 12 km^2.
   a. The notation 12 km^2 means 12 km $*$ km. On a card write down 12 km $*$ km.
   b. From the conversion table, write down the conversions you think will be helpful in this problem.
   c. Make *two* unit fraction cards for each conversion in part b.
   d. Use your unit fraction cards to determine how many square miles (mi $*$ mi) are in 12 km^2.

6. A recreational swimmer can swim 10 laps in 15 minutes. If one lap is 50 meters long, determine the rate that the person is swimming in miles per hour.
   a. We can write the ratio of 10 laps to 15 min as the fraction

   $$\frac{10 \text{ laps}}{15 \text{ min}} \quad \text{or} \quad \frac{15 \text{ min}}{10 \text{ laps}}$$

   Our goal is to determine the speed in miles per hour ($\frac{\text{miles}}{\text{hour}}$), which is a ratio of distance to time. We therefore need to start with a ratio of $\frac{\text{distance}}{\text{time}}$, so make a card with the ratio $\frac{10 \text{ laps}}{15 \text{ min}}$.

   b. From the conversion table, write down the conversions you think will be helpful in this problem. In addition, the relationship between the length of the pool and a lap can be considered a conversion *for this problem*.
   c. Make unit fraction cards for all of the conversions from part b.
   d. Use your unit fraction cards to determine the rate the person is swimming in miles per hour ($\frac{\text{miles}}{\text{hour}}$).

Discussion 1.4

Unit Conversion

In the activities we found that the process of multiplying **unit fractions** is the same as multiplying numerical fractions. Because unit fractions are equivalent to 1, multiplying by a unit fraction does not change the value of the original expression. Using these ideas, together with the general strategy that follows here, we can convert between different units of measurement.

Strategy

Converting Between Units

To convert a given quantity to an equivalent measurement with different units do the following.

1. Identify the units in the goal.
2. Start with the given quantity. If more than one piece of information is given, then we need to choose a quantity whose units have the same meaning as our goal. For example, if our goal is miles per hour, then we need to start with a measurement whose units are distance per unit of time such as $\frac{ft}{sec}$ or $\frac{laps}{min}$.
3. Multiply by unit fractions in a way that the units you are trying to eliminate cancel and the units you are aiming for remain. (If we are converting from square units to square units we will usually need two factors of each unit fraction. Similarly, if we are converting from cubic units to cubic units we will usually need three factors of each unit fraction.)
4. When the units cancel appropriately, leaving you with the units you desire, perform the multiplication and record your results.

Example 1

How many ounces are in 5 pounds (lb)?

Solution Because both ounces and pounds are American units, we will look at the American–American Conversions on our conversion table in the back of the book. From the table, we know that there are 16 ounces (oz) in 1 pound. If we multiply 5 lb by the unit fraction $\frac{16 \, oz}{1 \, lb}$, the pounds associated with the 5 will cancel the pounds in the denominator of the unit fraction, leaving the correct units of ounces. Multiplying a quantity by 1 does not change its value. Since the unit fraction $\frac{16 \, oz}{1 \, lb}$ is equal to 1, we can multiply 5 pounds by $\frac{16 \, oz}{1 \, lb}$ without changing the weight.

$$5 \, lb$$
$$= \frac{5 \, \cancel{lb}}{1} * \frac{16 \, oz}{1 \, \cancel{lb}}$$
$$= 80 \, oz$$

There are 80 oz in 5 lb.

Example 2

How many tablespoons are in a cup?

Solution Again both units of measurement are American, so from the American–American Conversions we obtain the following conversion values.

$$8 \text{ fluid ounces (fl oz)} = 1 \text{ cup (c)}$$
$$1 \text{ tablespoon (1 tbsp)} = 0.5 \text{ fluid ounces}$$

We need to start with the given value of 1 c. Next we will multiply by a unit fraction that will cancel the cups. If we multiply 1 c by the unit fraction $\frac{8 \text{ fl oz}}{1 \text{ c}}$, the cups cancel, leaving fluid ounces. Then we can multiply this by the unit fraction $\frac{1 \text{ tbsp}}{0.5 \text{ fl oz}}$, and the fluid ounces cancel, leaving tablespoons, which is the desired unit. This process can be seen as follows:

$$1 \text{ c}$$
$$= \frac{1 \text{ c}}{1} * \frac{8 \text{ fl oz}}{1 \text{ c}} * \frac{1 \text{ tbsp}}{0.5 \text{ fl oz}}$$
$$= \frac{1 * 8 * 1}{1 * 1 * 0.5} \text{ tbsp}$$
$$= 16 \text{ tbsp}$$

There are 16 tablespoons in 1 cup.

Example 3

How many square feet are in 2.43 acres?

Solution We use the conversions 1 mile^2 = 640 acres and 5280 ft = 1 mile. Recall that the notation 1 mile^2 means 1 mile * mile.

$$2.43 \text{ acres}$$
$$= 2.43 \text{ acres} * \frac{1 \text{ mile} * \text{mile}}{640 \text{ acres}} * \left(\frac{5280 \text{ ft}}{1 \text{ mile}}\right) * \left(\frac{5280 \text{ ft}}{1 \text{ mile}}\right)$$
$$= \frac{2.43 * 1 * 5280 * 5280}{640 * 1 * 1} \text{ ft}^2$$
$$\approx 106{,}000 \text{ ft}^2$$

There are about 106,000 square feet in 2.43 acres.

There are several things to notice in the preceding example. First, when we multiplied the given 2.43 acres by $\frac{1 \text{ mile}^2}{640 \text{ acres}}$ we obtained square miles. To cancel the square miles, we must multiply by two factors of $\frac{5280 \text{ ft}}{1 \text{ mile}}$ because mile^2 = mile * mile. Second, we rounded the results. The given information is an approximate measurement and contains three significant digits. Each conversion used was an exact conversion. We therefore rounded our answer to three significant digits.

Example 4

We know that there are 3 feet in 1 yard. Explain why there are 27 cubic feet (ft^3) in 1 cubic yard.

Solution A cubic yard can be visualized as a cube with dimensions 1 yd by 1 yd by 1 yd. This is the same as a cube with dimensions 3 ft by 3 ft by 3 ft.

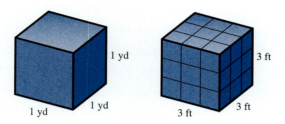

From the figure, we can see that there are

$$(3 \text{ ft}) * (3 \text{ ft}) * (3 \text{ ft}) = 27 \text{ ft}^3$$

in 1 yd^3.

We can also convert 1 cubic yard to cubic feet.

$$1 \text{ yd}^3$$
$$= 1 \text{ yd} * \text{yd} * \text{yd} * \left(\frac{3 \text{ ft}}{1 \text{ yd}}\right) * \left(\frac{3 \text{ ft}}{1 \text{ yd}}\right) * \left(\frac{3 \text{ ft}}{1 \text{ yd}}\right)$$
$$= 1 * 3 * 3 * 3 \text{ ft}^3$$
$$= 27 \text{ ft}^3$$

Example 5

How many milliliters are in 3 tablespoons?

Solution Because milliliters is a metric unit and tablespoons is an American unit we will need to find a volume conversion between the American units and metric units. Using the conversion table in the back of the book, we discover that the only American–metric volume conversion is 1.06 quarts ≈ 1 liter. In addition to this conversion, we may need the following conversion values.

1.06 quarts (qt) ≈ 1 liter (L)   8 fluid ounces = 1 cup
1000 milliliters (mL) = 1 liter   2 cups = 1 pint (pt)
1 tablespoon = 0.5 fluid ounce   2 pints = 1 quart

First, we start with the given 3 tbsp. Then, we multiply by unit fractions until we obtain milliliters.

$$= 3 \text{ tbsp}$$
$$\approx 3 \text{ tbsp} * \frac{0.5 \text{ fl oz}}{1 \text{ tbsp}} * \frac{1 \text{ c}}{8 \text{ fl oz}} * \frac{1 \text{ pt}}{2 \text{ c}} * \frac{1 \text{ qt}}{2 \text{ pt}} * \frac{1 \text{ L}}{1.06 \text{ qt}} * \frac{1000 \text{ mL}}{1 \text{ L}}$$
$$= \frac{3 * 0.5 * 1000}{8 * 2 * 2 * 1.06} \text{ mL}$$
$$\approx 44 \text{ mL}$$

Because 3 tablespoons is most likely measured more accurately than the nearest tablespoon, we will round our result to two significant digits. There are about 44 milliliters in 3 tablespoons.

Example 6

A recreational swimmer can swim 10 laps in 15 minutes. If one lap is 50 meters (m) long, determine the rate that the person is swimming in miles per hour.

Solution

In this problem we are given two different pieces of information. We know that

$$10 \text{ laps} = 15 \text{ minutes} \quad \text{and} \quad 1 \text{ lap} = 50 \text{ meters}$$

Each piece of the information given in this problem can be written as a ratio. For example the ratio of 10 laps to 15 minutes can be written as the fraction

$$\frac{10 \text{ laps}}{15 \text{ min}} \quad \text{or} \quad \frac{15 \text{ min}}{10 \text{ laps}}$$

Similarly, the ratio of 1 lap to 50 m can be written as the fraction

$$\frac{1 \text{ lap}}{50 \text{ m}} \quad \text{or} \quad \frac{50 \text{ m}}{1 \text{ lap}}$$

So, which information do we start with? One strategy is to start with the information whose units have the same meaning as our goal. Our goal is to determine the speed in miles per hour ($\frac{\text{miles}}{\text{hour}}$), which is a ratio of distance to time. We therefore need to start with a ratio of $\frac{\text{distance}}{\text{time}}$, so we use the ratio $\frac{10 \text{ laps}}{15 \text{ min}}$.

By beginning with laps per minute we have two units of measurement to convert. We need to convert the laps to miles and the minutes to hours. It does not matter which we start with. Both conversions can be done in a single expression. We first multiply by the unit fractions $\frac{50 \text{ m}}{1 \text{ lap}}$, $\frac{1 \text{ km}}{1000 \text{ m}}$, and $\frac{1 \text{ mile}}{1.609 \text{ km}}$, which converts the laps to miles, giving us units of miles per minute. Then, we complete the conversion by multiplying by $\frac{60 \text{ min}}{1 \text{ h}}$.

$$\frac{10 \text{ laps}}{15 \text{ min}}$$

$$= \frac{10 \text{ laps}}{15 \text{ min}} * \frac{50 \text{ m}}{1 \text{ lap}} * \frac{1 \text{ km}}{1000 \text{ m}} * \frac{1 \text{ mile}}{1.609 \text{ km}} * \frac{60 \text{ min}}{1 \text{ h}}$$

$$= \frac{10 * 50 * 60}{15 * 1000 * 1.609} \frac{\text{mile}}{\text{h}}$$

$$\approx 1.24 \frac{\text{miles}}{\text{h}}$$

The person is swimming at a rate of about $1\frac{1}{4}$ mph.

Notice that even though we converted two different types of units, the conversion was still done using a single expression.

So far, we have used unit analysis and unit fractions to convert between different units of measurement. Unit analysis can also be used as a problem-solving tool. Often in problem solving, we are unsure which mathematical operation to use. Should we multiply or divide two quantities, and if we divide, in which order do we perform the division? Analyzing the units in a problem can be helpful in deciding which operation to use.

Example 7

The 3000-Meter Competition. Marise just completed a 3000-meter preliminary race in 12 min 35 sec. The race was held on an indoor track and runners had to run 10 laps to complete the 3000-meter distance. She now wants to know how fast she has to run each lap in the finals competition so that her time is as good as in the preliminary race. Marise is uncertain about which order to perform the division. Which one is correct?

$$\frac{10 \text{ laps}}{12 \text{ min } 35 \text{ sec}} \quad \text{or} \quad \frac{12 \text{ min } 35 \text{ sec}}{10 \text{ laps}}$$

Solution The units in the first expression are in laps per length of time. This is not what Marise wants to know. In the second expression the resulting units will be length of time per lap; therefore, performing the division in the order $\frac{12 \text{ min } 35 \text{ sec}}{10 \text{ laps}}$ will give the length of time it will take to run one lap. To completely answer the question of how fast Marise should run, we will first need to convert 12 minutes and 35 seconds to seconds.

$$12 \text{ min} * \frac{60 \text{ sec}}{1 \text{ min}} + 35 \text{ sec} = 755 \text{ sec}$$

Then,

$$\frac{12 \text{ min } 35 \text{ sec}}{10 \text{ laps}} = \frac{755 \text{ sec}}{10 \text{ laps}} = \frac{75.5 \text{ sec}}{\text{lap}} = 75.5 \text{ sec per lap.}$$

Marise must run each lap in about 75.5 sec (or 1 min and 15.5 sec) to reach her goal.

In this section, we used **unit analysis** to convert between different units of measurement. In general, we start with the given quantity and multiply by **unit fractions** until we reach our desired units. If we are converting from square units to square units, we usually use two factors of each unit fraction. Similarly, if we are converting from cubic units to cubic units, we usually use three factors of each unit fraction.

Unit analysis can also be used as a problem-solving tool. If you are unsure whether to multiply or divide by a number, analyze the units. When the units work out correctly, you have most likely performed the correct operation.

Problem Set 1.4

For Problems 1–17, write a single expression using **unit fractions** to determine the answer to each question. Perform all computations in one step on your calculator. Round all answers reasonably.

1. How many feet are in 2.8 miles?

2. How many centimeters are in 4.7 kilometers?

3. How many feet are in 275 centimeters?

4. How many ounces are in 2.98 American tons?

5. How many grams are in 0.186 ounces?

6. How many kilograms are in 801 milligrams?

7. How many seconds are in 1 year?

8. How many milliliters are in 1.5 cups?

9. How many square feet are in 0.0313 mile^2?

10. Convert 7.5 pints per minute to cubic feet per hour.

11. Convert 21.6 cents per gram to dollars per American ton.

12. Convert 2032 cubic centimeters to cubic inches.

13. How many cubic inches are in a 3.0-liter engine?

14. Mercury is a liquid with a density of 13.6 grams per cubic centimeter (at room temperature). What is the density of mercury in pounds per cubic inch?

15. In the 1996 Summer Olympics, the United States men's 4 × 100-meter freestyle relay team won a gold medal. Each of the four team members, Josh Davis, Gary Hall Jr., Jon Olsen, and Brad Schumacher, swam 100 m. Their winning time was 3:15.41 min (3 min and 15.41 sec). What was the team's average speed in miles per hour?

16. Gail Devers, of the United States, earned the title of "world's fastest woman" in the 1996 Summer Olympics by winning the 100-m race. Her winning time was 10.94 sec. What was her average speed in miles per hour?

17. If water weighs 62.4 lb per cubic foot, what is the weight in ounces of one cubic inch of water?

18. If fencing costs $3.25 per foot, what will be the cost of fencing a rectangular lot that is 20 yd wide and 30 yd long?

19. The distance between two road control points on a level road is 380. ft. In an aerial photo of this road, the distance on the photo between these two road control points is about 0.58 in.
    a. How many feet does one photo inch represent?
    b. If two points on the photo are 3.5 in. apart, how far apart are the actual objects?
    c. How many photo inches represent an actual distance of 1 mile?

20. The distance between two road control points on a level road is 4520 ft. On an aerial photograph of this area, the distance between these two points is 2.5 in. If two objects in this photo are measured to be 11.2 in. apart, how many miles apart are they?

21. In a book on aerial photography, it is stated that the typical photograph scale is $\frac{1}{18,450}$. This means that 1 photo inch is equal to 18,450 ground inches or 1 photo centimeter is equal to 18,450 ground centimeters, or any similar relationship.
    a. In a photograph with this scale, how many photo inches represent 3 ground miles?
    b. If the distance between two points on the photo is 11 in., what is the ground distance between these two points?

22. Determine the volume of the cylinder in the illustration. Then convert your results to cubic meters using unit fractions. (*Note:* The volume of a cylinder is equal to the product of the area of the base and the height.)

23. a. Measure the following figure and write a single expression, including units, to determine the perimeter in centimeters. Then convert your results to inches.
    b. Measure the figure and write a single expression, including units, to determine the area in square centimeters. Then convert your results to square inches.

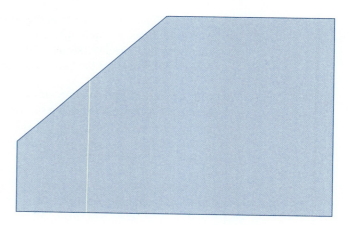

24. a. Measure the following figure and write a single expression to determine the perimeter in centimeters. Then convert your results to inches.

   b. Measure the figure and write a single expression to determine the area in square centimeters. Then convert your results to square inches.

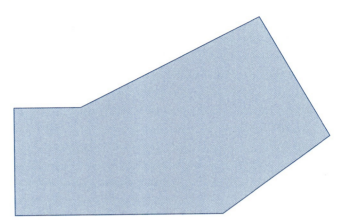

25. If one acre of land is in the shape of a square, what are the dimensions of this piece of land in feet?

26. The Aquarium

   a. A rectangular aquarium tank measures 4.30 m by 5.70 m by 9.40 m. Use unit fractions to determine the number of gallons of water necessary to fill the tank to a height 1 ft below the top.

   b. If two divers were in the tank feeding the fish, would the water overflow?

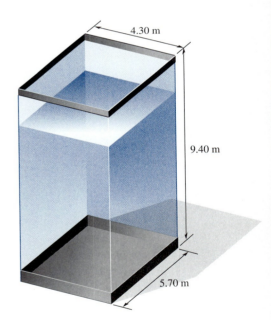

27. **Landscaping**
    a. In determining the length of the edging needed to go around the perimeter of garden area, a landscaper needed to evaluate the following expression. Determine both the numerical value and the units for this expression.

    $$4.2 \text{ ft} + 13.2 \text{ ft} + 4.8 \text{ ft} + 15.7 \text{ ft}$$

    b. The cost of the edging was $2.25 per foot. The following expression could be used to determine the total cost. Evaluate both the numerical value and the units for this expression.

    $$(4.2 \text{ ft} + 13.2 \text{ ft} + 4.8 \text{ ft} + 15.7 \text{ ft}) * \frac{\$2.25}{\text{ft}}$$

28. A student brought in the following two problems, which are already worked out. Both problems start with the measurement 28 ft^3, but the student does not understand why the first problem includes three factors of a given unit fraction but the second one does not. Explain how to help this student understand the similarities and differences between these two problems and their solutions.

    First Problem Convert 28 ft^3 to cubic inches.

    $$= 28 \text{ ft}^3$$
    $$= 28 \text{ ft}^3 * \frac{12 \text{ in.}}{1 \text{ ft}} * \frac{12 \text{ in.}}{1 \text{ ft}} * \frac{12 \text{ in.}}{1 \text{ ft}}$$
    $$\approx 48{,}000 \text{ in.}^3$$

    Second Problem Convert 28 ft^3 to gallons.

    $$= 28 \text{ ft}^3$$
    $$= 28 \text{ ft}^3 * \frac{1 \text{ gal.}}{7.481 \text{ ft}^3}$$
    $$\approx 3.7 \text{ gal}$$

Chapter 1 Summary

In this chapter, we learned a lot of new vocabulary that will help us read and interpret mathematical expressions.

Terms of an expression are the parts of a sum or difference. To identify the terms of an expression, we look for addition or subtraction symbols that are outside of grouping symbols. The expressions on each side of these are the terms of the expression. If there are no such symbols, then the expression is a single term.

To **evaluate** an expression means to find its value. To evaluate an expression showing all steps, we first find the value of each term. Then we add or subtract terms from left to right.

Factors are the parts of a product. At this point we have not used factors a great deal, but they will be important as we proceed through the course.

After we evaluated expressions, we proceeded to apply these skills to estimate expressions. In most real life situations that involve mathematics, the numbers are not easy to compute mentally. Therefore, we often estimate the value of an expression and then use a calculator to find the exact value. The estimate helps us determine whether or not we have entered the expression correctly in our calculator. For this reason the estimate must be performed without the use of the calculator.

To **estimate** the value of a numerical expression we round all numbers so that the arithmetic can be done easily. The way we round depends on the operations to be performed. In general, when adding or subtracting numbers, we usually round the numbers to the same place value. When multiplying numbers, we usually round to a single digit. When dividing numbers, we must round so the division is easy to perform. After we round the numbers, we perform the calculations in each step without using the calculator.

Exact numbers result from counting or numbers that are defined. All other numbers are **approximate** numbers. Because approximate numbers have some uncertainty, we must round our results reasonably when working with approximate numbers. In many applications, we can determine a reasonable way to round a result by considering the problem situation. For other situations we adopt some general guidelines for rounding approximate numbers.

> *Guidelines for Rounding Approximate Numbers*
> 1. When adding or subtracting approximate numbers, round results to the same place value as the least precise measurement.
> 2. When multiplying or dividing approximate numbers, round results to the same number of significant digits as the measurement with the least number of significant digits.

Throughout Chapter 1, we applied the geometrical concepts of perimeter and area to irregularly shaped figures. We found that we had to break the figure into shapes that we knew, measure appropriate dimensions, and use known formulas or definitions to determine the desired quantity. The guidelines for rounding approximate numbers helped us round our results reasonably.

In most applications, the numbers in a problem have units associated with them. It is therefore important to know how to use units correctly as well as to convert between different units of measurement. In Section 1.4, we learned a process for converting between different units of measurement. Beginning with the given measurement and its units, we multiply by unit fractions so as to cancel the units we are trying to eliminate and arrive at the desired units. When converting between square measurements or cubic measurements, we must remember that we need repeated factors of the unit fractions.

Finally, we saw throughout the chapter the importance of the language of mathematics. Understanding the meaning of the symbols of mathematics allows us to read expressions correctly. Being able to read mathematics correctly reinforces the correct meaning of the symbols and allows us to communicate with others.

Chapter Two

An Introduction to Algebra—Numerically, Graphically, and Algebraically

The focus of Chapter 1 was on numerical aspects of mathematics. In Chapter 2, we will learn how to look for patterns and summarize these patterns using the language of mathematics. We will learn how to model patterns and problem situations numerically with tables of values, algebraically with expressions and equations, and graphically. The advantages and disadvantages of each of these models will be explored. The connections between the models will be emphasized.

2.1 Using Block Patterns to Introduce Variables and Graphs

Activity Set 2.1

1. a. Construct each of the structures shown here using blocks.
   b. Construct the fourth and fifth structures that would extend the pattern.
   c. Describe, in words, how you would construct the 50th structure.
   d. Complete the first five rows in the following table.
   e. How many blocks would be in the 50th structure? Add this value to the table. (*Hint:* You may find it helpful to rearrange the blocks in each structure to determine a pattern for finding this value.)
   f. If you are given a structure number, describe how you would determine the number of blocks needed to build the structure.

Structure 1   Structure 2   Structure 3

Structure Number	Number of Blocks
1	
2	
3	
4	
5	
50	

2. a. Construct each of the structures shown here using blocks.
   b. Construct the fourth and fifth structures that would extend the pattern.
   c. Describe, in words, how you would construct the 50th structure.
   d. Complete the first five rows in the following table.
   e. How many blocks would be in the 50th structure? Add this value to the table. (*Hint:* You may find it helpful to rearrange the blocks in each structure to determine a pattern for finding this value.)
   f. If you are given a structure number, describe how you would determine the number of blocks needed to build the structure.

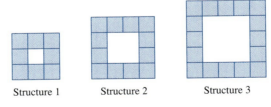

Structure 1   Structure 2   Structure 3

Structure Number	Number of Blocks
1	
2	
3	
4	
5	
50	

Discussion 2.1

Using Block Patterns to Introduce Variables and Graphs

In the activities, you looked at sequences of structures and generalized patterns. Patterns can be created from a variety of approaches. Some people may recognize patterns from geometrical shapes, and others may discern them from a sequence of numbers. The next three examples look at different approaches to generalizing a pattern for determining the number of blocks in Activity 2 in the activity set.

Example 1

First we look at the sequence of structures geometrically. It is usually helpful to build the next couple of structures in the sequence to see the pattern that is being established.

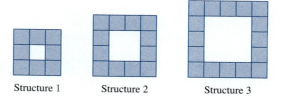

The fourth and fifth structure are added to the sequence.

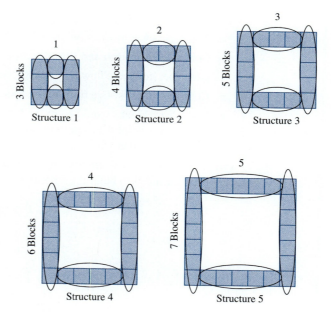

By looking at the sequence, you might observe that the first structure consists of two sets of 3 blocks and two sets of 1 block in a square formation. The second structure consists of two sets of 4 blocks and two sets of 2 blocks in a square formation. The third structure consists of two sets of 5 blocks and two sets of 3 blocks in a square formation. When this pattern is continued, the 50th structure will consist of two sets of 52 blocks and two sets of 50 blocks in a square formation.

Therefore, the number of blocks in

The first structure is	$2 * 3 + 2 * 1 = 8$ blocks
The second structure has	$2 * 4 + 2 * 2 = 12$ blocks
The third structure has	$2 * 5 + 2 * 3 = 16$ blocks
The fourth structure has	$2 * 6 + 2 * 4 = 20$ blocks
The fifth structure has	$2 * 7 + 2 * 5 = 24$ blocks

.

.

.

The 50th structure would have $\quad 2 * 52 + 2 * 50 = 204$ blocks

To generalize, any structure consists of two sets of two more than the structure number and two sets of the structure number in a square formation. Therefore, the number of blocks in any structure is equal to the sum of twice two more than the structure number and twice the structure number, that is,

number of blocks $= 2(\textit{structure no.} + 2) + 2(\textit{structure no.})$

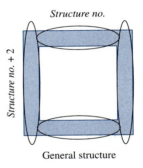

General structure

Rather than writing an equation using the phrases *number of blocks* and *structure no.*, we can let letters represent these quantities. If B represents the number of blocks and n represents the structure number, then the previous equation becomes

$B = 2(n + 2) + 2n$

However, before we use this formula for any structure number, we should make sure that it works for the first few structure numbers for which we know the number of blocks.

Let $n = 1$ in our equation. $B = 2(1 + 2) + 2 * 1$
$B = 8$

Let $n = 2$ in our equation. $B = 2(2 + 2) + 2 * 2$
$B = 12$

Let $n = 3$ in our equation. $B = 2(3 + 2) + 2 * 3$
$B = 16$

We can see that our equation gives the correct number of blocks for the first three structures. We can, therefore, be fairly confident that it will produce the correct number of blocks for any structure number.

> ### Definition
> Letters that represent numbers are called **literal symbols.** For example, B, n, and π are all literal symbols. Literal symbols that can represent different values are called **variables,** and literal symbols that represent values that cannot change are called **constants.**

We know that the area of a circle can be found by using the formula $A = \pi r^2$, where A represents the area and r represents the radius. In this formula, there are three literal symbols, A, π, and r. The letters A and r are variables, because they can represent different values, and π is the constant that is an irrational number and is approximately 3.14159 to the nearest hundred-thousandth.

Example 2

Let's look at a second approach to generalizing the pattern in Activity 2. You might see the sequence of structures as squares with a hole in the middle.

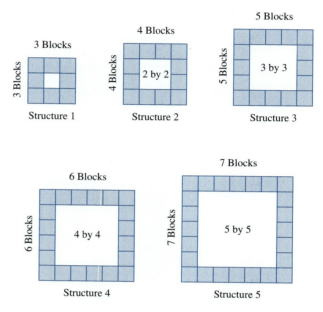

Seeing the pattern this way, you might observe that the first structure consists of a square that is 3 by 3 and a hole consisting of 1 block. The second structure consists of a square that is 4 by 4 and a square hole that is 2 by 2. The third structure consists of a square that is 5 by 5 and a square hole that is 3 by 3. When this pattern is continued, the 50th structure would consist of a square that is 52 by 52 and a square hole that is 50 by 50.

One way to determine the number of blocks in these structures is to determine the number of blocks in the large squares and subtract the number of blocks in the holes.

For example,

The first structure has	$3^2 - 1 = 8$ blocks
The second structure has	$4^2 - 2^2 = 12$ blocks
The third structure has	$5^2 - 3^2 = 16$ blocks
The fourth structure has	$6^2 - 4^2 = 20$ blocks
The fifth structure has	$7^2 - 5^2 = 24$ blocks
.	
.	
.	
The 50th structure would have	$52^2 - 50^2 = 204$ blocks

2.1 Using Block Patterns to Introduce Variables and Graphs

To generalize, any structure consists of a square whose side length is two more than the structure number and a square hole whose side length is the structure number. Therefore, the number of blocks in any structure is equal to the difference of the square of two more than the structure number and the square of the structure number, that is,

number of blocks = (*structure no.* + 2)^2 − (*structure no.*)^2

Structure no. + 2

Structure no.

General structure

If we let B represent the number of blocks and n represent the structure number, then the preceding equation becomes

$B = (n + 2)^2 - n^2$

Let's verify this formula for the first few structure numbers.

Let $n = 1$ in our equation. $B = (1 + 2)^2 - 1^2$
$B = 8$

Let $n = 2$ in our equation. $B = (2 + 2)^2 - 2^2$
$B = 12$

Let $n = 3$ in our equation. $B = (3 + 2)^2 - 3^2$
$B = 16$

These are the correct values for the first three structures.

Example 3

Now, let's look at the pattern in Activity 2 numerically. We complete a table of values for the number of blocks in each structure for the first five structures. Then we use these values to see if we can determine a pattern.

Structure Number	Number of Blocks
1	8
2	12
3	16
4	20
5	24
50	

Looking at the right-hand column, we can see that the number of blocks increases by 4 each time. Repeatedly adding 4 is the same as multiplying 4 by some number. Therefore, we try 4 times the structure number.

Structure Number	Number of Blocks	4 * Structure No.
1	8	4
2	12	8
3	16	12
4	20	16
5	24	20
50		

The expression *4 * structure no.* does not work; the result is always 4 less than what we want. If we add 4 to our original guess we should obtain the desired numbers.

Structure Number	Number of Blocks	4 * Structure No. + 4
1	8	8
2	12	12
3	16	16
4	20	20
5	24	24
50		

This expression seems to work. Therefore, the number of blocks in the 50th structure will be $4 * 50 + 4 = 204$. If we let B represent the number of blocks and n represent the structure number, then our equation is

$B = 4n + 4$

In Examples 1–3, we looked at three different ways to determine the number of blocks in any structure. In doing so, we found three different equations:

$B = 2(n + 2) + 2n$
$B = (n + 2)^2 - n^2$, and
$B = 4n + 4$

where B is the number of blocks and n is the structure number.

Because we verified each of these equations, we are confident they are correct. This means that the expressions

$2(n + 2) + 2n$

$(n + 2)^2 - n^2$

$4n + 4$

must be equivalent. In a later section we will be able to show algebraically that these expressions are equivalent.

An Introduction to Graphing

It is sometimes helpful to visualize a table of values. Before we can do this, we need to look at some basics of coordinate graphing.

We create a **coordinate graph** by drawing two perpendicular lines as shown in the figure. The horizontal line is called the **horizontal axis,** and the vertical line is called the **vertical axis.** The point where the axes intersect is called the **origin.**

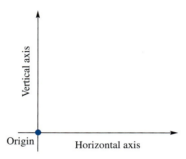

Both the horizontal and the vertical axes must be marked off in equal increments; however, the increments used for the horizontal axis may be different from those used for the vertical axis. The way the axes are marked off is referred to as the **scale** of the graph. The scale that is selected depends on the data being graphed.

On the following graph, the horizontal axis is marked off in increments of 1 unit as indicated by the tick marks and scale. The vertical axis is marked off in increments of 5 units. On a given axis, the distance between tick marks must be the same; however, the distance between tick marks on the vertical axis may differ from those on the horizontal axis.

To name the point shown in the graph, we must find the horizontal distance as measured from the vertical axis and the vertical distance as measured from the horizontal axis. The point is named by the **ordered pair,** (*horizontal distance, vertical distance*). On the following graph, the point is labeled as (7, 20) because the horizontal distance to the point is 7 units and the vertical distance to the point is 20 units. Notice that in an ordered pair, the horizontal distance is always given first and the vertical distance second.

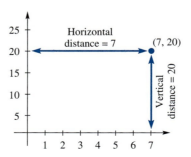

Example 4

Determine the ordered pair that names each of the points on the following graph. Assume that all points lie on integer coordinates.

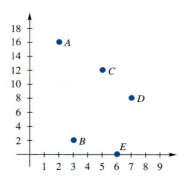

Solution

Point A is named (2, 16) because the horizontal distance to A is 2 units and the vertical distance to A is 16 units.

Point B is named (3, 2) because the horizontal distance to B is 3 units and the vertical distance to B is 2 units.

Point C is named by the ordered pair (5, 12), point D by the ordered pair (7, 8) and point E by the ordered pair (6, 0).

We can now return to the task of visualizing a table of values with a graph.

In general, table values are created to relate two types of information. Usually one piece of information depends on the other. For example, in the block patterns, the number of blocks in a structure depends on the structure number. For this reason, one of the variables is called the dependent variable and the other is called the independent variable. Again, in the block patterns, the number of blocks is the dependent variable and the structure number is the independent variable.

Definition

A **dependent variable** is a variable whose value depends on the other variable's value. The other variable is known as the **independent variable.**

To establish some consistency, the independent variable is given in the left column of a table, and the dependent variable is given in the right column. Then, each row in the table can be made into an ordered pair *(left column entry, right column entry)*.

Independent Variable	Dependent Variable

Or if a table is given horizontally, the independent variable is in the top row, and the dependent variable is in the bottom row. Each column in a table can be made into an ordered pair (*top row entry, bottom row entry*).

Independent Variable	
Dependent Variable	

Because we must know the value of the independent variable to determine the value of the dependent variable, we often refer to the independent variable as the **input.** Then the dependent variable is referred to as the **output.** In the block patterns, the structrure number is the input and the number of blocks is the output.

In the case of the block patterns, the number of blocks depends on the structure number. In the equations, n was the independent variable and B was the dependent variable.

On a graph, the independent variable goes on the horizontal axis and the dependent variable on the vertical axis. We use each row of the table to create an ordered pair called a **data point.** Then plotting the points results in a graph that represents the table data. We label the axes on the graph to correspond to the column headings in the table. Our scale is determined based on the range of values in each column of the table.

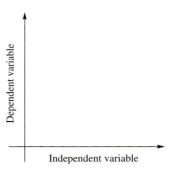

Example 5

Graph the first five data points from the table in Example 3.

Structure Number	Number of Blocks
1	8
2	12
3	16
4	20
5	24

Solution Because the left column data range from 1 to 5, we might let our horizontal axis go from 0 to 6 or 0 to 7, with increments of 1. Because the right column data range from 8 to 24, we might decide to scale our vertical axis from 0 to 25 or 30. We use increments of 5 on this axis because that is more convenient for this scale than increments of 1.

When constructing a graph we need to include labels for each axis. A label is a title indicating what the axis represents. Notice the labels on the horizontal and vertical axes of the following graph. The horizontal axis is labeled "Structure no." and the vertical axis is labeled "Number of blocks."

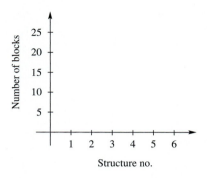

Each row in a two-column table represents an ordered pair. The first row in the table is represented by the ordered pair (1, 8). To plot this point, first go 1 unit along the horizontal axis from the origin. From there go up 8 units according to the vertical scale. Mark the point. On this graph we need to estimate a vertical distance of 8 because this number falls between our tick marks. To graph the entire table, we need to plot each ordered pair.

In this particular situation, no structure number 2.5 exists; therefore, the discrete points give a presentation of the problem situation. However, sometimes it is difficult to visualize a pattern by looking at discrete points. Therefore, we connect these points with a curve to see the pattern. In our example, this "curve" is a straight line. The following is a graph of the table of values connected with a curve.

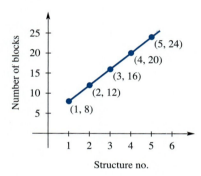

In the previous example, the data between the plotted points were not part of the problem situation, but there are many examples in which the data between points are meaningful in the problem situation.

Example 6

Graph the data from the following table.

Input	Output
1	3
2	8
3	15
4	24
5	35

Solution Recall, the left column represents the independent variable and is graphed on the horizontal axis. The right column represents the dependent variable and is graphed on the vertical axis.

With these data, the left column again ranges from 1 to 5. Our horizontal axis can then go from 0 to 5 or 6 in increments of 1. The right column data range from 3 to 35. We might again choose increments of 5 with the scale going from 0 to 40.

Other good choices are possible for the vertical scale, but increments of 5 work well because most of us count by 5's fairly easily. Increments of 1 on the vertical scale would be impractical.

Graphing the data points from the table with the graph scaled as described gives us the following graph. Again, we connect the points in the graph to see the pattern formed from these points. Notice that in this example, connecting the points in a smooth curve does not give us a straight line.

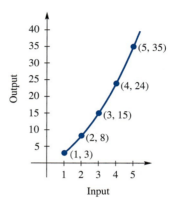

In this section we learned how to generate a **table** of values and to write an **algebraic equation** to fit a given pattern. Having written an equation, it is important to verify that it works for the given information. In writing the equations, we were introduced to **literal symbols** and the difference between **variables** and **constants.** We then learned how to use a **coordinate graph** to visualize the pattern. This gives us three ways to see the same data: numerically with a table, algebraically with an equation, and visually with a graph. We will revisit these three types of models throughout the course.

Problem Set 2.1

1. For each collection of structures, follow these directions:
   i. Draw the fourth and fifth structures that would extend the pattern.
   ii. Describe, in words, how you would construct the 50th structure.
   iii. How many blocks would be in the 50th structure?
   iv. Complete the following table.
   v. If n is the structure number, write an equation for the number of blocks needed to build the structure.

Structure number	1	2	3	4	5	50
Number of blocks						

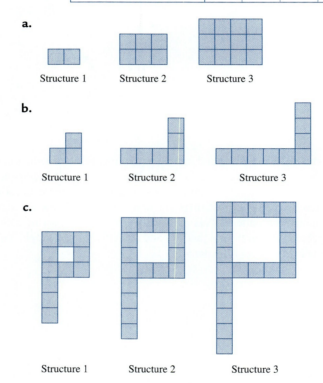

a.
Structure 1   Structure 2   Structure 3

b.
Structure 1   Structure 2   Structure 3

c.
Structure 1   Structure 2   Structure 3

2. For the collection of structures, follow these directions:
   i. Draw the fourth and fifth structures that would extend the pattern.
   ii. Make a table for the number of sticks for the first five structures.
   iii. Plot the first five points from the table on a graph.
   iv. If n is the structure number, write an equation for the number of sticks needed to build the structure.

   Structure 1   Structure 2   Structure 3

Structure number	1	2	3	4	5	50
Number of blocks						

3. For each collection of structures, follow these directions:
   i. Draw the fourth and fifth structures that would extend the pattern.
   ii. Describe, in words, how you would construct the 50th structure.
   iii. Make a table for the number of blocks for the first five structures.
   iv. Plot the first five points from your table on a graph.
   v. If n is the structure number, write an equation for the number of blocks needed to build the structure.

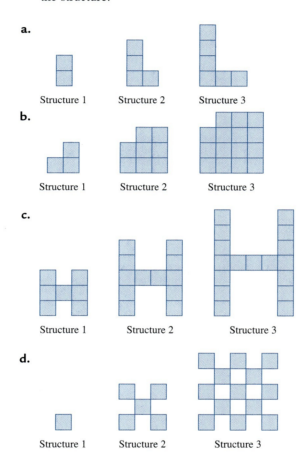

a.

Structure 1   Structure 2   Structure 3

b.

Structure 1   Structure 2   Structure 3

c.

Structure 1   Structure 2   Structure 3

d.

Structure 1   Structure 2   Structure 3

2.2 The Rectangular Coordinate System

Activity Set 2.2

1. The Eastside Bike Club planned a Sunday bike tour to the mountains. They met at Main Park at 8:00 A.M. and left promptly. Bill overslept and arrived at the park about 2 hours late. He knew that on group rides, the club averaged about 16 miles per hour (mph). Bill can easily average 23 mph. He began the trip at 10:00 A.M.

   a. Complete the following table.

Number of Hours After 10:00 A.M.	Distance Traveled by Bill (in miles)
0	
1	
2	
3	
4	

   b. Write an equation for the number of miles traveled by Bill in terms of the number of hours after 10:00 A.M.

   c. Plot the information from part a. Label the horizontal axis *Number of hours after 10:00 A.M.* and the vertical axis *Miles traveled*. Include an appropriate scale for this problem situation.

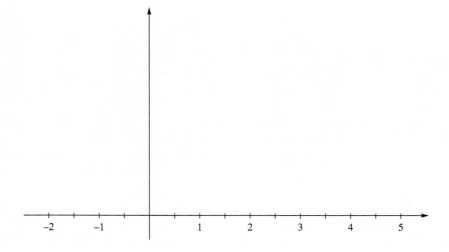

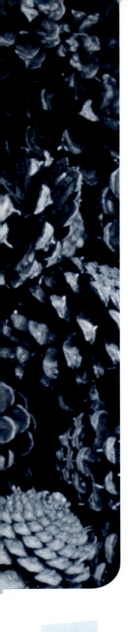

2.2 The Rectangular Coordinate System

d. The Eastside Bike Club left 2 hours *before* 10:00 A.M. How far had the club traveled when Bill began his ride? What point is this on the graph?

e. Since 8 A.M. is 2 hours before 10 A.M., we can think of 8:00 A.M. as −2. Complete the following table.

Number of Hours After 10:00 A.M.	Distance Traveled by Eastside Bike Club (in miles)
−2	
−1	
0	
1	
2	
3	
4	

f. Write an equation for the distance traveled by the Eastside Bike Club in terms of the number of hours after 10:00 A.M. Check your equation by substituting values from the table in part e.

g. Plot the information in the table from part e on the graph from part b. You may find it helpful to graph this information in a different color.

h. *Approximate the answers to the following questions using your graph.*

At what time had Bill traveled 50 miles?

At what time had the Eastside Bike Club traveled 50 miles?

How long will it take Bill to catch up with the club?

How many miles will he have ridden?

Discussion 2.2

In the previous section, you were introduced to graphing in a coordinate system. We looked only at ordered pairs in which both coordinates were positive numbers.

The portion of the coordinate graph that we looked at in the last section is called the first quadrant, or **Quadrant I.**

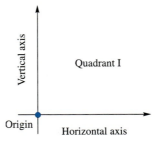

Activity 1 of this section described a situation in which the horizontal axis needed to be extended to include negative numbers. If we extend both the horizontal and the vertical axes to include negative values, we get the complete **rectangular coordinate system** shown in the following figure. This is sometimes referred to as the **Cartesian coordinate system.** The four quadrants are numbered as indicated. The positive and negative direction are indicated on each axis. The horizontal axis is often referred to as the x-axis and the vertical axis as the y-axis.

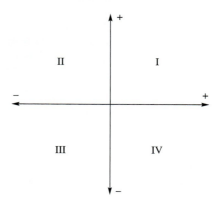

Example 1

Determine the ordered pair that names each of the points on the following graph.

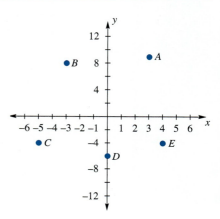

Solution We approximate the location of the points to the nearest integer coordinates. Point A is about (3, 9) because the horizontal distance to A is 3 units and the vertical distance to A is 9 units.

Point B is about ($^-$3, 8) because the horizontal distance is 3 units to the left (negative) and the vertical distance is 8 units up (positive).

Point C is about ($^-$5, $^-$4), point D is about (0, $^-$6), and point E is about (4, $^-$4).

Example 2

Target Heart Rates. For aerobic exercise to be effective, you must maintain a heart rate, in beats per minute (bpm), between a given minimum level and a given maximum level. The appropriate level is related to age. The following ordered pairs represent ages and the appropriate maximum target heart rate for each.

(30 yr, 159 bpm), (54 yr, 137 bpm), (72 yr, 121 bpm), (46 yr, 145 bpm), (80 yr, 114 bpm)

a. Graph the given ordered pairs on a rectangular coordinate graph. Connect the points with a smooth curve.

b. Use the graph to determine the maximum target heart rate for a person who is 40 years old.

c. At what age should a person's maximum target heart rate be 132 beats per minute?

Solution a. In selecting scales for the graph, we might notice that all the values are positive, so we only need the first quadrant. We can also see that the largest age we are given is 80 years and the highest heart rate is 159 bpm. On the horizontal axis we go from 0 to 90 in increments of 10. On the vertical axis we go from 0 to 160 in increments of 20.

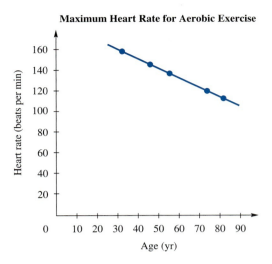

Although nothing is wrong with this graph, it could be better. Our points are all clustered in the upper right-hand part of the first quadrant. Because none of the heart rates are below 100 bpm, we can start the vertical scale at 100. We can still see all of the points, and our graph is easier to read. To start at a value other than zero, we use a special mark to show the reader that a break occurs in the axis.

> **Breaks in Axes**
>
> A break can be inserted *only* at the beginning of a scale. Putting a break in the middle of a scale may distort the graph.

The following figures show two different ways of denoting a break in the vertical axis. Drawing a break on the vertical axis as seen in the next figure is an easy method to denote a vertical break. On the other hand, drawing the break so that it looks like someone has cut across the graph horizontally and then put it back together is not as easy to draw, but clearly demonstrates what a cut in the vertical axis means.

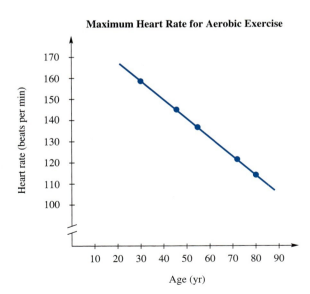

b. To determine the appropriate maximum target heart rate for a person 40 years old, we find 40 on the horizontal axis and draw a vertical line to the graph. From the point where the vertical line meets the graph, draw a horizontal line to the vertical axis. This point on the axis gives us the heart rate for age 40. Drawing in the vertical and horizontal lines is a way of showing how we reached our conclusion. We conclude that a 40-year-old should have a maximum target heart rate of approximately 148 bpm.

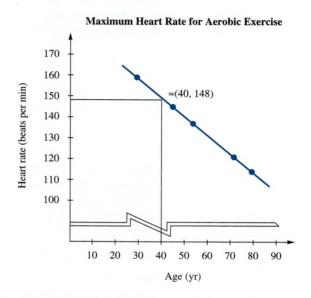

c. To determine the age of a person whose maximum target heart rate is 132 bpm, we use a similar technique. This time we start on the vertical axis at 132 and find the corresponding value on the horizontal axis. We conclude that 132 bpm is the maximum target heart rate of a 61-year-old.

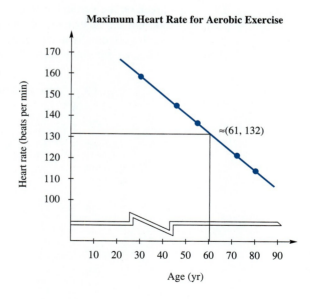

In this section we extended the **coordinate graphing system** to include both positive and negative values for both the horizontal and vertical axes. This resulted in a graph with four **quadrants.**

Many real-world applications are graphed only in quadrant I; however, some situations, such as the ones we saw in Activity 1, require us to use negative coordinates.

Much of the data we encounter in everyday experience is presented to us in graphical format. As you read newspapers or magazines over the next few days, look for graphical presentations. Examine the data carefully and determine if they are presented in a way that makes them easily understood.

In presenting data graphically, it is very important that we select appropriate **scales.** In applications, it is also important to **label the axes** for easy understanding of the information displayed.

Problem Set 2.2

1. Determine the ordered pair that names each of the points on the following graph. Assume that the coordinates of each point are integers.

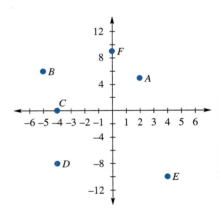

2. For each of the following tables, draw a coordinate system and plot the points. Scale your graph appropriately for the data. Do not connect the points.

a.

Input	Output
−6	24
7	62
2	12
−2	−10
5	−22
1	5

b.

Input	Output
−15	6
10	12
−25	3
34	6
20	−10
6	−8

3. For each part, graph the collection of ordered pairs. Scale each graph appropriately for the data. Do not connect the points.

   a. (−12, 6), (17, −5), (3, −4), (9, 13), (−5, 5)
   b. (−15, 475), (15, 325), (−4, 420), (8, 360), (0, 400), (3, 385)

4. **Cigarette Consumption.** The following graph shows the per capita cigarette consumption in the United States for persons who are 18 years of age or older. Use the graph to estimate the answers to the questions.

   a. What is the first year shown on the graph? What is the last year?
   b. In what year did the per capita consumption peak? What was the per capita consumption that year?
   c. In what year did the consumption return to the level it had been in 1950?
   d. In 1966, the U.S. Surgeon General's health warnings began appearing on cigarette packages. Does the graph seem to indicate that the warnings have had some effect? Explain.

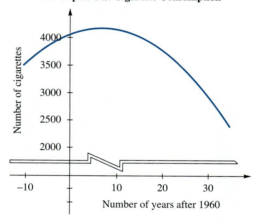

5. **Total Miles Driven in the United States.** The following graph approximates the number of miles driven in the United States for each year after 1950. Use the graph to estimate the answers to the questions.

   a. How many miles were driven in 1950? 1955? 1972?
   b. In what year were approximately 800 billion miles driven? 950,000,000,000 miles driven?
   c. Assuming the model works reasonably well for years before 1950, interpret the meaning of the point $(-3, 360)$.

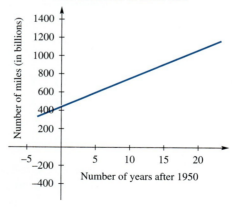

6. Newspaper Circulation. The following graph shows circulations for both morning and evening newspapers. Use the graph to answer the questions.
   a. In what year was the circulation for morning papers about the same as that for evening papers?
   b. What was the circulation for morning newspapers in 1989?
   c. In what year was the circulation of evening papers approximately 25,000,000?
   d. Interpret the meaning of the point (8, 40). Interpret the meaning of the point (−1, 33.5).
   e. In 1986, was the circulation greater for morning or evening papers? How much greater was it?

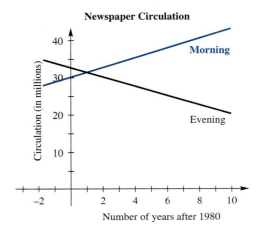

7. Portland Rainfall. The following graph shows the rainfall for each month of 1998 in Portland, Oregon.
   a. What month had the highest rainfall? Approximately what was the rainfall that month?
   b. From the graph estimate the rainfall for each month of 1998.
   c. What was the average rainfall per month for 1998?
   d. In how many months was the rainfall less than 3 in.?

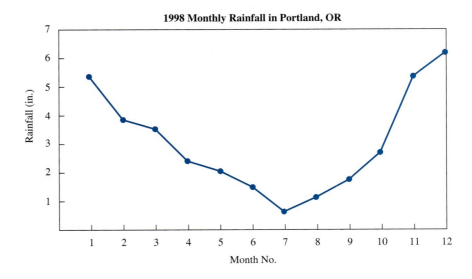

8. **Blood Alcohol Levels.** The following graph shows the blood alcohol level (BAL) in a person who has quickly drunk two alcoholic beverages on an empty stomach. The body metabolizes alcohol at varying rates. Levels rise and fall with time as the alcohol is first absorbed into the bloodstream, and then the effects wear off. Use the following graph to answer the questions.

   a. When does the BAL first reach 0.05%? When does it fall below 0.05%?

   b. If a person is considered intoxicated when the BAL is above 0.10%, when is this person considered intoxicated?

   c. In several states, a person is considered intoxicated when the BAL is above 0.08%. When is this person considered intoxicated if he or she is in a state that has adopted this new standard?

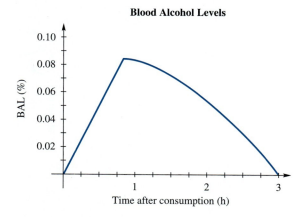

9. **Dave's Dine-Inn Restaurant.** Dave noticed a drop in profits during the 12 months since January 1999. The data is represented in the following bar graph. Dave is concerned about what will happen to his business over the next year if this trend continues. He extended the graph for the next 12 months following the pattern he observed. Use the graph to answer the following questions.

   a. What was Dave's monthly profit in January 1999? December 1999?

   b. When was Dave's monthly profit about $2250? $1400?

   c. What would you predict the monthly profit to be in January 2000? April 2000? December 2000?

   d. When do you predict that Dave's profit will fall below zero? (*Note:* When profit falls below zero it is referred to as a negative profit or as a loss.)

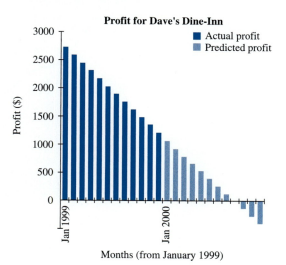

2.3 Graphing Equations

Discussion 2.3

In Section 2.1 we modeled block patterns with a table of values, with a graph, and with an equation. All three of these models are related. From a table of values we can draw a graph and sometimes write an equation, from a graph we can make a table and sometimes write an equation, and from an equation we can make a table and draw a graph. Throughout the remainder of the text we will revisit the relationships between the models. In this section we focus on drawing the graph of an equation.

Suppose we start with the equation $x + 2 = 5$. When an equation contains a variable, we often want to know what value(s) of this variable make the equation true. Such values are called **solutions** to the equation. In our example $x + 2 = 5$, we can see that if $x = 3$, then the equation is true; therefore, $x = 3$ is the solution to the equation. To **solve** an equation means to find all of the solutions.

Example 1

Identify the solutions to each equation by inspection.

a. $m + 5 = 15$
b. $k - 3 = 20$

Solution

a. We want to know what number we need to add to 5 to obtain 15. We can see that 10 works because $10 + 5 = 15$. Therefore, the solution to the equation $m + 5 = 15$ is $m = 10$.
b. Similarly, the solution to $k - 3 = 20$ is $k = 23$ because $23 - 3 = 20$.

As you might imagine, not all equations can be solved by inspection. In Chapter 3, we will discuss an algebraic process for solving some equations.

Next let's consider an equation that contains two different variables, for example, $y = x + 2$. To find the solution to an equation in two variables, we need to find a value for x and a value for y, that together make the equation true. For example, if $x = 3$ and $y = 5$, then the equation $y = x + 2$ is true. Does this equation have other solutions? What if $x = 1$? Then $y = 1 + 2$, or $y = 3$; therefore, $x = 1$ and $y = 3$ is another solution to the equation.

One way to identify solutions to an equation that contains two variables is to pick a value for one of the variables and determine the value of the other. In our equation $y = x + 2$, we pick values for x and determine the corresponding values for y. We record our results in a table.

x	$y = x + 2$
-2	$-2 + 2 = 0$
-1	$-1 + 2 = 1$
0	$0 + 2 = 2$
1	$1 + 2 = 3$
2	$2 + 2 = 4$
3	$3 + 2 = 5$

Therefore, $x = {}^-2$ and $y = 0$ is a solution, as well as $x = {}^-1$ and $y = 1$, is a solution, and so on. Because each solution is a value of x and y, we often list the solutions as ordered pairs. For example, we say $({}^-2, 0)$ is a solution, $({}^-1, 1)$ is a solution, and so on. As you can imagine from this table, we can continue this process indefinitely. Because the equation $y = x + 2$ has an infinite number of solutions, each of which is an ordered pair, it is not possible to list all of them. Instead, we graph the solutions by plotting the ordered pairs from our table and connecting the points with a smooth curve, which is shown on the graph.

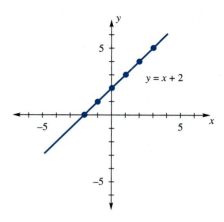

As you can see, the points we plotted for this equation appear to lie in a straight line, and in fact they do.

The line drawn is a representation of *all* of the solutions to the equation $y = x + 2$. For example, if we pick any point that lies on this line, it should be a solution to the equation. Let's try it. The point $(1.5, 3.5)$ appears to lie on the line. Let's see if it is a solution.

$3.5 = 1.5 + 2$

$3.5 = 3.5$

We can see that $(1.5, 3.5)$ is a solution.

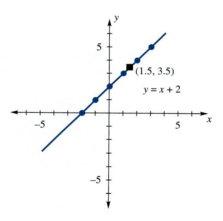

To recap:

- Equations in two variables have an infinite number of solutions.
- Each solution to an equation in two variables is an ordered pair.
- We visually represent the solutions to an equation in two variables by graphing the equation. At this point, we create the graph by plotting some of the ordered-pair solutions and connecting the points with a smooth curve.

Therefore, if we need to graph an equation in two variables, we must first find some solutions to the equation. One way to do this is to make a table of values. When making a table of values, it is best to choose both positive and negative values for one of the variables because we do not know what the graph is going to look like.

Example 2

Draw a graph of the solutions for each of the following equations by plotting points.

a. $y = 4 - 3x$ b. $y = x(x + 2)$

Solution a. To plot points, we first make a table. In making our table, we pick values for x that allow us to easily determine the values for y.

x	y = 4 − 3x
−3	4 − 3(−3) = 13
−2	4 − 3(−2) = 10
−1	7
0	4
1	1
2	−2
3	−5

Now we can plot these points. The points appear to lie in a straight line, so we draw a line through them. The line we draw represents all of the ordered pairs that are solutions to the equation $y = 4 - 3x$. We assume when drawing the line that it will extend in both directions so that it represents all of the points that solve $y = 4 - 3x$.

We can graph this equation in a calculator to gain confidence that our hand-drawn graph is correct. To do this, we should set our calculator window to match our hand-drawn graph.

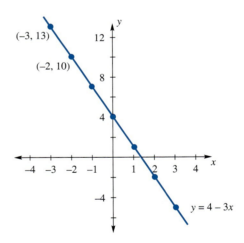

b. Similarly, we make a table of values for $y = x(x + 2)$ and plot these points. Remember that it is a good idea to pick both positive and negative numbers for x.

x	y = x(x + 2)
−3	−3(−3 + 2) = 3
−2	−2(−2 + 2) = 0
−1	−1
0	0
1	3
2	8
3	15

Next, we plot these points and draw the graph. These points clearly do not lie in a straight line. It is a good idea to check the shape of this graph using your graphing calculator before trying to draw the graph with these few points. Set the window the same as the graph you have drawn by hand.

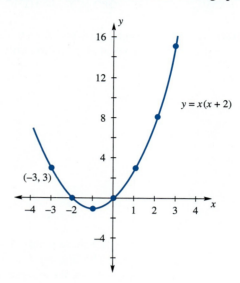

In parts a and b of this example, we used a table to find several points on the graph of the given equations. We then connected the points with a straight line or a smooth curve. We do not know enough about the graphs of these equations to be certain that we have graphed them completely. (A complete graph of an equation would include all places where the graph changes direction.) As you study more mathematics you will learn how the graphs of these and other equations should look. For now, when you are asked to graph an equation, be sure to pick several values for the input variable: some that are positive and some that are negative. This gives you the best chance of seeing how the complete graph looks.

Problem Set 2.3

1. Identify the solution to each equation by inspection. Check your solution.
   a. $x + 6 = 18$ b. $p - 2 = 7$ c. $6 + x = 9$

2. List three solutions to each equation.
   a. $y = x - 5$ b. $y = 10 - x^2$ c. $y = 2\sqrt{x}$

3. a. Complete the following table for the equation $y = 15 - x$.
   b. Graph the equation $y = 15 - x$ by plotting the points from your table.

x	$y = 15 - x$
-3	
-2	
-1	
0	
1	
2	
3	

4. a. Complete the following table for the equation.
   $$y = \frac{x + 3}{2}$$
   b. Graph the equation by plotting the points from your table.
   $$y = \frac{x + 3}{2}$$

x	$y = \dfrac{x+3}{2}$
-5	
-3	
-1	
0	
1	
3	
5	

5. a. Make a table of values for the equation $y = (x - 5)(x + 5)$.
   b. Draw a graph of the equation $y = (x - 5)(x + 5)$.

6. Draw a graph that represents the solutions to the equation $y = x^3$.

2.4 Modeling with Tables, Graphs, and Equations

Activity Set 2.4

1. Following is the graph of the cost of a phone call to Philadelphia, Pennsylvania, from Portland, Oregon. Use the graph to answer the following questions.

    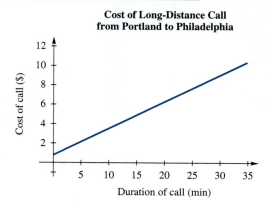

    a. How much does a call from Portland to Philadelphia cost that lasts 5 minutes? 20 minutes? 17 minutes?

    b. If you want to keep the cost of the call below $6.00, how long should you talk?

2. Following is a table used by a telephone company for the cost of a phone call to Philadelphia, Pennsylvania, from Portland, Oregon. Use the table to answer the questions.

    a. How much does a call from Portland to Philadelphia cost that lasts 5 minutes? 20 minutes? 17 minutes?

    b. If you want to keep the cost of the call below $6.00, how long should you talk?

Length of Call (min)	Total Cost of Call ($)
1	1.07
5	2.15
10	3.50
15	4.85
20	6.20
25	7.55
30	8.90
35	10.25

3. The equation $C = 0.27(m - 1) + 1.07$ gives us the cost of a phone call in dollars C to Philadelphia, Pennsylvania, from Portland, Oregon, in terms of the length of the call in minutes m. Use the equation to answer the following questions.

    a. How much does a call from Portland to Philadelphia cost that lasts 5 minutes? 20 minutes? 17 minutes?

    b. If you want to keep the cost of the call below $6.00, how long should you talk? Explain how you made your decision.

4. In Activities 1, 2, and 3 you answered the same questions using three different mathematical models: a graphical model (the graph), a numerical model (the table), and an algebraic model (the equation). Describe the advantages and disadvantages of using each model to answer the questions.

Discussion 2.4

In this section, we will look at mathematical modeling. A mathematical model can be a numerical table of values, a graph, or an algebraic equation that provides information about a problem situation. We will see how these models can be used to answer questions in the following examples.

Example 1

Windchill Temperatures. When the wind blows and the temperature is cold, we perceive the temperature to be colder than it actually reads. Wind lowers the body temperature by blowing away heat from the surface of your skin. To quantify this effect, the windchill factor was devised. The following table shows the equivalent windchill temperature for various wind speeds if the actual temperature is 30° Fahrenheit (°F). The *equivalent windchill temperature* is how cold your skin feels if exposed to the wind. (*Note:* Winds in excess of 40 mph produce little additional chilling effect.)

Use the table to answer the following questions based on a winter day when the actual temperature is 30°F.

   a. What is the equivalent windchill temperature if the winds are 35 mph?
   b. What is the equivalent windchill temperature if the winds are 11 mph?
   c. How fast are the winds blowing if the equivalent windchill temperature is 0°F?

EQUIVALENT WINDCHILL TEMPERATURES (FOR 30°F)

Wind Speed (mph)	Temperature (°F)
0	30
5	27
10	16
15	9
20	4
25	1
30	−2
35	−4
40	−5

Solution

a. Reading from the table, we see that if the wind is 35 mph, the equivalent windchill temperature is −4°F.

b. We see that 11 mph is not in our table of values. It falls between 10 mph and 15 mph, and is closer to 10 mph. Because the equivalent windchill temperature falls between 16°F and 9°F and is closer to 16°F, we estimate it to be 15°F.

c. At 25 mph the equivalent windchill temperature is 1°F and at 30 mph the equivalent windchill temperature is −2°F, so we can estimate that 27 mph winds result in a temperature that feels like 0°F to exposed skin.

Example 2

Heat and Humidity. When humidity combines with high temperature, the body's ability to keep the temperature under control through perspiration is reduced. Humidity refers to the amount of moisture in the air. On a warm day, the higher the humidity, the warmer it feels. *Apparent temperature* is how hot a person at a particular temperature and humidity feels. The following graph shows the relationship between apparent temperatures and relative humidity when the actual temperature is 85°F.

Use the graph to answer the following questions:

a. When the temperature is 85°F and the relative humidity is 50%, what is the apparent temperature?

b. At what relative humidity does a day at 85°F feel like 100°F?

c. When the temperature is 85°F and the relative humidity is 0%, what is the apparent temperature?

Solution When reading from a graph it is helpful to sketch in vertical and horizontal lines to assist in answering the questions. A ruler can help make estimates from a graph more accurate.

a. To find the apparent temperature when the relative humidity is 50%, draw a vertical line at 50% humidity on the horizontal axis to the point where the line intersects the graph. Then draw a horizontal line over to the vertical axis.

The apparent temperature is about 88°F.

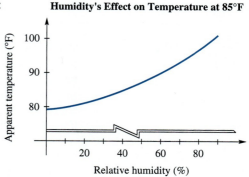

b. To find the relative humidity when the apparent temperature is 100°F, draw a horizontal line at 100°F on the vertical axis to the point where the line intersects the graph. Then draw a vertical line down to the horizontal axis.

The relative humidity is approximately 85%.

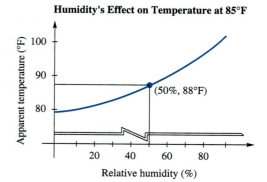

c. The apparent temperature when the relative humidity is 0% corresponds to the point (0, 78) on the graph. Therefore, the apparent temperature is about 78°F.

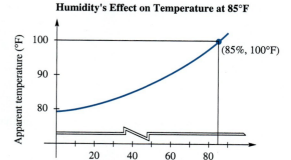

Example 3

Converting Degrees. A well-known formula is used to convert temperatures from degrees Fahrenheit (°F) to degrees Celsius (°C). The formula is

$$C = \frac{5}{9}(F - 32)$$

where C = degrees Celsius and
F = degrees Fahrenheit.

Use the formula to convert 70°F to Celsius.

Solution Substitute $F = 70$ into the formula.

$$C = \frac{5}{9}(70 - 32)$$

$$C \approx 21$$

We see that 70°F is approximately 21°C.

In the first three examples and in the activities, we saw how numerical, graphical, and algebraic models can be used to relate information. Each of the models has advantages and disadvantages. Often we need more than one model to help us understand a problem situation.

In each of the previous examples, we looked at the relationship between two different quantities. When we work with relationships in two variables, it is helpful to decide which variable depends on the other. For example, in the formula for the area of a circle, $A = \pi r^2$, the area clearly **depends** on the radius. Therefore, we call the area A the dependent variable and the radius r the independent variable. Knowing which variable is dependent and which is independent is helpful when you are modeling relationships between two variables.

Definition

A **dependent variable** is a variable whose value depends on the other variable's value. The other variable is known as the **independent variable.**

Example 4

In each of the following situations, decide which variable is the dependent variable and which is the independent variable.

a. The lake you are camping at this summer rents kayaks for $5 an hour. The cost of a rented kayak is C, and the number of hours it was rented is H.

b. A table for the amount of medication to prescribe lists the weight of the child in pounds and the amount of medication in milligrams. The weight of a child is W, and the amount of medication a pediatrician prescribes is M.

c. A community needs to predict the amount of water that will be available throughout the summer. One formula for this prediction uses the winter's snowfall S and the amount of spring water run-off W.

Solution
a. The cost of a rented kayak depends on how many hours you rent it for. So, C is the dependent variable and H is the independent variable.
b. The amount of medication a pediatrician prescribes depends on the weight of the child. So, M is the dependent variable and W is the independent variable.
c. The amount of spring run-off depends on the number of inches of snowfall. So, W is the dependent variable and S is the independent variable.

In a table we usually report the independent variable first because the other variable depends on this value.

Independent Variable	Dependent Variable

Or the table might be oriented horizontally.

Independent Variable	
Dependent Variable	

When graphing, the independent variable is graphed on the horizontal axis and the dependent variable is graphed on the vertical axis.

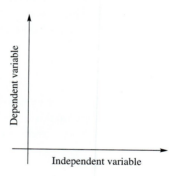

Example 5

Weather Balloon — A Numerical Model. A weather balloon can register the temperature at each meter above sea level. The temperature at sea level registers 60.0°F. After the balloon is released, the data show a relationship between the temperature and the height of the balloon above sea level. The temperature drops 0.16°F for every meter the balloon rises.

a. What would you define as the dependent variable? What would you define as the independent variable?
b. Make a table of several possible heights above sea level and the corresponding temperature.
c. Determine the temperature when the balloon is 65 meters above sea level.
d. Determine the height where the temperature falls below freezing.

Solution

a. In this problem, the temperature *depends* on the height of the balloon, the height does *not* depend on the temperature; therefore, the temperature is the dependent variable and height is the independent variable.

b. Because height is the independent variable, we can make our table by picking some values for the height h and use these to compute the temperature. We begin at $h = 1$ meter above sea level. If the temperature drops 0.16°F per meter above sea level, then we can find the temperature 1 meter above sea level by subtracting 0.16°F from 60.0°F. To determine the temperature at $h = 2$ meters above sea level, we need to subtract 0.16°F from 60.0°F twice. This can be simplified by subtracting twice 0.16°F from 60.0°F. We can use this same process to find the temperature at $h = 3$ meters, 4 meters, and so on. The computations are shown in the following table. In the table, the height above sea level is the independent variable; therefore, the height is in the left column.

Height Above Sea Level (m)	Temperature (°F)
0	60.0
1	$60.0 - 0.16 \approx 59.8$
2	$60.0 - 2 * 0.16 \approx 59.7$
3	$60.0 - 3 * 0.16 \approx 59.5$

Looking at these values, we see that the temperature does not drop very much. We follow the same process to determine more values for our table, but we should choose larger values for the height. Remember, the questions in parts c and d? Our goal is to make a table that will help us answer these questions. Thinking about the questions that need to be answered helps us decide what numbers to select for our table. We know that 65 meters is a value we need. We also know that we need to pick values for h such that the temperature falls below freezing.

Height Above Sea Level (m)	Temperature (°F)
0	60.0
20	$60.0 - 20 * 0.16 = 56.8$
40	$60.0 - 40 * 0.16 = 53.6$
60	$60.0 - 60 * 0.16 = 50.4$
65	$60.0 - 65 * 0.16 = 49.6$
80	$60.0 - 80 * 0.16 = 47.2$
100	$60.0 - 100 * 0.16 = 44.0$

c. Knowing that we needed the temperature for 65 meters above sea level, we include this value in the table. Therefore, we know that at 65 meters above sea level, the temperature is 49.6°F.

d. Freezing occurs at 32°F. Even though we extended our table to 100 meters above sea level, we still do not have temperatures this low. This means we need to expand the table further.

Height Above Sea Level (m)	Temperature (°F)
100	60.0 − 100 ∗ 0.16 = 44.0
150	60.0 − 150 ∗ 0.16 = 36.0
200	60.0 − 200 ∗ 0.16 = 28.0

Freezing, 32°F, falls between 36°F and 28°F in the table, so at a height somewhere between 150 and 200 meters the temperature falls below freezing. If we want to find a more precise answer we could try several more values between 150 and 200 meters. At this point, the process is somewhat tedious even with the help of a calculator.

Example 6

Weather Balloon Revisited Graphically. A weather balloon registers the temperature at each meter above sea level. From the data it is determined that the temperature drops 0.16°F for every meter the balloon rises. It registers 60.0°F at sea level. Sketch a graph of this problem situation. *Use the graph* to answer the following questions:

a. What is the temperature when the balloon is 65 meters above sea level?
b. At what height does the temperature fall below freezing?

Solution From the previous example, we have a table of values that we can plot to visualize the relationship between the height above sea level and the temperature. Height is on the horizontal axis because it is the independent variable. To draw this graph, we use the tables generated in the previous example and plot the points.

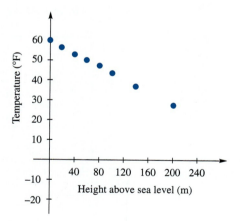

From the graph we again see that as the height of the balloon increases the temperature decreases. We can also see that the points appear to lie in a straight line, and in fact they do. We can connect these points to obtain a graph that is more useful in answering the questions. We sketch in horizontal and vertical lines to help us answer the questions.

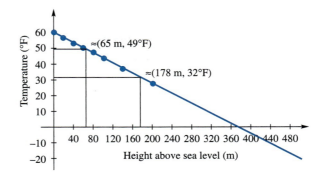

a. At 65 meters above sea level, the temperature is about 49°F.

b. To determine at what height the temperature falls below freezing, we use 32°F on the vertical axis. At approximately 178 meters the temperature drops below freezing.

Example 7

Weather Balloon Revisited Algebraically. A weather balloon registers the temperature at each meter above sea level. From the data we determine that the temperature drops 0.16°F for every meter the balloon rose. It registers 60.0°F at sea level.

a. Write an equation to model this problem situation.

b. *Use your equation* to determine the temperature when the balloon is 65 meters above sea level.

c. *Use your equation* to determine the height where the temperature falls below freezing.

Solution   a. To model an application with an equation, we must first define our variables.

Let

T = temperature in degrees Fahrenheit, and

h = height above sea level in meters

To model a situation with an equation, we write

Dependent variable = Expression written in terms of the independent variable

Recall that temperature T is the dependent variable, and height h is the independent variable. So, our equation should be

T = *Expression written in terms of* h.

The following table is from Example 5. Is it helpful in finding a pattern for writing this equation?

Height Above Sea Level (m)	Temperature (°F)
0	60.0
20	56.8
40	53.6
60	50.4
65	49.6
80	47.2

For most of us, it is very difficult to see the pattern from this table.

The next table, includes *the expressions* as well as the results we computed earlier. Using this next table, it is easier to determine the pattern and then to write an equation from the pattern. What pattern do you see? What numbers are constant (stay the same)? What numbers change?

Height Above Sea Level (m)	Temperature (°F)
0	60.0
20	$60.0 - 20 * 0.16 = 56.8$
40	$60.0 - 40 * 0.16 = 53.6$
60	$60.0 - 60 * 0.16 = 50.4$
80	$60.0 - 80 * 0.16 = 47.2$

In each of the expressions in the second column, we see a difference between 60.0 and the product of a number and 0.16. In the first term, 60.0 is constant. In the second term, 0.16 is constant. The number that varies in the second term is the same as the number in the "height" column. From these observations we can say that the temperature is the difference of 60.0 and the product of the height and 0.16. Then the equation is

$T = 60.0 - h * 0.16.$

We usually write numerical coefficients first, so we rewrite the equation as

$T = 60.0 - 0.16h$

Strategy

Using a Table to Create an Equation

When using a table to discover a pattern, be sure to write *the expressions* you use to determine the values, not just the results.

To verify our equation we choose a value for h and use the equation to determine T. We select a value from the table that is known, such as (40, 53.6).

$53.6 \stackrel{?}{=} 60.0 - 0.16 * 40$

$53.6 = 53.6$ ✔

Now we can use the equation to answer the questions.

b. To determine the temperature at 65 meters, we substitute 65 for h.

$T = 60.0 - 0.16h$

$T = 60.0 - 0.16 * 65$

$T = 49.6$

The temperature is about 49.6°F when the balloon is 65 meters high.

c. To determine the height at which the temperature is freezing, we substitute 32 for T.

$T = 60.0 - 0.16h$

$32 = 60.0 - 0.16h$

Not knowing h is more difficult, but we know from the table in Example 5 that the height is between 150 and 200 feet. We can build a table using our equation and a calculator with a table feature to help us answer the question.

Height Above Sea Level (m) h	Temperature (°F) $T = 60.0 - 0.16h$
150	36.0
160	34.4
170	32.8
180	31.2

The temperature drops below freezing at a height between 170 and 180 meters. We might estimate this to be about 175 meters. If we need a more accurate answer, we can build a table of values between 170 and 180.

Height Above Sea Level (m) h	Temperature (°F) $T = 60.0 - 0.16h$
171	32.6
172	32.5
173	32.3
174	32.2
175	32.0
176	31.8

The temperature reaches freezing at 175 meters above sea level. At any height above 175 meters, the temperature is below freezing.

We have now seen three ways to model a problem situation. We can use a **numerical model,** which often consists of a table of values. A table provides concrete values from which to analyze the behavior of the situation. We can use a **graphical model,** which consists of one or more graphs. A graph provides a visual model from which to analyze the behavior. We can use an **algebraic model,** which consists of an equation. Algebra provides us with a mathematical language to describe patterns. We can also see that these three models are intimately related. The graph is a visualization of the table. The equation is a generalization of the table. The algebra is a mathematical description of the graph. What are the advantages and disadvantages to each of these models?

In problem situations with two variables, one is called the **dependent variable** since its value depends on the other variable. The other variable is called the **independent variable.** In problems like the weather balloon, it is clear which variable depends on the other. In other situations it may not be obvious. If knowing either of the variables allows you to determine the other one, you can just assign one of the variables as the dependent variable and the other as the independent, and then proceed.

In the problem set in this section, we ask you to model some problems using all three techniques so that you can become familiar with the advantages and disadvantages of each. However, sometimes you must decide which model or models to use. More than one correct decision is usually possible, and the advantages and disadvantages you discover in the problem set will help prepare you to make this decision.

Problem Set 2.4

1. In each of the following tables, observe the pattern and use it to complete the table. For the last row in each table, write an expression in terms of the variable that generalizes the pattern.

a.
w	A
1	1 + 4
2	2 + 4
3	3 + 4
4	
w	

b.
t	S
1	5 * 1
2	5 * 2
3	5 * 3
4	
t	

c.
x	y
1	5 * 3
2	5 * 4
3	5 * 5
4	
x	

d.
p	q
1	3
2	3 * 3
3	3 * 3 * 3
4	
p	

e.
T	S
2	$2^2 - 5 * 2$
4	$4^2 - 5 * 4$
6	$6^2 - 5 * 6$
8	
T	

f.
n	P
2	$2^2 - 5 * 1$
4	$4^2 - 5 * 3$
6	$6^2 - 5 * 5$
8	
n	

2. The following two graphs depict the data in the table. Are the graphs correct? Explain.

x	y
0	0
1	1
2	4
3	9
4	16
5	25

a.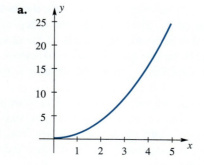

b.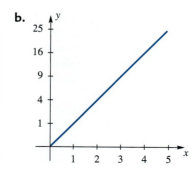

3. The following two graphs depict data in the table. Are the graphs correct? Explain.

x	y
0	0
1	2
2	4
3	6
4	8
5	10

a.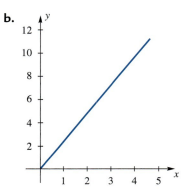

b.

4. The following two graphs depict the data in the table. Are the graphs correct? Explain.

x	y
0	1
1	2
2	4
3	8
4	16
5	32

a.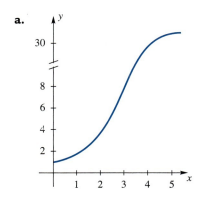

b.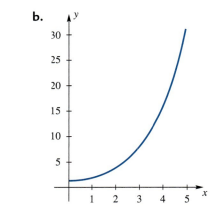

5. In each of the following situations, decide which variable is the dependent variable and which is the independent variable.
   a. You plan to put aside $35 each week into savings. The amount in savings is S and the number of weeks is W.
   b. The population of Argentina is increasing by about 86,000 people per year. If we know the population this year, we can predict the population P of Argentina for the number of years y from this year.
   c. A local bank uses a formula to relate the amount of time T a customer stands in line and the number of customers N standing in front of them.

6. **A Registered Letter.** A registered letter mailed from San Francisco to Chicago costs $5.09 for the first ounce and $1.20 for each additional ounce.
   a. How much does it cost to send a registered letter from San Francisco to Chicago weighing 1 ounce? 2 ounces? 4 ounces?
   b. Does the cost of sending the letter depend on its weight or does the weight depend on the cost? Given your response, what would you define as the dependent variable? What would you define as the independent variable?
   c. Make a table of at least five values for the cost of sending a registered letter for several weights (under 1 pound).
   d. Draw a graph of the cost of mailing in terms of the weight in ounces. The horizontal axis represents the independent variable and the vertical axis represents the dependent variable. Label your axes and include a scale for each axis.
   e. Write an equation for the cost of mailing in terms of the weight in ounces. Verify your equation.
   f. What is the cost of a letter weighing 3 ounces? 1 pound?
   g. How heavy a letter can you send for $15.00?

7. **The Capital Hotel** in Washington, D.C., charges a 13% tax plus a $1.50 per day room charge. Room rates are usually quoted before the tax and room charge are added. (The tax only applies to the room rate.)
   a. Your room rate for a conference is $90 per night. How much does it actually cost for one night? What is your actual cost at a quoted rate of $120 per night? $190 per night?
   b. What would you define as the dependent variable? What would you define as the independent variable?
   c. Make a table of several possible room rates and determine the total cost per night. Your table should include at least five points and range from $70 to $200 per night.
   d. Draw a graph of the total cost in terms of the room rate.
   e. Write an equation for the total cost per night in terms of the room rate. Verify your equation.
   f. Your room rate for a vacation is $186 per night. How much does it actually cost for the night?
   g. You need to stay for three nights and the room rate is $145 per night. How much does it actually cost for the three nights?

8. **New York Car Rental.** On a trip to New York you find that a midsize car rents for $41 a day with unlimited mileage. You can return the car full of gas or the rental company will charge $2.87 per gallon to refill the tank when you return the car. Because gas in New York is averaging $1.50 per gallon, you decide to fill the tank before you return the car. You need to rent the car for four days. Several places might be of interest to you during your four-day trip. If you assume the car you are driving gets 23 miles to the gallon, determine the cost to drive the rental 100 miles, 200 miles, and 400 miles. How far can you drive if you budgeted $200 for the total cost of the car rental over the four-day trip?

9. **The Car Rental Comparison.** In this problem we look at two different car rental companies, Cars-4-U and Beaters-R-Us.
   a. The Cars-4-U car company rents a compact car for $25 a day and 23¢ a mile.
      i. Complete the second column in the following table.
      ii. Draw a graph of the cost to rent a car for one day in terms of the number of miles traveled.
      iii. Write an equation for the cost to rent a car for one day in terms of the number of miles traveled. Verify your equation.
   b. The Beaters-R-Us car company does not charge a daily fee. They rent compact cars for 35¢ a mile.
      i. Complete the third column in the following table.
      ii. On the same axes as part a, draw a graph of the cost to rent a car for one day in terms of the number of miles traveled. It might be helpful to use a different color.
      iii. Write an equation for the cost to rent a car for one day in terms of the number of miles traveled. Verify your equation.
   c. What is the cost to rent a car from each company if you plan on driving 75 miles in one day? What is the difference?
   d. For each company, how far can you drive for $40.00?
   e. At what point are the costs of the two companies equal?
   f. If you had to drive 100 miles, which company would you use? Why?
   g. List at least one other consideration, besides cost, that may influence your choice.

Miles	Cost to Rent a Car from Cars-4-U for One Day	Cost to Rent a Car from Beaters-R-Us for One Day
0		
50		
100		
150		
200		
250		

10. **Dog Pen.** You just moved into a new home and want to build a dog pen for Max, your rather large shaggy dog. You have 70 feet of fencing and plan on building a rectangular pen. (A square is a special kind of rectangle.)

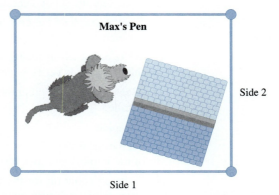

Side 1

a. Draw several rectangular shapes that use all 70 feet of fencing. What is the area of each of these?
b. If side 1 of the pen is 5 feet, what is the length of side 2? What is the area of this pen?
c. If the length of side 1 of the pen is 10 feet, what is the length of side 2? What is the area of this pen?
d. Complete a table similar to the following one.
e. Draw a graph of the area in terms of the length of side 1.
f. How should you build the pen so that Max has the maximum area?

Length of Side 1 (ft)	Length of Side 2 (ft)	Area (ft²)
5		
10		
15		
20		
25		
30		

11. Using the information for Problem 10, "Dog Pen," do the following:
    a. Write an equation for the length of side 2 in terms of the length of side 1.
    b. Write an equation for the area of the pen in terms of the length of side 1.

2.5 Evaluating Expressions with Units

Activity Set 2.5

1. **a.** Draw a rectangle whose area is 3 cm². (A square is a special kind of rectangle.)
   b. Draw a rectangle whose area is (3 cm)².

Discussion 2.5

We saw units play an important role in geometry applications and in unit conversions. In both of these situations we often know what kind of units we are starting with and what units we expect for our result. However, in many other applications the resulting units are not as obvious. These applications often arise in formulas from other disciplines, including chemistry, physics, and biology. In these instances we must analyze the units mathematically.

In this section, we will look at expressions that include numerical values and units. We will not have a context for these expressions, so we will use our understanding of units and mathematics to determine the numerical value and the units associated with the results.

Before we begin, we must have a clear understanding of how units work with exponents. In the activities, you considered two different expressions involving units and squares. The first expression, 3 cm², can be read as "three square centimeters." A square centimeter is the area equivalent to the area of a square whose side lengths are 1 cm. Therefore, a rectangle whose side lengths are 1 cm and 3 cm has an area of 3 cm².

Now let's compare this with the expression (3 cm)². This can be read as "the square of three centimeters." By the definition of exponents, (3 cm)² is equivalent to 3 cm ∗ 3 cm. The area of a square whose side lengths are 3 cm produces an area of 3 cm ∗ 3 cm = 9 cm².

In the first expression, 3 cm², we are working with square centimeters. In other words, just the units are being squared. In the second expression, (3 cm)², the exponent is applying to the measurement 3 cm and this is equivalent to 3 cm ∗ 3 cm = 3² cm². From this we can see that both the numerical value and the units are squared.

Example 1

The following expression arose from a geometry application. Determine the numerical value and the units of the result.

54.1 cm (25.3 cm + 14.8 cm)

Solution First let's analyze the units.

cm(cm + cm)	Inside the parentheses we have centimeters plus centimeters, which results in centimeters.
= cm(cm)	Then we have a product of centimeters and centimeters, which results in square centimeters.
= cm²	

The units of the result are square centimeters. Next, we can find the numerical value and combine this with the units for our final result.

$$54.1 \text{ cm } (25.3 \text{ cm} + 14.8 \text{ cm})$$
$$= 54.1(25.3 + 14.8) \text{ cm}^2 \quad \text{We know the units result in square centimeters.}$$
$$\approx 2170 \text{ cm}^2$$

Notice that the result was rounded to three significant digits.

Example 2

The following expression arose from an application. Determine the numerical value and the units of the result.

$$35 \frac{\text{miles}}{\text{h}} * 0.5 \text{ h} + 55 \frac{\text{miles}}{\text{h}} * 2.5 \text{ h}$$

Solution First let's analyze the units. In each term, we have $\frac{\text{miles}}{\text{h}} * \text{h}$. The hours will cancel resulting in miles. Then we have miles plus miles, which results in miles.

$$\frac{\text{miles}}{\cancel{\text{h}}} * \cancel{\text{h}} + \frac{\text{miles}}{\cancel{\text{h}}} * \cancel{\text{h}}$$
$$= \text{miles} + \text{miles}$$
$$= \text{miles}$$

The units of the result are miles. Therefore,

$$35 \frac{\text{miles}}{\text{h}} * 0.5 \text{ h} + 55 \frac{\text{miles}}{\text{h}} * 2.5 \text{ h}$$
$$= (35 * 0.5 + 55 * 2.5) \text{ miles} \quad \text{We know the resulting units are miles.}$$
$$= 155 \text{ miles}$$
$$\approx 160 \text{ miles}$$

Example 3

The following expression arose from a geometry application. Determine the numerical value and the units of the result.

$$\frac{2\pi(4.5 \text{ ft})^3}{3} + \pi(4.5 \text{ ft})^2 * 12 \text{ ft}$$

Solution First let's analyze the units.

$$(\text{ft})^3 + (\text{ft})^2 * \text{ft}$$
$$= \text{ft}^3 + \text{ft}^3$$
$$= \text{ft}^3$$

The units of the result are cubic feet. Next, we can find the numerical value and combine this with the units.

$$\frac{2\pi(4.5 \text{ ft})^3}{3} + \pi(4.5 \text{ ft})^2 * 12 \text{ ft}$$

$$= \left(\frac{2\pi * 4.5^3}{3} + \pi * 4.5^2 * 12\right) \text{ ft}^3$$

$$\approx 950 \text{ ft}^3$$

NOTE: To correctly evaluate the preceding expression, cube 4.5 in the first term and square 4.5 in the second term.

Example 4

The following expression arose at a plant nursery. Determine the numerical value and the units of the result.

$$\frac{110 \text{ ft} * 62 \text{ ft}}{\dfrac{550 \text{ ft}^2}{1 \text{ lb}}}$$

Solution First let's analyze the units.

$$\frac{\text{ft} * \text{ft}}{\dfrac{\text{ft}^2}{\text{lb}}}$$

$$= \frac{\text{ft}^2}{\dfrac{\text{ft}^2}{\text{lb}}} \qquad \text{In the numerator we have square feet.}$$

Here we have a complex fraction, that is, a fraction within a fraction. To simplify a complex fraction we need to remember some properties of division. Recall that dividing by a number is equivalent to multiplying by the reciprocal of that number. For example,

$$\frac{7}{\dfrac{1}{2}}$$

$$= 7 \div \frac{1}{2}$$

$$= 7 * \frac{2}{1} \qquad \text{Multiply by the reciprocal of the divisor.}$$

$$= 14$$

We can use this property of division to determine the units of our result.

$$= \frac{\text{ft}^2}{\frac{\text{ft}^2}{\text{lb}}}$$

$$= \text{ft}^2 \div \frac{\text{ft}^2}{\text{lb}}$$

$$= \text{ft}^2 * \frac{\text{lb}}{\text{ft}^2} \qquad \text{Multiply by the reciprocal of the divisor.}$$

$$= \text{lb} \qquad \text{The ft}^2 \text{ cancel, resulting in pounds.}$$

The units of the result are pounds; therefore,

$$\frac{110 \text{ ft} * 62 \text{ ft}}{\frac{550 \text{ ft}^2}{1 \text{ lb}}}$$

$$= \frac{110 * 62}{550} \text{ lb}$$

$$\approx 12 \text{ lb}$$

NOTE: To correctly evaluate the expression, you must *not* square 550. The expression 550 ft² represents a measurement of 550 square feet. Only the units are squared.

Example 5

The following expression arose in a physics class. Determine the numerical value and the units of the result.

$$-9.8 \, \frac{\text{m}}{\text{sec}^2} * 2.0 \text{ sec} + 35.0 \, \frac{\text{m}}{\text{sec}}$$

Solution First let's analyze the units.

$$\frac{\text{m}}{\text{sec}^2} * \text{sec} + \frac{\text{m}}{\text{sec}}$$

$$\frac{\text{m}}{\text{sec} * \text{sec}} * \text{sec} + \frac{\text{m}}{\text{sec}} \qquad \text{In the first term, one factor of seconds cancels.}$$

$$= \frac{\text{m}}{\text{sec}} + \frac{\text{m}}{\text{sec}}$$

$$= \frac{\text{m}}{\text{sec}}$$

The units of the result are meters per second; therefore,

$$-9.8 \, \frac{\text{m}}{\text{sec}^2} * 2.0 \text{ sec} + 35.0 \, \frac{\text{m}}{\text{sec}}$$

$$= 15.4 \, \frac{\text{m}}{\text{sec}}$$

In some applications, people determine the units of the result from the context of the problem. For example, if the application involves the amount of fertilizer necessary for a given lawn, the units for the answer are either *pounds* or possibly *bags*, depending on what information is given. However, for a solution to be read and understood the units must be included and used correctly. In addition, as we can see from the previous examples, sometimes the units of the result are not as obvious and, therefore, must be analyzed mathematically.

2.5 Evaluating Expressions with Units

Problem Set 2.5

1. For each of the following expressions, determine the numerical value and units of the result.

   a. $2 * 825 \text{ ft} + 140 \text{ ft}$

   b. $25 \text{ in.} (100 \text{ in.} - 76 \text{ in.})$

   c. $88 \dfrac{\text{ft}}{\text{sec}} * 50 \text{ sec}$

   d. $\pi (5.0 \text{ in.})^2$

   e. $\pi (36 \text{ ft}^2)$

   f. $\dfrac{12.3 \text{ cm}(7.4 \text{ cm} + 6.7 \text{ cm})}{2} + \pi (3.4 \text{ cm})^2$

   g. $\pi * 4.5 \text{ ft}[4.5 \text{ ft} * 16.0 \text{ ft} + \dfrac{4}{3}\pi * (4.5 \text{ ft})^2]$

   h. $273 \text{ K} * \dfrac{2.75 \text{ L}}{1.00 \text{ L}} - 273 \text{ K}$ (K represents the kelvin, which is a temperature measurement.)

   i. $(7.40 \text{ m})^2 + 100.0 \text{ m}^2$

   j. $\dfrac{46.2 \text{ kg} * \dfrac{1000 \text{ g}}{\text{kg}}}{20.0 \text{ cm} * 7.00 \text{ cm} * 30.0 \text{ cm}}$

   k. $2.34 \dfrac{\text{g}}{\text{L}} * \dfrac{22.4 \text{ L}}{\text{mol}}$

   l. $\dfrac{0.25\pi * (24 \text{ in.})^2 * 30. \text{ in.}}{231 \dfrac{\text{in.}^3}{\text{gal}}}$

   m. $\dfrac{49 \text{ g}}{98 \dfrac{\text{g}}{\text{mol}} * 2.0 \text{ L}}$

   n. $\dfrac{25 \dfrac{\text{ft}}{\text{sec}} - 12 \dfrac{\text{ft}}{\text{sec}}}{2.5 \text{ sec}}$

   o. $\dfrac{200.59 \dfrac{\text{g}}{\text{mol}}}{6.0221 * 10^{23} \dfrac{\text{molecules}}{\text{mol}}}$

   p. $\dfrac{215 \text{ g}}{2.40 \dfrac{\text{g}}{\text{cm}^3}}$

2. **Foreign Exchange.** The following are wholesale interbank transfer rates for Monday, May 12, 1997, 7:30 A.M., from U.S. Bank of Oregon. Retail rates for individuals vary from bank to bank.

To help you understand how to read the table, let's look at the first row. The first column tells us that one Australian dollar is approximately equal to 0.7791 U.S. dollar. The second column tells us that 1.2835 Australian dollars is approximately equal to 1 U.S. dollar. Use the table and unit conversions to answer the following questions.

  a. You are planning a trip to Mexico. You plan to take about $500 (U.S.). You exchange this for new pesos; how many new pesos do you receive?

  b. Chris and Sean are currently in Japan. After staying in Japan for three weeks, they have about 85,000 yen left. On their return trip to Canada they decide to stop in London. They exchange their yen for pounds. How many pounds do they get?

  c. When Chris and Sean return to Canada, they have about 440 Canadian dollars left, and their parents want to know how much money they spent in London. Determine how much money they spent during their stay in London.

	Value ($U.S.)	Units per dollar		Value ($U.S.)	Units per dollar
Australia, dollar	0.7791	1.2835	Israel, shekel	0.2951	3.3887
ds Austria, shilling	0.0840	11.9100	Italy, lira	0.000595	1679.9500
Brazil, real	0.9378	1.0663	Japan, yen	0.008440	118.4000
Canada, dollar	0.7186	1.3916	Mexico, new peso	0.1262	7.9270
Costa Rica, colon	0.0044	228.9300	New Zealand, dollar	0.6950	1.4388
France, franc	0.1753	5.7060	Philippines, peso	0.0379	26.3600
Germany, mark	0.5908	1.6925	Portugal, escudo	0.0059	170.4000
Great Britain, pound sterling	1.6162	0.6187	Singapore, dollar	0.6970	1.4348
Greece, drachma	0.0037	270.3900	Spain, peseta	0.0070	143.1200
Hong Kong, dollar	0.1291	7.7432	Switzerland, franc	0.7008	1.4270
India, rupee	0.0279	35.8100	Taiwan, dollar	0.0361	27.7300
Ireland, punt	1.5176	0.6589			

3. Write a single expression, including units, to determine the area of the following figure.

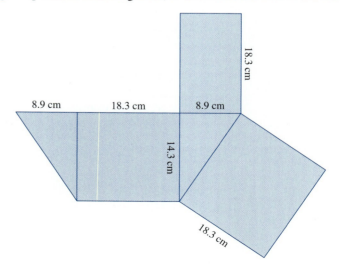

4. In a mathematics class students needed to determine the area of the following rectangle as part of a problem. Two answers predominated. About half of the class said the area was approximately 54 ft², and the other half said it was about 650 ft². Without computing the exact answer, determine which answer is reasonable. Explain how you made your decision.

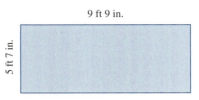

9 ft 9 in.

5 ft 7 in.

5. Without computing the exact area, determine which of the following answers is a reasonable approximation for the area of the circle. Explain how you made your decision.
   a. 32 ft²
   b. 84 ft
   c. 1006 ft²
   d. 32 ft
   e. 84 ft²
   f. 12,000 in.²
   g. 390 in.²

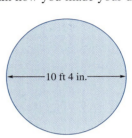

10 ft 4 in.

6. You are asked to evaluate the expression

$$\frac{\pi * (2.73 \text{ cm})^3}{3}$$

Identify all of the following calculator entries that correctly evaluate the numerical value of this expression. More than one response may be correct.
   a. (π * 2.73)^3/3
   b. π * 2.73^3/3
   c. π * (2.73)^3/3
   d. (π * 2.73^3)/3
   e. π * (2.73^3)/3
   f. (π * (2.73))^3/3

7. You are asked to evaluate the expression

$$42 \text{ ft}^2 * \frac{\$31.50}{10 \text{ ft}^2}$$

Identify all of the following calculator entries that correctly evaluate the numerical value of this expression.
   a. 42^2 * (31.50/10²)
   b. 42^2 * 31.50/10²
   c. 42 * 31.50/10
   d. (42 * 31.50)/10

8. Which of the following expressions are equivalent?
   a. π * 5 cm²
   b. π * (5 cm)²
   c. π * 25 cm
   d. π * 25 cm²

9. You are asked to evaluate the expression

$$(4.2 \text{ ft})^3 * \frac{1 \text{ gal}}{7.481 \text{ ft}^3}$$

Identify all of the following expressions that correctly evaluate the numerical value of this expression.

a. 4.2^3/7.481^3

b. 4.2/7.481

c. 4.2^3/7.481

d. 4.2/7.481^3

10. You are asked to evaluate the expression

$$\frac{32 \text{ ft}^2 + 27 \text{ ft}^2 + 18 \text{ ft}^2}{9 \frac{\text{ft}^2}{\text{yd}^2}} * \frac{\$22.50}{\text{yd}^2}$$

Identify all of the following calculator entries that correctly evaluate the numerical value of this expression.

a. (32^2 + 27^2 + 18^2)/9^2 * 22.50

b. ((32^2 + 27^2 + 18^2)/9^2) * 22.50

c. (32 + 27 + 18)/9 * 22.50

d. 32^2 + 27^2 + 18^2/9^2 * 22.50

e. 32 + 27 + 18/9 * 22.50

f. ((32 + 27 + 18)/9) * 22.50

g. (32^2 + 27^2 + 18^2)/9 * 22.50

h. (32 + 27 + 18/9) * 22.50

2.6 Formula Evaluation

Activity Set 2.6

1. Using the formula for the area of a rectangle, area = length * width, find the area of the square. Substitute the numerical values and the units into your formula.

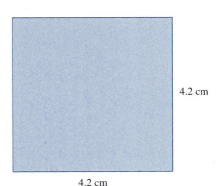

4.2 cm

4.2 cm

2. Using the formula for the area of a square, area = side², find the area of the square. Substitute the numerical values and the units into your formula.

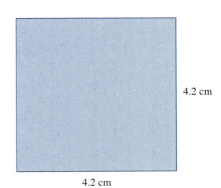

4.2 cm

4.2 cm

Discussion 2.6

In the activity set, you found the area of a square whose side length is 4.2 cm in two ways. Using the formula area = length * width, we obtain

$A = 4.2 \text{ cm} * 4.2 \text{ cm}$

$\approx 18 \text{ cm}^2$

Notice that we rounded the result to two significant digits and that our units are square centimeters.

Alternatively, because the figure is a square we can use the area formula for a square, which is area = side². To correctly substitute the side length of 4.2 cm into this formula we need to include parentheses.

$A = (4.2 \text{ cm})^2$

$= 4.2^2 \text{ cm}^2$

$\approx 18 \text{ cm}^2$

Using either formula, we obtain the correct value for the area. The area is about 18 cm².

What happens if we do not use parentheses when substituting 4.2 cm into the formula area = side²? Without parentheses the formula might look like the examples in the following table.

INCORRECT SUBSTITUTIONS FOR $s = 4.2$ cm INTO $A = s^2$	
Square Units Only	**Square Number Only**
$A = 4.2$ cm^2	$A = 4.2^2$ cm
$= 4.2$ cm^2	≈ 18 cm

In the first column only the units are squared, so we obtain the correct units but the incorrect numerical value. In the second column only the numerical value is squared, so we obtain the correct numerical value but the units are incorrect. Therefore, it is important to include parentheses when substituting a measurement into a formula.

> **Substituting Measurements into Formulas**
> To correctly substitute a measurement into an expression or formula involving a power, the measurement must be enclosed in parentheses.

Mathematical formulas can be used to model many situations. Electricians can determine the amount of electricity available in a circuit design. Design engineers can determine the pressure needed to produce a desired height for various nozzles of a fountain to create a visually pleasing effect. Police officers can calculate the speed of a car prior to an emergency stop by measuring the length of the skid marks. In each of these examples, the variables in formulas represent measured physical quantities that include units. It is important to understand how to use units in formulas.

Formulas come in two types. The first type of formula we will call a **general formula.** In a general formula, the variables are defined but the units for the variables are not designated. When substituting values into a general formula, the units must be substituted along with the numerical values to determine the units of the result.

Example 1

Use the following formula to answer the questions.

$$D = r * t$$

where r is rate,

t is time, and

D is distance traveled.

a. How far can you travel if you are driving 60 miles per hour for $3\frac{1}{2}$ hours?

b. How far does a lap swimmer actually swim in 35 minutes if she can swim 41 meters per minute?

c. How far can you travel if you are driving 60 miles per hour for 20 minutes?

Solution

a. Notice in the definition of the variables that t represents time, but the units for time are not specified. This means that we can substitute seconds, minutes, or hours into the formula. The units for the rate and distance are also not specified. This means that this is a general formula and we need to substitute both the numerical values and the units for the rate and time into the formula. The rate of the car is 60 miles per hour, or $60 \frac{\text{miles}}{\text{hour}}$. The time is $3\frac{1}{2}$ hours.

$$D = 60 \frac{\text{miles}}{\cancel{\text{h}}} * 3.5 \cancel{\text{h}}$$

$$= 210 \text{ miles}$$

The distance traveled is 210 miles. Notice that the units in the result are appropriate.

b. To answer the next question we substitute 41 meters per minute for the rate and 35 minutes for time.

$$D = 41 \frac{\text{m}}{\cancel{\text{min}}} * 35 \cancel{\text{min}}$$

$$= 1435 \text{ m}$$

Swimmers usually state their distance to the nearest length of a pool. If we assume that this is a standard 25-meter pool, we would say that the distance traveled by the swimmer was about 1425 meters.

c. Next, we need to determine how far you have driven if you drove 60 miles per hour for 20 minutes. Can we still use the formula? Let's look at what happens if we substitute the values and the units into the formula.

$$D = 60 \frac{\text{miles}}{\text{h}} * 20 \text{ min}$$

$$= 1200 \frac{\text{miles} * \text{min}}{\text{h}}$$

We can see that the units did not cancel and are not useful. We first need to convert 20 minutes to hours so that the units cancel.

$$20 \text{ min} = \frac{1}{3} \text{ h}$$

Using this result, we can now find the distance.

$$D = 60 \frac{\text{miles}}{\cancel{\text{h}}} * \frac{1}{3} \cancel{\text{h}}$$

$$= 20 \text{ miles}$$

The distance traveled is 20 miles.

> In **general formulas** the numerical values and the units are substituted into the formula, and the units of the result depend on these units.

The other type of formula is called a **unit-specific formula.** In a unit-specific formula, the formula defines the variables *and* defines the units to be used for all of the variables in the formula. In this type of formula, the numerical values are substituted into the formula, but not the units. The units of the result are specified in the definition of the variables for the formula.

Example 2

The following formula is used for determining the amount of calories burned when a person walks at a normal pace for 30 minutes.

$$C = 1.105w - 2.9$$

where w is the weight of the person in pounds, and

C is the number of calories burned.

a. Is this a general formula or a unit-specific formula?

b. Determine the number of calories burned if a 120-pound person walks for 30 minutes.

Solution

a. Notice that the units for each variable are specified. For example, weight *must be given in pounds*. This means that this is a unit-specific formula. If a person's weight is given in kilograms or some unit other than pounds, we first need to convert the weight to pounds *before* substituting the value into the formula.

b. To determine the number of calories burned by a 120-pound person walking for 30 minutes, we first check that the units match those defined by the formula. Because our units for weight are in pounds, we substitute 120 into the formula but not the units.

$$C = 1.105 * 120 - 2.9$$
$$\approx 130$$

Because C is defined to be number of calories burned, we say that this person burned about 130 calories during the walk.

Notice that we could not have substituted pounds into the formula to arrive at a reasonable answer.

$$C = 1.105 * 120 \text{ lb} - 2.9$$
$$= 132.6 \text{ lb} - 2.9$$

The first term is in units of pounds and the second term has no units. Because these units do not match, we cannot combine these terms.

Example 3

Vital Capacity. A person's *vital capacity* is the amount of air that can be exhaled after one deep breath. An estimate of the vital capacity can be obtained by the formula

$$V = 0.041h - 0.018A - 2.69$$

where V is the vital capacity in liters

h is the person's height in centimeters, and

A is the person's age in years.

a. Is this a general formula or a unit-specific formula?

b. Determine the vital capacity for Professor W who is 157.5 cm tall and 33 years old.

c. Determine the vital capacity for Professor S who is 5 ft 9 in. tall and 51 years old.

Solution

a. The formula defines what each variable represents and the units required in this formula; therefore, it is a unit-specific formula.

b. Because this is a unit-specific formula, we substitute the appropriate numerical values without units into the formula. We cannot ignore the units, but we use them in a different way. The height of Professor W is in centimeters and the age is in years. These units match the defined variable units so we can substitute the given values into the formula.

$$V = 0.041(157.5) - 0.018(33) - 2.69$$
$$\approx 3.2$$

Professor W's vital capacity should be about 3.2 liters. Notice that the unit of the result is liters, which was defined by the original formula.

c. We cannot substitute 5 ft 9 in. into the formula for h. The formula states that h must be in centimeters. So, first we must convert 5 ft 9 in. to centimeters.

$$5 \text{ ft } 9 \text{ in.}$$
$$= 5 \text{ ft} * \frac{12 \text{ in.}}{1 \text{ ft}} + 9 \text{ in.}$$
$$= 69 \text{ in.}$$

Next convert inches to centimeters to match the units specified in the formula.

$$69 \text{ in.} * \frac{2.54 \text{ cm}}{1 \text{ in.}}$$
$$\approx 175 \text{ cm}$$

Substituting into the formula, we have

$$V = 0.041(175) - 0.018(51) - 2.69$$
$$\approx 3.6$$

Professor S's vital capacity should be about 3.6 liters.

> In **unit-specific formulas,** first check that the units of the given measurements match the units defined by the formula. If they do not, you must first convert the measurements so that the units match. Once the units match those defined by the formula, the numerical values are substituted into the formula. The units are *not* substituted. The final answer must include the appropriate unit as defined by the formula.

Some formulas include variables with subscripts. For example, a trapezoid is a four-sided figure with two sides parallel as seen in the following figure. The area of a trapezoid is given as

$$A = \frac{h(b_1 + b_2)}{2}$$

The variables b_1 and b_2 are used in the formula to represent the lengths of the parallel sides of the trapezoid. The parallel sides are referred to as the bases of the trapezoid. The variables b_1 and b_2 represent two *different values*. The numbers 1 and 2 are called **subscripts.** We read the variables b_1 and b_2 as "b sub 1" and "b sub 2," respectively. The variables b_1 and b_2 represent different values in the formula just like L and W do in the formula for the area of a rectangle ($A = lw$). It is important not to confuse a subscripted variable such as b_2 with b^2 (which represents a product where b is used as a factor two times).

Example 4

a. Is the formula for the area of a trapezoid a general formula or a unit-specific formula?
b. Find the area of a trapezoid in which $b_1 = 3.5$ cm, $b_2 = 5.3$ cm, and $h = 2.8$ cm.
c. Find the area of a trapezoid in which $b_1 = 15$ ft, $b_2 = 26$ ft, and $h = 18$ in.

Solution

a. The variables are defined, but the units associated with the variables are not specified; therefore, the area of a trapezoid is a general formula.

b. Because this is a general formula, we must substitute both the numerical values and the units into the formula. Then the units must be analyzed in the formula to determine the units of the result. The variables b_1, b_2, and h in the formula represent lengths that we substitute into the formula along with their units.

$$A = \frac{2.8 \text{ cm } (3.5 \text{ cm} + 5.3 \text{ cm})}{2}$$

$$\approx 12 \text{ cm}^2$$

The area is approximately 12 square centimeters.

c. Looking at the measurements for part b, we see that some of the measurements are in feet and some are in inches. We want the area to be in square feet or square inches (not inch-feet!). We can convert all of the measurements to feet or to inches. We will use feet in this example.

$$18 \text{ in.} * \frac{1 \text{ ft}}{12 \text{ in.}} = 1.5 \text{ ft}$$

Substitute the numerical values and units into the formula.

$$A = \frac{1.5 \text{ ft } (15 \text{ ft} + 26 \text{ ft})}{2}$$

$$\approx 31 \text{ ft}^2$$

The area is approximately 31 square feet.

In this section, we saw some examples of how formulas are used in applications of mathematics. When we evaluate a formula the numerical result and the appropriate units of the result are required. In **general formulas,** the variables can represent measurements, such as length, weight, and time, but the units for these are not specified. In a general formula, the numerical values and the units are substituted into the formula. The units of the result are determined from this substitution. In **unit-specific formulas,** the units are specified when the variables are defined. Care must be taken that the units of the values you are given match those specified in the formula. Once this is accomplished, only the numerical values are substituted into the formula. The units of the result are specified in the definition of the variables. When evaluating a formula, we must first determine which of these types of formulas we are working with, then substitute and evaluate accordingly.

Problem Set 2.6

1. Determine whether the following formulas are general or unit-specific.

   a. $D = r * t$
      where D is the distance traveled,
      r is the speed of the object, and
      t is the time the object has been traveling.

   b. $E = IR$
      where E is the voltage in volts,
      I is the current in amperes, and
      R is the resistance in ohms.

   c. $V = {}^-gt + v_0$
      where V is the speed of the object,
      g is the acceleration due to gravity,
      t is the length of time since the object was thrown, and
      v_0 is the initial speed of the object.

   d. $C = \dfrac{\pi DN}{12}$
      where C is the cutting speed of a lathe in feet per minute (fpm),
      D is the diameter in inches, and
      N is the revolutions per minute (rpm).

   e. $L = \pi D + 2C$
      where L is the length of the pulley belt,
      C is the center to center distance, and
      D is the diameter of pulley.

   f. $E = \dfrac{I - P}{I}$
      where E is the engine efficiency,
      I is the heat input, and
      P is the heat output.

2. The surface area S of a silo formed by a cylinder with a hemisphere on the top is given in the following formula. Find the surface area if the radius is 6.2 feet and the height is 24 feet.

   $$S = \pi r(2h + 2r)$$

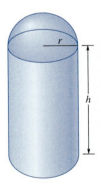

3. The cost to carpet a rectangular room is given by the formula

   $$C = l * w * c$$

   where C is the total cost to carpet,
   l is the length of the room,
   w is the width of the room, and
   c is the unit cost of the carpet.

   Determine the cost to carpet a room that is 13.5 feet long and 10 feet wide with carpet that costs $15.50 per square yard.

4. The speed of an object thrown into the air is given by the formula

$$V = {}^-gt + v_0$$

where V is the speed of the object,

g is acceleration due to gravity,

t is the length of time since the object was thrown, and

v_0 is the initial speed of the object.

   a. Find V given $g = 9.8$ m/sec^2, $t = 2.5$ sec, and $v_0 = 53$ m/sec.
   b. Find V given $g = 32$ ft/sec^2, $t = 5.0$ sec, and $v_0 = 125$ ft/sec.

5. The capacity, in gallons, of cylindrical tanks can be found using the formula

$$C = 0.0034 D^2 L$$

where C is capacity in gallons,

D is diameter of tank in inches, and

L is length of tank in inches.

   a. What is the capacity of a tank, in gallons, if the diameter is 10.0 inches and the length is 30.0 inches?
   b. What is the capacity of a tank in a filling station that measures 6 feet in length and 4 feet in diameter?
   c. A drum measures 4 feet in height and 25 inches in diameter. Is this a 25-, 50-, 100-, 200-, or 500-gallon drum?

6. The speed of a car can be determined by the formula

$$V = \frac{C * W}{R * 168}$$

where V is the speed of the car in miles per hour,

C is the crankshaft speed in revolutions per minute (rpm),

W is the tire radius in inches, and

R is the rear axle ratio.

Find the speed of the car if the crankshaft is running 1500 rpm, the tire radius is 32 cm, and the rear axle ratio is 3.50.

7. The air resistance on an airplane can be determined by the formula

$$AR = 0.0025 * V^2 * FA * DC$$

where AR is the air resistance in pounds,

V is the speed in miles per hour,

FA is the frontal area in square feet, and

DC is the drag coefficient.

When an airplane is flying at the speed of 528 kilometers per hour, what is the air resistance on the plane if the frontal area is 5.7 square meters, and the drag coefficient is 5.5?

8. The velocity, in meters per second, of sound in air is approximately given by the formula

$$v = \sqrt{\frac{1.410p}{d}}$$

where v is velocity in meters per second (m/sec),

   p is the pressure in pascals (Pa), and

   d is the density in kilograms per cubic meter (kg/m³).

Find v given $p = 1.01 * 10^5$ Pa, and $d = 1.29$ kg/m³.

9. **Cardiac Output.** The volume of blood pumped by the heart per unit of time is called the **cardiac output.** The cardiac output C of a person is equal to the product of the person's heart rate H and the volume of blood pumped per heart beat (stroke volume) S. Determine a person's cardiac output if the heart rate is 68 beats per minute and the stroke volume is 75 milliliters per beat.

10. **Doppler Shift.** Have you ever noticed the change in pitch in a train's whistle when the train approaches you and when it's moving away? The pitch seems higher as the sound gets closer to you and drops as the train moves away. To a person riding on the train, that same whistle has a single pitch. The change in the pitch of the train whistle to the observer on the side of the tracks can be explained by a phenomenon called the Doppler shift. The pitch that we hear is determined by the frequency of the sound wave, that is, the greater the frequency, the higher the pitch. We can determine the frequency heard by the observer using the formula

$$f = \frac{Fc}{c - v}$$

where F is the frequency of the sound to someone riding on the train,

   c is the speed of sound,

   v is the speed of the train, and

   f is the frequency of the sound heard by the observer.

(*Note:* F and f represent different variables in this formula.)

Find f if $F = 420$ vibrations per second, $c = 1{,}120$ feet per second, and $v = 95$ feet per second. Is the train moving toward the observer or away?

11. **Maximum Heart Rate.** You can determine your maximum heart rate in beats per minute by subtracting one half your age in years from 210, then subtract 1% of your body weight in pounds, and then add 4 if you are male or add 0 if you are female.

   a. Write out these two formulas. Identify the variables.
   b. Are these unit-specific or general formulas?
   c. Use the appropriate formula to determine the maximum heart rate for a 50-year-old male who weighs 175 pounds.
   d. Compare your answer from part c to your maximum heart rate.

12. The Olympians. The average speed of a person who has traveled 200 meters is approximately given by the formula

$$S = \frac{450}{t}$$

where S is the person's average speed in miles per hour, and

t is the amount of time, in seconds, it took to travel the 200 meters.

a. In the 1996 Summer Olympics, Michael Johnson, of the United States, won the 200-meter race in the world record time of 19.32 seconds. Use the preceding formula to determine his average speed in miles per hour.

b. Use unit fractions to determine where the number 450 in the formula came from. (*Hint:* Start with $\frac{200 \text{ m}}{t \text{ sec}}$ and convert this to miles per hour. Use 1.6 km $\approx$ 1 mi.)

c. Write a formula to determine the speed, in miles per hour, of a person who has traveled d meters in t seconds. Use your formula to determine the average speed of each of the following Olympians.

In swimming:

   Amy Van Dyken, of the United States, won the gold medal in the women's 50-meter freestyle with a time of 24.87 seconds.
   Aleksandr Popov, of Russia, won the gold medal in the men's 50-meter freestyle with a time of 22.13 seconds.
   Brad Bridgewater, of the United States, won the gold medal in the men's 200-meter backstroke with a time of 1:58.54 (1 minute and 58.54 seconds).
   Brooke Bennett, of the United States, won the gold medal in the women's 800-meter freestyle with a time of 8:27.89 (8 minutes and 27.89 seconds).

In rowing:

   Both the U.S. men and women's team won a silver medal in the coxless four.
   Each race is 2000 meters long.
   The men's time was 5:56.68, and the women's time was 6:31.86.

In track and field:

   Donovan Bailey, of Canada, won the gold medal in the men's 100-meter race with a time of 9.84 seconds.
   Haile Gebrselassie, of Ethiopia, won a gold medal in the men's 10,000-meter race with a time of 27:07.34.
   Fatuma Roba, of Ethiopia, won the women's marathon with a time of 2 hours, 26 minutes, and 5 seconds. A marathon is 26 miles, 385 yards.

d. Explain why it was beneficial to write a formula to determine the speed of these Olympians.

13. The formula for the surface area of a right circular cone is given as

    $SA = \pi r^2 + \pi r s$

    where SA is surface area,

       r is radius, and

       s is slant height.

Determine the surface area for a cone whose radius is $3\frac{3}{4}$ inches and slant height is $11\frac{5}{8}$ inches.

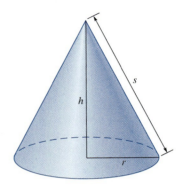

Chapter 2 Summary

In this chapter we began our study of algebra. We learned that letters that represent numbers are called **literal symbols.** If a literal symbol can take on different values it is called a **variable.** If a literal symbol represents a value that does not change, like π, it is called a **constant.**

We used variables to write equations relating two different kinds of information. In doing this, we start by identifying the independent and dependent variables. The **dependent variable** is the variable whose value depends on the other variable's value. The other variable is the **independent variable.**

In **creating a table of values,** the independent variable is represented first. This means that the independent variable either occurs in the leftmost column or in the top row, depending on the orientation of the table.

In **creating a graph,** the independent variable is graphed on the horizontal axis and the dependent variable on the vertical axis. The horizontal axis must be scaled in equal increments; likewise, the vertical axis must be scaled in equal increments. However, the scale used on the horizontal axis may be different from that used on the vertical axis. The scale should be clearly indicated on each axis. Each axis should include a label indicating what it represents. A break may only be used at the beginning of the axis and must be clearly indicated.

To **graph an equation,** we make a table of values, plot the points from the table, and connect the points with a smooth curve. At this point, we do not know how the graph of a given equation will look, so we like to choose several values for a table. Even when we are finished we cannot be sure that we have a "complete" graph. We will begin to study the graphs of specific equations later in this course.

In this chapter we looked at relationships in tables, on graphs, and through equations. Each view of a relationship has advantages and disadvantages. Perhaps one of the more challenging tasks is to create an equation from a problem situation. In Section 2.4, we saw that it is often helpful to use a table to create an equation. We begin by using the problem situation to make a table. In the table, we *write the expressions* used to determine the values, not just the results. These expressions can then be generalized into an equation.

We concluded the chapter by looking at formula evaluation. We saw that formulas are of two types: unit-specific and general. In a **unit-specific formula,** the formula defines the variables and the units to be used with the variables. Therefore, to evaluate a unit-specific formula, we substitute the numerical value of a variable, assuming it has units that match those defined by the formula. If the units do not match those defined in the formula, we must first convert the units and then substitute the new numerical value. The units on the final answer are defined by the formula.

In a **general formula,** the variables are defined but the units are not specified. To use a general formula, we substitute the numerical value and the units, making sure the units are compatible. The units of the result are determined by analyzing the units in the formula.

Throughout this chapter we saw that problem situations can be modeled in three ways: with a table of values, with a graph, or with an equation. All three of the models are related. How you enter a problem depends on how the information is given to you and what representation you are most comfortable with.

Chapter Three

Working with Algebraic Expressions and Equations

In Chapter 2, we modeled many patterns and problems algebraically with equations. In the process, we found that different people might model the same situation with apparently different equations. However, by substituting numbers into the different equations, these equations appeared to be equivalent. In Chapter 3, we will learn how to show algebraically that the different equations are in fact equivalent.

At the end of this chapter, we will be able to simplify expressions by combining like terms, multiplying factors, and applying the distributive property. Then, we will combine these techniques with some properties of equality to solve linear equations and linear literal equations.

3.1 Simplification of Algebraic Expressions

Often we think of algebra as symbolic and not geometrical; however, we can represent algebra geometrically. In this activity we use various algebra pieces to represent algebraic expressions.

We represent algebra pieces using an area model. The following figure shows picture representations of three of the pieces. The piece whose dimensions are 1×1 represents 1 unit because $1 \times 1 = 1$. The variable x represents an arbitrary number. The piece whose dimensions are $1 \times x$ represents the arbitrary number x because $1 \times x = x$. Similarly, the piece whose dimensions are $1 \times y$ represents the different arbitrary number y because $1 \times y = y$.

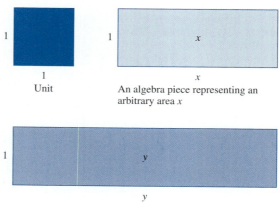

An algebra piece representing an arbitrary area y

With these pieces we can represent algebraic expressions. For example, $2x + 1$ can be represented by two x pieces and 1 unit as shown in the following diagram.

The square of x, x^2, can be represented by a square with side length x. Similarly, y^2 can be represented by a square with side length y.

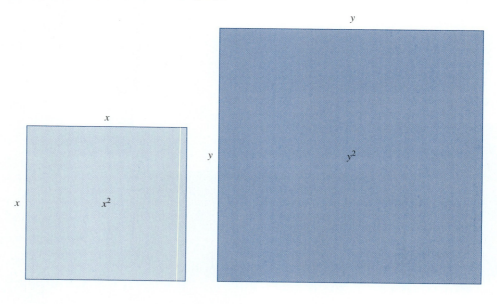

The product of *x* and *y* can be represented by a rectangle whose width is *x* and length is *y*.

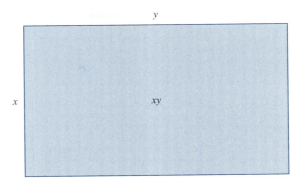

1. The expression $3x + 2 + x + 1$ can be represented with the algebra pieces as shown in the following figure.

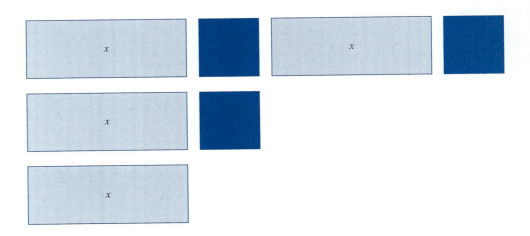

   a. How many terms does the expression $3x + 2 + x + 1$ have?
   b. Rearrange the algebra pieces and rewrite $3x + 2 + x + 1$ as an equivalent expression with only two terms.
   c. Pick a number for *x*, other than 0, 1, or 2. Verify that your expression in part b is correct by substituting the number you picked for *x* into $3x + 2 + x + 1$ *and* into your result for part b. You should get the same value for both expressions. Did you?

2. For expressions a and b follow these directions:
   i. Identify the number of terms in the original expression.
   ii. Represent the *original* expression using algebra pieces.
   iii. Rearrange the algebra pieces to rewrite the expression with the fewest possible terms.
   iv. Verify your result to part iii by choosing a value (other than 0, 1, or 2) for each variable and substituting these values into the original expression and into your result for part iii.
   a. $4x + 2x^2 + x^2 + 1$ b. $3y + 2xy + x + xy + 2y$

3. In Activities 1 and 2 you simplified expressions. You found that "like" terms are combined. Describe *in detail* what makes terms "alike."

4. The expression ⁻x is read as "the opposite of x." The expression ⁻1x is read as "negative one times x." Complete the following table. Show your substitutions.

x	⁻x	⁻1x
3		
2		
1		
0		
⁻1		
⁻2		

Based on your results from the table, what is the relationship between the expressions ⁻x and ⁻1x?

Discussion 3.1

Algebraic Vocabulary

In Section 1.1, we defined the terms of a numerical expression as the parts of a sum or difference. If the expression did not consist of a sum or difference, then the entire expression was a single term. This same definition works for algebraic expressions. The technique of identifying the terms of an expression by finding all the addition and subtraction symbols that are not in grouping symbols also works with algebraic expressions.

Example 1

Identify the terms of the algebraic expression $3k - 2k^2 + k(k + 8)$.

Solution To identify the terms, we first find the addition and subtraction symbols that are not in grouping symbols. Those are pointed at in the following expression.

$$3k - 2k^2 + k(k + 8)$$

These symbols separate the expression into the three terms that are underlined as shown.

$$\underline{3k} - \underline{2k^2} + \underline{k(k + 8)}$$

We can see that the three terms of the expression are $3k$, $2k^2$, and $k(k + 8)$.

Example 2

Identify the factors of each of the following expressions.

    a. $3km$ b. $-2k^2$ c. $k(k + 8)$

Solution Recall that in a given product, the quantities being multiplied are called the factors.

a. The factors of $3km$ are 3, k, and m. (*Note:* These are the *simple factors* of $3km$. Products formed by combining these are also factors, that is, $3k$, $3m$, km, and $3km$ are also factors of $3km$. Assume in this text that when we ask for the factors of a term, we only want a list of the simple factors.)

b. The factors of $-2k^2$ are -2, k, and k. (Recall that k^2 means $k * k$.) We could also say that the factors are -2 and k^2.

c. The factors of $k(k + 8)$ are k and $k + 8$.

We need some additional vocabulary to make it easier for us to talk about our processes and strategies as we learn to simplify algebraic expressions and solve equations.

> **Definition**
> In a single term, the numerical factor is known as the **numerical coefficient** of the term. If there is no numerical factor, the numerical coefficient is 1.

> **Definition**
> In a single term, the **coefficient** of a specified factor consists of the product of all of the other factors in the term. If there are no other factors, the coefficient is 1.

Example 3

For the expression $-5mpt^2$ do the following.

    a. Identify the numerical coefficient.
    b. Identify the coefficient of t^2.
    c. Identify the coefficient of p.
    d. Identify the coefficient of m.

Solution

a. The numerical coefficient of $-5mpt^2$ is -5.

b. We identify all of the factors of the expression other than t^2. These are -5, m, and p; therefore, the coefficient of t^2 is $-5mp$.

c. The coefficient of p in the expression $-5mpt^2$ is $-5mt^2$.

d. The coefficient of m in the expression $-5mpt^2$ is $-5pt^2$.

Using the algebra pieces in Activities 1–3, you may have discovered that certain terms of an algebraic expression can be combined. Those terms that can be combined are called like terms.

> **Definition**
> Terms that have the same literal factors are called **like terms**.

Example 4

Which of the following are like terms?

$$4t \quad t \quad 4t^2 \quad 4 \quad 5t \quad \frac{t}{3} \quad 4tp$$

Solution Like terms have the same literal factors. In other words, like terms have the same variables, and the variables are raised to the same powers.

The terms, $4t$, t, $5t$, and $\frac{t}{3}$ each have the variable t and it occurs as a factor exactly once; therefore, $4t$, t, $5t$, and $\frac{t}{3}$ are like terms. The term $4t^2$ has the variable t but it is a factor twice, 4 does not have a factor of t, and $4tp$ has a factor of p, which none of the other terms have.

> **Combining Like Terms**
>
> To combine like terms, add or subtract the numerical coefficients of the terms and keep the same literal factors.

Example 5

For the expression $3m^2 - \frac{5}{3}mp + \frac{p}{4} + m^2$ do the following.

a. Underline the terms.

b. Identify the numerical coefficient of each term.

c. Combine any like terms in the expression.

Solution a. The expression has four terms. Underlining the terms gives us

$$\underline{3m^2} - \underline{\frac{5}{3}mp} + \underline{\frac{p}{4}} + \underline{m^2}$$

b. The numerical coefficient of the first term is 3.

The numerical coefficient of the second term can be considered $\frac{5}{3}$ or $-\frac{5}{3}$, depending on whether you think of the subtraction as addition of the opposite.

The expression $\frac{p}{4}$ can be written as $\frac{1}{4} * p$. Therefore, the numerical coefficient of the third term is $\frac{1}{4}$.

Because the term m^2 is the same as $1m^2$, the numerical coefficient of the fourth term is 1.

c. The terms $3m^2$ and m^2 are like terms because they have the same literal factor of m^2. Adding the numerical coefficients of 3 and 1, we get $4m^2$. The expression can then be simplified to $4m^2 - \frac{5}{3}mp + \frac{p}{4}$.

Simplifying Algebraic Expressions

In this section, we will begin simplifying algebraic expressions. To do this it is important that we understand and apply the vocabulary correctly. Consider the following examples.

Example 6

a. List the factors of $5 * 7x$.

b. Which of the following expressions have the same factors as $5 * 7x$?

$7 * 5x \qquad 7x \qquad 5x * 7x \qquad 5x * 7 \qquad 35x$

Solution

a. The factors of $5 * 7x$ are 5, 7, and x.

b. Let's list the factors of each expression.

Expression	Factors
$7 * 5x$	7, 5, and x
$7x$	7 and x
$5x * 7x$	5, 7, x, and x
$5x * 7$	5, 7, and x
$35x$	35 and x or 5, 7, and x (because $5 * 7 = 35$)

Because $5 * 7x$, $7 * 5x$, $5x * 7$, and $35x$ all have the same factors, they must be equivalent. The simplest version of these is $35x$; therefore, $5 * 7x$, $7 * 5x$, and $5x * 7$ all simplify to $35x$. Notice that, in essence, we are rearranging the factors so that the numerical factors can be multiplied.

Example 7

a. List the factors of $5x * 7x$.

b. Which of the following expressions have the same factors as $5x * 7x$?

$35x \qquad 35x * 7x \qquad 35 * x * x \qquad 35x^2 \qquad 5 * 7 * x * x$

Solution

a. The factors of $5x * 7x$ are 5, 7, x, and x.

b. Let's list the factors of each expression.

Expression	Factors
$35x$	35 and x or 7, 5, and x (because $5 * 7 = 35$)
$35x * 7x$	35, 7, x, and x or 5, 7, 7, x, and x
$35 * x * x$	35, x, and x or 5, 7, x, and x
$35x^2$	35, x, and x or 5, 7, x, and x
$5 * 7 * x * x$	5, 7, x, and x

Because, $5x * 7x$, $35 * x * x$, $35x^2$, $5 * 7 * x * x$ all have the same factors, they must be equivalent. The simplest version of these is $35x^2$. Therefore, $5x * 7x$, $35 * x * x$, and $5 * 7 * x * x$ all simplify to $35x^2$. Notice again, we multiply the numerical factors. This time we also write repeated variable factors using exponents.

Example 8

Simplify each expression.

   a. $8m * 3$ b. $9k * 4m$ c. $4p * p$ d. $mt * m$ e. $^-2x * 9x$

Solution

a. The expression $8m * 3$ has factors 8, 3, and m. Therefore, it can be rewritten as $24m$.

$$8m * 3 = 24m$$

b. The expression $9k * 4m$ has factors 9, 4, k, and m; therefore, $9k * 4m$ simplifies to $36km$.

c. The expression $4p * p$ has 4 as a factor and p as a factor twice; therefore, $4p * p$ simplifies to $4p^2$.

d. The expression $mt * m$ has two factors of m and one factor of t; therefore, $mt * m$ simplifies to m^2t.

e. The factors of $^-2x * 9x$ are $^-2$, 9, x, and x; therefore, $^-2x * 9x$ simplifies to $^-18x^2$.

In Example 5b, we noted that m^2 is the same as $1m^2$. In Activity 4, you saw that *the opposite of x* is equivalent to *negative 1 times x*. That is, $^-x = ^-1 * x$. Writing the 1 or $^-1$ reminds us what the numerical coefficient of the term is. This can be important in combining like terms and in other simplifying that we will encounter.

Example 9

Simplify each expression, and verify your results numerically.

   a. $7k + 4k * 2p + 6kp + k$
   b. $14t^2 - 2 * 6t + t * 3t + 10t$
   c. $^-MT + 8M^2 - M * 4T + 1$

Solution

a. The expression $7k + 4k * 2p + 6kp + k$ contains four terms. The second term can be simplified, so we begin there.

$$7k + \underline{4k * 2p} + 6kp + k$$

$= 7k + 8kp + 6kp + 1k$ The second term has factors 4, 2, k, and p; therefore, $4k * 2p$ can be rewritten as $8kp$.

$= 8k + 14kp$ The first and last terms are alike because they contain the variable k. The middle terms are also alike because they contain the variables k and p.

To numerically verify our result we choose a value for each variable and evaluate the original and ending expressions to see if they result in the same value.

VERIFY   Let $k = 3$ and $p = 4$

$$7k + 4k * 2p + 6kp + k \stackrel{?}{=} 8k + 14kp$$

$$7 * 3 + 4 * 3 * 2 * 4 + 6 * 3 * 4 + 3 \stackrel{?}{=} 8 * 3 + 14 * 3 * 4$$ Substitute the chosen values.

$$192 = 192 \quad ✓$$ Evaluate each side using your calculator.

We conclude that $7k + 4k * 2p + 6kp + k$ simplifies to $8k + 14kp$.

In selecting a value, we avoid 0, 1, and 2. These special values can make expressions appear to be equal when they are not because of properties like $0 * n = 0$ and $1 * n = n$. In addition, $2 + 2$, $2 * 2$, and 2^2 are all equal to 4.

b. The expressions that we have simplified so far all involved terms connected by addition rather than subtraction.

> **Definition**
>
> Subtraction is addition of the opposite. That is, $a - b = a + {}^-b$. In words, *a minus b is equal to a plus the opposite of b*.

In simplifying expressions involving subtraction, it is helpful to use the preceding definition to rewrite the subtraction as addition of the opposite.

The expression $14t^2 - 2*6t + t*3t + 10t$ has four terms. Because this expression contains subtraction, we begin by rewriting the subtraction as addition of the opposite.

$\underline{14t^2} - \underline{2*6t} + \underline{t*3t} + \underline{10t}$	Underline the terms.
$= 14t^2 + {}^-2*6t + t*3t + 10t$	Rewrite the subtraction as addition of the opposite in the second term.
$= 14t^2 + {}^-12t + 3t^2 + 10t$	Simplify the middle two terms.
$= 17t^2 + {}^-2t$	The $14t^2$ and $3t^2$ are like terms. We add their coefficients to get $17t^2$. The ^-12t and $10t$ are like terms. We add their coefficients to get ^-2t.
$= 17t^2 - 2t$	The result can be rewritten as subtraction.

VERIFY Let $t = 5$. Notice that our choice for t is a number that does not occur in the original problem. This allows us to distinguish the numbers that are substituted from the other numerical factors and terms.

$$14t^2 - 2*6t + t*3t + 10t \stackrel{?}{=} 17t^2 - 2t$$
$$14*5^2 - 2*6*5 + 5*3*5 + 10*5 \stackrel{?}{=} 17*5^2 - 2*5$$
$$415 = 415 \quad ✔$$

We conclude that $14t^2 - 2*6t + t*3t + 10t$ simplifies to $17t^2 - 2t$.

c.

$\underline{{}^-MT} + \underline{8M^2} - \underline{M*4T} + \underline{1}$	Underline the terms.
$^-1MT + 8M^2 + {}^-1M*4T + 1$	Rewrite subtraction as addition of the opposite. Notice that the opposite of M, ^-M, is written as ^-1M.
$^-1MT + 8M^2 + {}^-4MT + 1$	Simplify the third term.
$^-5MT + 8M^2 + 1$	Combine like terms.

VERIFY Let $M = 3$ and $T = 6$.
$$^-MT + 8M^2 - M*4T + 1 \stackrel{?}{=} {}^-5MT + 8M^2 + 1$$
$$^-3*6 + 8*3^2 - 3*4*6 + 1 \stackrel{?}{=} {}^-5*3*6 + 8*3^2 + 1$$
$$^-17 = {}^-17 \quad ✔$$

We conclude that $^-MT + 8M^2 - M*4T + 1$ simplifies to $^-5MT + 8M^2 + 1$.

In this section, we looked at simplifying algebraic expressions. Before we began simplifying, we revisited the definitions of **terms** and **factors**. **Coefficients** and **numerical coefficients** were defined. We learned that **like terms,** terms having the same literal factors, can be combined by adding or subtracting their numerical coefficients.

Problem Set 3.1

1. For each of the following expressions, list the factors.
   a. $3mnk$ b. $\frac{4}{5}xy^2$ c. $5T(T-6)$ d. a^2b^3

2. Identify the numerical coefficient of each expression in Problem 1.

3. An expression has the factors 2, 5, x, x, y, and z. Identify all of the following that fit this description.
   a. $10x^2yz$ b. $(2xy)(5xz)$ c. $10x * xyz$ d. $25xy * xz$ e. $5z * x * 2xy$

4. For the expression $-25M^2PT$, do the following.
   a. Identify the numerical coefficient.
   b. Identify the coefficient of T.
   c. Identify the coefficient of P.
   d. Identify the coefficient of M^2.

5. a. What is the numerical coefficient of $\frac{2x}{3}$?
   b. What is the numerical coefficient of $\frac{x}{2}$?
   c. What is the numerical coefficient of $\frac{-x}{5}$?
   d. What is the numerical coefficient of x in the expression $10 + \frac{x}{4}$?

6. Answer parts a–d using the following expression.

   $$\frac{5xy^2z}{3}$$

   a. Identify the numerical coefficient.
   b. Identify the coefficient of z.
   c. Identify the coefficient of x.
   d. Identify the coefficient of y^2.

7. Answer parts a–e using the following expression.

   $$-7kt^2 + \frac{2k}{5} - t(t+2) + 3k(4-t)$$

   a. Identify the number of terms.
   b. Identify the numerical coefficient of each term.
   c. Identify the coefficient of k in the first term.
   d. Identify the coefficient of k in the fourth term.
   e. Identify the factors of the third term.

8. Which of the following terms are like terms?
   a. $8pw$
   b. $-w$
   c. $2w * 3p$
   d. $4w * w$
   e. $\frac{w}{2}$
   f. $4w * p$
   g. $-7w^2p$

9. To simplify each of the following expressions, first rewrite each subtraction as adding the opposite, and then combine like terms. Verify your results by numerical substitution.
   a. $12A + 3 + 7A + 5$
   b. $12A - 3 - 7A + 5$
   c. $-5P + M + 3P - 4M$
   d. $3k - 2k^2 - k^2 + 8k$
   e. $3.2R - 3.8RQ + 6.7RQ - 12.3$
   f. $1.2 * 10^5T + 3.5 * 10^5T^2 - 3.2 * 10^5T$

10. Simplify each expression.
    a. $4 * 5t$
    b. $8R * {^-3}$
    c. $4x * 6p$
    d. $^-12m * 2m$
    e. $4g(^-10g)$
    f. $z * 3z$
    g. $^-p(kp)$
    h. $2x^2 * 3px$
    i. $10m(^-6k^2m)$

11. Simplify each expression, and verify your results numerically.
    a. $12RT + 2 * 4R + 3T * 2R + R$
    b. $9x * 2x - 2 * 4x^2 + 4x * 3$
    c. $^-m(mp) + 5 * 3m - 3m + 4pm^2$
    d. $4h(^-10h) + 5h^2 * 8 + 1$
    e. $5 * 4w - 5 * wz + 2z * 4w - 2z * wz$
    f. $^-8(4x) + {^-8}(5) + 10$

12. The following is a student's work, showing how she simplified $10p + 3m * 6h - 4p + mh$ to $6p + 19mh$. Explain each step in the process.

    $10p + 3m * 6h - 4p + mh$
    $= 10p + 3m * 6h + {^-4p} + 1mh$
    $= 10p + 18mh + {^-4p} + 1mh$
    $= 6p + 19mh$

13. Write formulas for the perimeter and the area of the following figure.

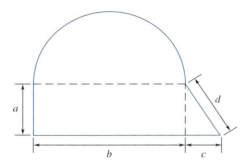

3.2 The Distributive Property

Activity Set 3.2

1. Find the area of the following figure in two different ways according to the directions in parts a and b.
   a. Write the length of the long side of the figure as a sum. Then write the area of the figure as a product of the length and the width.
   b. Find the area of regions I and II. Write the area of the figure as a sum of these two regions.
   c. Because the expressions you wrote in parts a and b both represent the area of the whole rectangle, they must be equal. Write an equation expressing this equality.

2. Follow the directions in Activity 1 to find the area of the following figure in two different ways.

3. Follow the directions in Activity 1 to find the area of the following figure in two different ways.

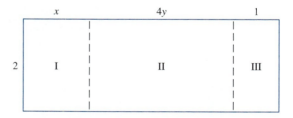

4. Draw a divided rectangle model, similar to the preceding figures, to illustrate the product of 4 and the sum of $2x$ and $3y$. Rewrite the product $4(2x + 3y)$ as a sum. Verify your result by numerical substitution.

5. a. Write an expression for the area of the following figure by adding the lengths on the long side and then multiply the length and the width.
   b. Write an expression for the area of the following figure by determining the areas of regions I and II and then adding the results.
   c. Describe a rule to determine the product $a(b + c)$.

Discussion 3.2

In Activity 1, we saw that the expression $2(x + 3)$ was equivalent to $2x + 6$. In general, we found that the product of a single term and the sum of terms could be rewritten as a sum. In algebraic notation, $a(b + c) = ab + ac$. This property is called the **distributive property**.

> **The Distributive Property**
> For any real numbers a, b, and c, $a(b + c) = ab + ac$
>
>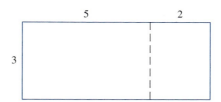

We can illustrate the distributive property geometrically using rectangles as we did in the activities. Consider the following rectangle.

We can find the area of this rectangle in two different ways. One way is to determine the dimensions of the entire rectangle by adding the lengths on the long side and then multiply the length and the width. This leads to the expression $3(5 + 2)$, where the addition is done first. Alternatively, we can find the area of the two smaller rectangles and then add the results to determine the area of the entire rectangle. This leads to the expression $3 * 5 + 3 * 2$, where the multiplication is done first. Because both of these expressions represent the area of the same rectangle, we know they are equivalent; therefore, $3(5 + 2) = 3 * 5 + 3 * 2$.

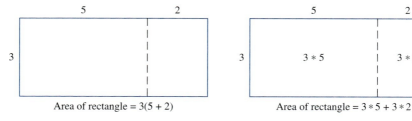

Area of rectangle = $3(5 + 2)$ Area of rectangle = $3 * 5 + 3 * 2$

The process of multiplying the factor of three times each of the terms of the sum is known as distributing multiplication over addition. This process is frequently used in simplifying algebraic expressions.

Similarly, the area of the following rectangle can be found with either the expression $a(b + c)$ or $ab + ac$. Therefore, $a(b + c) = ab + ac$. When we refer to the distributive property, we mean the process of distributing multiplication over addition or subtraction.

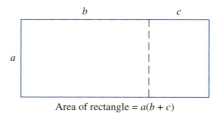
Area of rectangle = $a(b + c)$

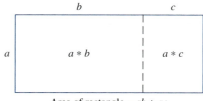
Area of rectangle = $ab + ac$

The next example shows how this property is used in simplifying algebraic expressions.

Example 1

Simplify each of the following expressions, and verify your results numerically.

a. $4(n + 5) + 6n$
b. $9x + 3x(2x + 7)$

Solution

a. The expression $4(n + 5) + 6n$ has two terms: $4(n + 5)$ and $6n$. We first apply the distributive property to the first term.

$\underline{4(n + 5)} + \underline{6n}$ Underline the terms.

$= 4(n + 5) + 6n$ Distribute the 4 over the sum of n and 5.
$= 4 * n + 4 * 5 + 6n$
$= 4n + 20 + 6n$ Simplify each term.
$= 10n + 20$ Combine like terms.

To numerically verify our result we choose a value for the variable n and evaluate the beginning and ending expressions to see if they result in the same value.

VERIFY   Let $n = 3$

$4(n + 5) + 6n \stackrel{?}{=} 10n + 20$
$4(3 + 5) + 6 * 3 \stackrel{?}{=} 10 * 3 + 20$
$50 = 50$ ✔

We conclude that $4(n + 5) + 6n$ simplifies to $10n + 20$.

b. The expression $9x + 3x(2x + 7)$ has two terms. The first term is $9x$, and the second term is $3x(2x + 7)$. Notice that because the $3x$ is not a term, it cannot be combined with the $9x$.

$\underline{9x} + \underline{3x(2x + 7)}$

$= 9x + 3x(2x + 7)$ Distribute the $3x$ over the sum $2x + 7$.
$= 9x + 3x * 2x + 3x * 7$
$= 9x + 6x^2 + 21x$ Simplify each term by multiplying.
$= 30x + 6x^2$ Combine like terms.

VERIFY   Let $x = 4$.

$9x + 3x(2x + 7) \stackrel{?}{=} 30x + 6x^2$
$9 * 4 + 3 * 4(2 * 4 + 7) \stackrel{?}{=} 30 * 4 + 6 * 4^2$
$216 = 216$ ✔

We conclude that $9x + 3x(2x + 7)$ simplifies to $30x + 6x^2$.

Example 2

Simplify each expression. Verify your results numerically.

a. $4k(3k - 5)$
b. $2m - 3(m + 1)$

Solution a. We first rewrite the subtraction as addition of the opposite and then apply the distributive property.

$$4k(3k - 5)$$
$$= 4k(3k + {}^-5) \quad \text{Rewrite the subtraction as adding the opposite.}$$
$$= 4k(3k + {}^-5) \quad \text{Distribute the } 4k \text{ over the sum.}$$
$$= 4k * 3k + 4k * {}^-5$$
$$= 12k^2 + {}^-20k \quad \text{Simplify each term.}$$
$$= 12k^2 - 20k \quad \text{Rewrite addition of the opposite as subtraction.}$$

VERIFY Let $k = 4$.

$$4k(3k - 5) \stackrel{?}{=} 12k^2 - 20k$$
$$4 * 4(3 * 4 - 5) \stackrel{?}{=} 12 * 4^2 - 20 * 4$$
$$112 = 112 \quad ✔$$

We conclude that $4k(3k - 5)$ simplifies to $12k^2 - 20k$.

b. The two terms of the expression $2m - 3(m + 1)$ are connected by subtraction. We rewrite the subtraction as addition of the opposite and distribute.

$$2m - 3(m + 1)$$
$$= 2m + {}^-3(m + 1) \quad \text{Rewrite subtraction as adding the opposite.}$$
$$= 2m + {}^-3(m + 1) \quad \text{Distribute the } {}^-3 \text{ over the } m + 1.$$
$$= 2m + {}^-3 * m + {}^-3 * 1$$
$$= 2m + {}^-3m + {}^-3 \quad \text{Simplify each term.}$$
$$= {}^-1m + {}^-3 \quad \text{Combine like terms.}$$
$$= {}^-m - 3 \quad \text{Rewrite addition of the opposite as subtraction.}$$

VERIFY Let $m = 5$.

$$2m - 3(m + 1) \stackrel{?}{=} {}^-m - 3$$
$$2 * 5 - 3(5 + 1) \stackrel{?}{=} {}^-5 - 3$$
$${}^-8 = {}^-8 \quad ✔$$

We conclude that $2m - 3(m + 1)$ simplifies to ${}^-m - 3$.

As we begin to simplify more complex expressions, it is helpful to have a strategy to follow.

Strategy *Simplifying Algebraic Expressions*

1. Identify the terms of the expression.
2. Rewrite subtraction as addition of the opposite as needed.
3. Apply the distributive property as appropriate.
4. Simplify each term individually.
5. Combine like terms as possible.
6. Verify your result.

The following examples should help you to see how to apply the preceding strategy.

Example 3

Simplify each expression, and verify your result using numerical substitution.

a. $2(3x + y) + 3(3y + x)$
b. $12(3K - P) - 4(K^2 + 15P)$
c. $3T^2 - (T - T^2)$
d. $4(m^2 + 3m) - 5m(4 - m)$

Solution

a. The expression $2(3x + y) + 3(3y + x)$ has two terms. We first simplify each term using the distributive property.

$$2(3x + y) + 3(3y + x)$$

$= 2(3x + y) + 3(3y + x)$ Distribute the 2 and the 3 over addition.
$= 2 * 3x + 2 * y + 3 * 3y + 3 * x$
$= 6x + 2y + 9y + 3x$
$= 9x + 11y$ Combine like terms.

VERIFY   Let $x = 4$ and $y = 5$.

$$2(3x + y) + 3(3y + x) \stackrel{?}{=} 9x + 11y$$
$$2(3 * 4 + 5) + 3(3 * 5 + 4) \stackrel{?}{=} 9 * 4 + 11 * 5$$
$$91 = 91 \quad ✔$$

We conclude that $2(3x + y) + 3(3y + x)$ simplifies to $9x + 11y$.

b. The expression $12(3K - P) - 4(K^2 + 15P)$ has two terms. First we rewrite each subtraction as adding the opposite and then begin to simplify each term using the distributive property.

$$12(3K - P) - 4(K^2 + 15P)$$

$= 12(3K + {}^-1P) + {}^-4(K^2 + 15P)$ Rewrite subtraction as adding the opposite.

$= 12(3K + {}^-1P) + {}^-4(K^2 + 15P)$ Distribute the 12 and the $^-4$ over addition.

$= 12 * 3K + 12 * {}^-1P + {}^-4 * K^2 + {}^-4 * 15P$
$= 36K + {}^-12P + {}^-4K^2 + {}^-60P$
$= 36K + {}^-72P + {}^-4K^2$ Combine like terms.
$= 36K - 72P - 4K^2$ Rewrite adding the opposite as subtraction.

VERIFY   Let $K = 7$ and $P = 3$.

$$12(3K - P) - 4(K^2 + 15P) \stackrel{?}{=} 36K - 72P - 4K^2$$
$$12(3 * 7 - 3) - 4(7^2 + 15 * 3) \stackrel{?}{=} 36 * 7 - 72 * 3 - 4 * 7^2$$
$$^-160 = {}^-160 \quad ✔$$

We conclude that $12(3K - P) - 4(K^2 + 15P)$ simplifies to $36K - 72P - 4K^2$.

c. The expression $3T^2 - (T - T^2)$ has two terms. First we rewrite each subtraction as adding the opposite and then begin to simplify each term using the distributive property.

$$3T^2 - (T - T^2)$$

$= 3T^2 + {}^-1(1T + {}^-1T^2)$ Rewrite subtraction as adding the opposite.

$= 3T^2 + {}^-1(1T + {}^-1T^2)$ Distribute the $^-1$ over addition.
$= 3T^2 + {}^-1T + 1T^2$
$= 4T^2 + {}^-1T$ Combine like terms.
$= 4T^2 - T$ Rewrite adding the opposite as subtraction.

VERIFY Let $T = 5$

$$3T^2 - (T - T^2) \stackrel{?}{=} 4T^2 - T$$
$$3 * 5^2 - (5 - 5^2) \stackrel{?}{=} 4 * 5^2 - 5$$
$$95 = 95 \quad ✔$$

We conclude that $3T^2 - (T - T^2)$ simplifies to $4T^2 - T$.

d. The expression $4(m^2 + 3m) - 5m(4 - m)$ has two terms. First we rewrite each subtraction as adding the opposite and then begin to simplify each term using the distributive property.

$$\underline{4(m^2 + 3m)} - \underline{5m(4 - M)}$$
$$= 4(m^2 + 3m) + {}^-5m(4 + {}^-1m) \qquad \text{Rewrite subtraction as addition of the opposite.}$$
$$= 4(m^2 + 3m) + {}^-5m(4 + {}^-1m) \qquad \text{Distribute the 4 and the } {}^-5m \text{ over addition.}$$
$$= 4 * m^2 + 4 * 3m + {}^-5m * 4 + {}^-5m * {}^-1m$$
$$= 4m^2 + 12m + {}^-20m + 5m^2 \qquad \text{Simplify each term.}$$
$$= 9m^2 - 8m \qquad \text{Combine like terms, and rewrite addition of the opposite as subtraction.}$$

VERIFY Let $m = 7$

$$4(m^2 + 3m) - 5m(4 - m) \stackrel{?}{=} 9m^2 - 8m$$
$$4(7^2 + 3 * 7) - 5 * 7(4 - 7) \stackrel{?}{=} 9 * 7^2 - 8 * 7$$
$$385 = 385 \quad ✔$$

We conclude that $4(m^2 + 3m) - 5m(4 - m)$ simplifies to $9m^2 - 8m$.

In this section, we looked at the **distributive property** and applied it to simplifying expressions. We saw that it is often helpful to rewrite subtraction as addition of the opposite when we are simplifying expressions.

Problem Set 3.2

1. Draw a divided rectangular model, similar to ones in the activity set, to show that the product $2(3x + y)$ is equivalent to $6x + 2y$.

2. Simplify each expression. You may find it helpful to rewrite subtraction as adding the opposite. Verify your results by numerical substitution.
   a. $3(5x + 2y)$
   b. $2(3b + 4a + 5)$
   c. $^-2(4x + 3a + 1)$
   d. $^-2(4x + 3a - 1)$
   e. $5(2x^2 - 3xy)$
   f. $^-5(2x^2 - 3xy)$
   g. $1.5(3.2k + 6.0p)$
   h. $^-8T + 16(4T + 5)$
   i. $4M - 3(M + 2MN)$
   j. $7.3(5.1w - 1.0c)$
   k. $4 - 2(Q^2 - Q)$
   l. $^-43(2T - 5) + 13(205 - T)$

3. Simplify each expression. You may find it helpful to rewrite subtraction as adding the opposite. Verify your results by numerical substitution.
   a. $4y(x + 3)$
   b. $12x(x + 3y)$
   c. $3T(15P + 41)$
   d. $51k(32k - 46m + 10)$
   e. $18Q(23R + 15 - 5T)$
   f. $2A(14 - A)$
   g. $^-M(3M + 14 + MN)$
   h. $^-12G(34 - 57G)$

4. Simplify each expression. You may find it helpful to rewrite subtraction as adding the opposite. Verify your results by numerical substitution.
   a. $12x^2 + 5x(x + 2)$
   b. $-4T(2T + 7) - 9T$
   c. $3mp(4 - m) + 2m^2(m + 6p)$
   d. $12(2x^2 + 3x - 4) - 3x(2x)$
   e. $7.3 * 10^3 RT - 8.1 * 10^3 R^2 + 9.0 * 10^4 RT$
   f. $2.4 * 10^3(3x - 1) + 4.1 * 10^3$

5. For each of the following expressions, determine the numerical value and units of the result.
   a. $2\left(1.5m * 10.0m + \dfrac{10.0m(1.5m + 3.8m)}{2} + 3.8m * 5.0m\right)$

   b. $\dfrac{1}{2}\left(\dfrac{4}{3}\pi(10.0 \text{ ft})^3\right) + \pi(10.0 \text{ ft})^2(17.5 \text{ ft})$

   c. $\dfrac{4080 \text{ gal}}{\dfrac{7.481 \text{ gal}}{1 \text{ ft}^3}}$

   d. $5.5 \text{ ft} * 6.0 \text{ ft} * 10.0 \text{ ft} + \pi(25 \text{ ft}^2)(3.5 \text{ ft})$

6. **Body Fat.** Exercise technicians use the following formula to estimate the ratio of body fat to total body weight for men.

   $F = 0.49W + 0.45P - 6.36R + 8.71$

   where F is the percent of body fat,

   W is the waist measurement in centimeters,

   P is the thickness of the skin fold above the pectoral muscle in millimeters, and

   R is the wrist diameter in centimeters

   a. Compute the percent body fat for a male whose waist measures 87.3 cm, skin fold thickness is 6.2 mm, and wrist diameter is 6.5 cm. Round your result to the nearest tenth of a percent.

   b. Compute the percent body fat for a male whose waist measures 34 inches, skin fold thickness is 0.5 inches, and wrist diameter is 2.5 inches.

7. Explain what is done in each of the following steps.

   $3m * 4p - 2m(10m - p)$
   $= 3m * 4p + {}^-2m(10m + {}^-1p)$
   $= 12mp + {}^-2m * 10m + {}^-2m * {}^-1p$
   $= 12mp + {}^-20m^2 + 2mp$
   $= 14mp + {}^-20m^2$
   $= 14mp - 20m^2$

3.3 Solving Linear Equations

Activity Set 3.3

1. The scale shown is balanced, which means that the objects on the left side of the scale weigh the same as those on the right. Each weight weighs 1 gram. We want to determine the weight of one of the bottles on the left side of the scale.

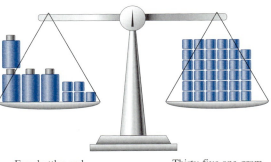

Four bottles and five one-gram weights

Thirty-five one-gram weights

   a. If you remove five 1-gram weights from the left side of the scale, what would you need to do to the right side of the scale to keep it balanced? Draw the scale that results after removing these weights.

   b. The bottles are equal in weight. How much does each bottle weigh? Explain how you determined this.

   c. An algebraic model of the original scale can be written as $4w + 5 = 35$, where w is the weight of one bottle in grams. Check your solution from part b by numerical substitution into this equation.

2. Write an equation that models the balance scale in the following figure. Determine the weight of one bottle. Explain how you reached your conclusion.

3. Write an equation that models the balance scale in the following figure. Determine the weight of one bottle. Explain how you reached your conclusion.

Discussion 3.3

Solving Linear Equations

In this section, we will solve linear equations in one variable. A **linear equation** in one variable is an equation that can be simplified to the form $ax + b = c$, where x is the variable and a, b, and c represent constants. For example, $2x + 3 = 10$ and $-0.5x + 4 = -8$ are linear equations. Solving an equation in one variable means finding all values of the variable that make the equation true. Consider the equation $x + 2 = 5$. We can see that if $x = 3$, the equation is true. Therefore, the solution to $x + 2 = 5$ is $x = 3$. We are able to solve this equation by inspection; however, as equations become more complicated we will need a strategy to solve them.

In the activities, we looked at balance scales that contained bottles and 1-gram weights. We were able to determine the weight of each bottle by manipulating the objects. Let's see how to model this process algebraically.

The balanced scale in the first activity can be modeled by the equation $4w + 5 = 35$, where w is the weight of one bottle in grams. The first thing we did was remove five 1-gram weights from each side of the scale. The scale still balances because we are removing the same amount of weight from each side. This is equivalent to subtracting 5 from each side of our equation. This left us with four bottles on the left-hand side and thirty 1-gram weights on the right-hand side, or equivalently the equation $4w = 30$.

$4w + 5 = 35$

$4w + 5 - 5 = 35 - 5$ Subtract 5 from each side of the equation.

$4w = 30$

Next, since we know that four bottles weigh 30 g, we know that one bottle weighs $\frac{30 \text{ g}}{4} = 7.5$ g. In our equation, this is equivalent to dividing each side of the equation by 4.

$4w = 30$

$\dfrac{4w}{4} = \dfrac{30}{4}$ Divide both sides of the equation by 4, the coefficient of w.

$w = 7.5$

In solving the equation $4w + 5 = 35$, we perform two main operations. First, we subtract a given amount from both sides of the equation. We choose to subtract 5 because that leaves us only bottles on the left-hand side. Second, we divide both sides of the equation by a given amount. We choose 4 because we have four bottles on the left and want to know the weight of one bottle. This provides us with the two main properties we need to solve linear equations.

Properties of Equality

1. Adding or subtracting the same amount on both sides of an equation does not change the equation's solution.
2. Multiplying or dividing both sides of an equation by the same nonzero number does not change the equation's solution.

In solving equations, we do not always have a physical model of the situation. The properties allow us to look at the equations more abstractly.

The equation $5w + 3 = 24$ represents the balance scale in Activity 2. We approach solving this equation without using the balance scale directly.

3.3 Solving Linear Equations

Recall that solving an equation means finding all the values of the variable that make the equation true. This means that we need to isolate the variable.

The equation $5w + 3 = 24$ tells us that w is first multiplied by 5. Then 3 is added to this result. The final result is 24. We need to undo both the multiplication and the division. To decide which we need to do first, we think back to the scale model. In Activity 2, we first take three 1-gram weights from each side of the scale. This isolates the bottles on one side of the equation. In looking at the equation $5w + 3 = 24$, we can see that subtracting 3 from each side *isolates the term containing the variable*. Notice that to get rid of the addition of 3, we need to subtract 3. Subtraction is the inverse operation for addition.

$$5w + 3 = 24$$
$$5w + 3 - 3 = 24 - 3 \quad \text{Subtract 3 from each side to undo the addition.}$$
$$5w = 21$$

We now see that w is being multiplied by 5. To undo this multiplication, we divide because division is the inverse operation for multiplication.

$$5w = 21$$
$$\frac{5w}{5} = \frac{21}{5} \quad \text{Divide both sides by 5 to undo the multiplication.}$$
$$w = 4.2$$

We examined two methods for using the properties of equality to solve linear equations: the balance scale model and the process of using inverse operations.

Let's look at several examples of solving linear equations. Pay attention to the role of inverse operations as you follow the solutions in the next example.

Example 1

Solve each linear equation, and check your solution.

a. $3m + 6 = 27$ b. $^-2x - 4 = 9$ c. $7 - 5p = {}^-12$

Solution

a. To solve the equation $3m + 6 = 27$, we need to isolate the variable m. The first step in isolating the variable m is to isolate the *term* that contains the variable m. The term that contains the variable is $3m$. Therefore, we need to isolate $3m$. Because 6 is added to $3m$, we subtract 6 from both sides of the equation.

$$3m + 6 = 27$$
$$3m + 6 - 6 = 27 - 6 \quad \text{Subtract 6 from both sides.}$$
$$3m = 21$$

Next, on the left-hand side, we have 3 times m. Therefore, to isolate m we divide both sides by 3.

$$\frac{3m}{3} = \frac{21}{3} \quad \text{Divide both sides by 3.}$$
$$m = 7$$

CHECK   To check our solution, we substitute 7 for m into the original equation.

$$3 * 7 + 6 \stackrel{?}{=} 27$$
$$27 = 27 \quad \checkmark$$

The solution to $3m + 6 = 27$ is $m = 7$.

b. To solve $-2x - 4 = 9$, we will first isolate the term containing the variable, which is $-2x$. Since 4 is subtracted from the $-2x$ we will add 4 to both sides of the equation.

$$-2x - 4 = 9$$
$$-2x - 4 + 4 = 9 + 4 \quad \text{Add 4 to both sides of the equation.}$$
$$-2x = 13$$

On the left, we now have -2 times x, so we divide both sides by -2.

$$\frac{-2x}{-2} = \frac{13}{-2} \quad \text{Divide both sides by } -2.$$
$$x = -6.5$$

CHECK To check, substitute -6.5 for x into the original equation.

$$-2 * -6.5 - 4 \stackrel{?}{=} 9$$
$$9 = 9 \quad \checkmark$$

The solution to $-2x - 4 = 9$ is $x = -6.5$.

c. In the equation $7 - 5p = -12$, what is the coefficient of p? Is it 5 or -5? To help us decide we rewrite the subtraction on the left as adding the opposite. The equation then becomes

$$7 + {-5p} = -12.$$

From this we can see that the coefficient of the variable is -5. The 7 is being added to the $-5p$; therefore, we subtract 7 from both sides of the equation. Then, once $-5p$ is isolated we divide by -5, the coefficient of p.

$$7 + {-5p} = -12$$
$$7 + {-5p} - 7 = -12 - 7 \quad \text{Subtract 7 from both sides of the equation.}$$
$$-5p = -19$$
$$\frac{-5p}{-5} = \frac{-19}{-5} \quad \text{Divide both sides by } -5.$$
$$p = 3.8$$

CHECK To check, substitute 3.8 for p in the original equation.

$$7 - 5 * 3.8 \stackrel{?}{=} -12$$
$$-12 = -12 \quad \checkmark$$

The solution to $7 - 5p = -12$ is $p = 3.8$.

The balanced scale in Activity 3 can be modeled by the equation $5w + 1 = 2w + 7$, where w is the weight of one bottle in grams. This equation looks different from the ones we looked at in Example 1. In this equation, variables occur on both sides of the equation.

In finding the weight of one bottle in Activity 3, we first remove two bottles from each side of the scale so that bottles appear on only one side of the scale. This is equivalent to subtracting $2w$ from each side of the equation. We then have the equation $3w + 1 = 7$. Notice that all the terms containing variables are now on one side. Next we remove a 1-gram weight from each side of the scale. This is equivalent to subtracting 1 from each side of the equation. We then have the equation $3w = 6$. If three bottles weigh 6 grams, then we know that one bottle must weigh 2 grams. The algebraic process is shown next.

$$5w + 1 = 2w + 7$$
$$5w + 1 - 2w = 2w + 7 - 2w \quad \text{Subtract } 2w \text{ from both sides.}$$
$$3w + 1 = 7$$
$$3w - 1 = 7 - 1 \quad \text{Subtract 1 from each side.}$$
$$3w = 6$$
$$\frac{3w}{3} = \frac{6}{3} \quad \text{Divide both sides by 3.}$$
$$w = 2$$

In this example, in which variable quantities occurred on both sides of the equation, our first step was to collect all of the terms containing the variable onto one side of the equation. The next step was to collect all terms not containing the variable on the opposite side of the equation. We can see that the properties of equality work whether the amounts we are adding or subtracting are numerical quantities or variable quantities.

In the equations in Example 1, both sides of the equation were completely simplified, and the variable occurred only on one side. We saw in the equation from Activity 3 that the variable may occur on both sides of the equation. We will also be given equations in which one or both sides may not be simplified. A strategy for solving any linear equation in one variable is shown next.

Strategy

Solving Linear Equations in One Variable

1. Simplify both sides of the equation by applying the distributive property or combining like terms (or both). If the equation contains fractions, you may want to multiply both sides of the equation by the least common denominator to eliminate the fractions.
2. Add or subtract terms from both sides of the equation to collect all terms containing the variable on one side and all other terms on the opposite side. Again combine like terms if possible.
3. Divide both sides by the coefficient of the variable.
4. Check your solution in the original equation.

Example 2

Solve each equation, and check your solution.

a. $14 = 2(3x - 5)$ b. $2(3 - m) = {}^-8m - 9$ c. $\frac{p}{5} - \frac{2p}{3} = 4$

Solution   a.
$$14 = 2(3x - 5)$$
$$14 = 6x - 10 \quad \text{Distribute the 2 over the } 3x - 5.$$
$$14 + 10 = 6x - 10 + 10 \quad \text{Add 10 to both sides to isolate the } 6x.$$
$$24 = 6x \quad \text{Simplify both sides of the equation.}$$
$$\frac{24}{6} = \frac{6x}{6} \quad \text{Divide both sides of the equation by 6, the coefficient of } x.$$
$$4 = x$$

CHECK   To check, substitute 4 for *x* into the original equation.

$$14 = 2(3x - 5)$$
$$14 \stackrel{?}{=} 2(3 * 4 - 5)$$
$$14 = 14 \quad ✔$$

The solution to $14 = 2(3x - 5)$ is $x = 4$.

b.
$2(3 - m) = {}^-8m - 9$	
$6 - 2m = {}^-8m - 9$	Distribute the 2 over the $3 - m$.
$6 - 2m + 2m = {}^-8m - 9 + 2m$	Add $2m$ to both sides to collect all terms containing m on the right-hand side.
$6 = {}^-6m - 9$	Simplify both sides of the equation.
$6 + 9 = {}^-6m - 9 + 9$	Add 9 to both sides to isolate ^-6m.
$15 = {}^-6m$	Simplify both sides of the equation.
$\dfrac{15}{^-6} = \dfrac{^-6m}{^-6}$	Divide both sides by $^-6$, the coefficient of m.
$^-2.5 = m$	

CHECK   To check, substitute $^-2.5$ for m into the original equation.

$$2(3 - m) = {}^-8m - 9$$
$$2(3 - {}^-2.5) \stackrel{?}{=} {}^-8 * {}^-2.5 - 9$$
$$11 = 11 \quad ✔$$

The solution to $2(3 - m) = -8m - 9$ is $m = -2.5$.

c. The equation $\dfrac{p}{5} - \dfrac{2p}{3} = 4$ contains fractions. Usually, the simplest way to solve equations involving fractions is to multiply through by a common denominator to eliminate the fractions. In this case, 15 is a common denominator because both 3 and 5 are factors of 15. Therefore, we multiply both sides of the equation by 15.

$\dfrac{p}{5} - \dfrac{2p}{3} = 4$	
$\mathbf{15}\left(\dfrac{p}{5} - \dfrac{2p}{3}\right) = 4 * \mathbf{15}$	Multiply both sides of the equation by 15.
$15 * \dfrac{p}{5} - 15 * \dfrac{2p}{3} = 60$	Distribute 15 over the sum on the left-hand side.
$3p - 10p = 60$	Simplify the terms on the left-hand side.
$^-7p = 60$	Combine like terms.
$\dfrac{^-7p}{^-7} = \dfrac{60}{^-7}$	Divide both sides by $^-7$, the coefficient of p.
$p = -\dfrac{60}{7} \quad$ or $\quad p \approx {}^-8.57$	

CHECK   To check, substitute $-\dfrac{60}{7}$ for p into the original equation.

$$\dfrac{p}{5} - \dfrac{2p}{3} = 4$$
$$\dfrac{-\dfrac{60}{7}}{5} - \dfrac{2 * -\dfrac{60}{7}}{3} \stackrel{?}{=} 4$$
$$4 = 4 \quad ✔$$

What happens if we check our result using our approximate solution of −8.57 instead of the exact solution? We substitute −8.57 for p into the original equation.

$$\frac{-8.57}{5} - \frac{2 * -8.57}{3} \stackrel{?}{\approx} 4$$

$$3.999 \approx 4 \quad ✔$$

Notice, when using our approximate solution to check, the value of the left side of the equation does not equal the right-hand side, but the values are very close. When we check or verify a solution with approximate numbers, the check may be approximately equal rather than exactly equal.

Example 3

If Rae can run 3 miles in $25\frac{1}{2}$ minutes, how long will it take for her to finish a local race that is 4.2 miles long?

Solution To solve this problem, we assume that Rae runs the race at the same rate that she does her 3-mile runs. This means that the ratio of time to distance is constant.

Because we want to determine the time, we let T represent the time (in minutes) to complete the local race. Then,

$$\frac{25.5 \text{ min}}{3 \text{ miles}} = \frac{T}{4.2 \text{ miles}}$$

$$\mathbf{4.2 \text{ miles}} * \frac{25.5 \text{ min}}{3 \text{ miles}} = \frac{T}{4.2 \text{ miles}} * \mathbf{4.2 \text{ miles}} \quad \text{To solve for } T, \text{ multiply both sides by 4.2 miles.}$$

$$35.7 \text{ min} = T$$

To check our solution, we need to decide whether the answer is reasonable. We know it takes her longer to run 4.2 miles than it takes to run 3 miles. Because 35.7 minutes is longer than 25.5 minutes, we assume that our answer is reasonable. Therefore, we conclude that it takes Rae about 35.7 minutes to complete the race.

Checking a proportion by analyzing if the answer is reasonable is a better method than substituting the solution back into the original equation. Because we set up the original equation, it's possible that we set it up wrong. Substituting back into it cannot show us that we set it up wrong.

In the previous example, we set up the proportion

$$\frac{25.5 \text{ min}}{3 \text{ miles}} = \frac{T}{4.2 \text{ miles}}$$

If we had set up the original equation as

$$\frac{3 \text{ miles}}{25.5 \text{ min}} = \frac{4.2 \text{ miles}}{T}$$

the variable is in the denominator. This equation is more difficult to solve; therefore, when setting up a proportional equation, *make sure the variable is in the numerator.*

Notice the units in the proportion

$$\frac{25.5 \text{ min}}{3 \text{ miles}} = \frac{T}{4.2 \text{ miles}}$$

The units are not all the same; however, the units in the ratio on the left-hand side of the equation are equivalent to those on the right-hand side. The equations we have solved so far have all had a single solution. That is, we found one number that made the original equation true. This type of equation is called a conditional equation. Occasionally, we have linear equations that are true for all values of the variable or, at the other extreme, true for no values of the variable.

> **Definition**
>
> An equation that is true for all values of the variable is called an **identity.** An equation that is true for no values of the variable is called a **contradiction.** An equation that is true for some specific value(s) of the variable is called a **conditional equation.**

Example 4   Identify each of the following equations as an identity, contradiction, or conditional equation.

a. $3(x + 2) - x = 5 + 2x$
b. $3(x + 2) = x + 2(3 + x)$

Solution   a.

$$3(x + 2) - x = 5 + 2x$$
$$3x + 6 - x = 5 + 2x \quad \text{Distribute the 3 over the } x + 2.$$
$$2x + 6 = 2x + 5 \quad \text{Combine like terms.}$$
$$2x + 6 - \mathbf{2x} = 2x + 5 - \mathbf{2x} \quad \text{Subtract } 2x \text{ from both sides.}$$
$$6 = 5 \quad ✗ \quad \text{Simplify both sides.}$$

Because we know that $6 \neq 5$, we must conclude that the equation is not true for any value of the variable. No matter what value we substitute for x, we get a false statement. Therefore, the equation $3(x + 2) - x = 5 + 2x$ is a contradiction.

b.
$$3(x + 2) = x + 2(3 + x)$$
$$3x + 6 = x + 6 + 2x \quad \text{Distribute on both sides.}$$
$$3x + 6 = 3x + 6 \quad \text{Combine like terms.}$$
$$3x + 6 - \mathbf{3x} = 3x + 6 - \mathbf{3x} \quad \text{Subtract } 3x \text{ from both sides.}$$
$$6 = 6 \quad \text{Simplify both sides.}$$

Because $6 = 6$ is always true, our equation appears to be true no matter what value we give x. Let's try a few values to see if this is the case. Let's substitute $x = 5$ into the equation.

$$3(x + 2) = x + 2(3 + x)$$
$$3(5 + 2) \stackrel{?}{=} 5 + 2(3 + 5)$$
$$21 = 21 \quad ✓$$

Let's substitute $x = 8$ into the equation.

$$3(8 + 2) \stackrel{?}{=} 8 + 2(3 + 8)$$
$$30 = 30 \quad ✓$$

It appears that our conclusion is true; any value of x satisfies the equation. You might try a couple more values to convince yourself.

Because the equation is true for all values of the variable, the equation $3(x + 2) = x + 2(3 + x)$ is an identity.

In general, we expect to get one solution when we solve a linear equation. The last example shows us that some circumstances result in either an identity or a contradiction. Later you will be able to see graphically why this occurs.

In this section, we looked at techniques for algebraically solving linear equations in one variable. Our first step is to simplify both sides of the equation or if the equation contains fractions, to multiply both sides by a common denominator. Second, we collect and isolate all of the terms that contain the variable. And finally, we divide both sides by the coefficient of the variable and check our result. Usually this results in a single solution. If following this technique results in a statement that is either always true or always false, we have an identity or a contradiction rather than a conditional equation.

Problem Set 3.3

1. Solve the following equations. Show each step in the process. Check your solutions by numerical substitution into the original equations.

   a. $a - 13 = 27$
   b. $2x + 18 = 12$
   c. $1.5 + 2p = 7.6$
   d. $2(m - 5) = 39$
   e. $3y - 5 = 22$
   f. $-5Q = 9$
   g. $2x - 4 = 18$
   h. $3x + 20 = 5$
   i. $3x - 2 = -5$
   j. $30w + 15 = -15$
   k. $15 - 2x = 5x + 7$
   l. $3(5 + 2x) = 7$
   m. $3x + 2 = 2(x + 1)$
   n. $3 + x = 24 - x$
   o. $2(x - 1) = x - 3$
   p. $3(2n - 4) + 15 = 18$
   q. $5x - 4(x + 7) = -27$
   r. $3(x - 1) + x = 2(x - 1)$

2. Solve the following equations. Show each step in the process. Check your solutions by numerical substitution into the original equations.

   a. $\dfrac{x}{5} = -8$
   b. $\dfrac{2x}{3} = 12 - x$
   c. $\dfrac{x}{3} = 10$
   d. $\dfrac{7}{12} + r = \dfrac{1}{2}$
   e. $\dfrac{-3}{4}T = -12$
   f. $\dfrac{1}{3} - 2b = 8$
   g. $\dfrac{x}{0.81} = -2.79$
   h. $\dfrac{x}{2} = x - 4$
   i. $\dfrac{M}{6} - \dfrac{M}{3} = 5$
   j. $12 = \dfrac{w + 5}{3}$
   k. $\dfrac{13}{5} = \dfrac{3(x + 4)}{2}$

3. The following solution seems to check but is incorrect. Find the errors.

   Solution:
   $3(m - 6) = 15$
   $3m - 6 = 15$
   $3m - 6 + 6 = 15 + 6$
   $3m = 21$
   $m = 7$

   CHECK
   $3(7 - 6) \stackrel{?}{=} 15$
   $3 * 7 - 6 \stackrel{?}{=} 15$
   $21 - 6 + 6 \stackrel{?}{=} 15 + 6$
   $21 = 21$ ✔

4. Solve the following equations. Show each step in the process. Check your solutions.
   a. $-3(y + 5) = -12 + 2(y - 3)$
   b. $4(2x + 1) = 3(x - 3) - 7$
   c. $7p + 5 = 17 - p$
   d. $\dfrac{x}{2} + 2 = \dfrac{x}{5} + 5$
   e. $1.26x - 3.65 = 8.75 - 5.14x$
   f. $\dfrac{x}{3} + \dfrac{x}{4} = \dfrac{7}{12}$
   g. $4.6(k + 1.2) = 2.3(k - 2.1)$

5. Solve the following equations. Show each step in the process. Express answers as specified. Check your solutions.
   a. $5.3 = 3.2x + 7.1$ nearest hundredth
   b. $21(7m - 3) = 5.6 - 3.6m$ two significant digits
   c. $\dfrac{5T - 16.4}{3.45} = 2T$ three significant digits
   d. $5.7 * 10^4 P - 3.6 * 10^6 = 4.2 * 10^6$ two significant digits

6. **Scales on a Map.** The scale on a map is given as 3 inches = 5 miles. On the map the distance between two towns measures $8\frac{1}{4}$ inches. What is the actual distance between these two towns?

7. **Adjusting the Recipe.** A recipe for four people requires $2\frac{1}{2}$ cups of liquid. How many cups of liquid would you use if you were preparing the recipe for three people?

8. **Election Time.** If it takes 15 volunteers 1 hour to stuff 750 letters of support for a candidate in an upcoming election, how many volunteer hours will it take to stuff 2500 letters?

9. **Marc's Car Trip.** Marc has been monitoring his gas usage in anticipation of an upcoming trip. He found that his car used 9 gallons of gas in 250 miles. How many gallons of gas can Marc expect to use on a 1200-mile trip? At this rate, how much will Marc spend on gas for the trip if the price of gas averages around $1.50 per gallon?

10. **Copy Machine.** Your office leases a copy machine. The lease includes a maintenance program. You are charged $305.98 per month plus an additional $0.04 per copy.
    a. Complete the following table.

Copies per Month	Cost of Copier per Month
250	
500	
750	
1000	
1500	

    b. Write an equation for the cost of the copier per month in terms of the number of copies printed.
    c. The bill for October was $361.30. Algebraically determine the number of copies made in October by solving your equation.

11. a. i. Solve the equation $2(2n + 1) + n = 5n + 2$ for n. If you are not sure of the solution go on to parts ii and iii.

   ii. Substitute $n = 3$ into the equation in part i. Is the statement true for $n = 3$? Substitute $n = 4, -5$, and 101 into the equation in part i. Is the statement true for these values?

   iii. What is the solution to $2(2n + 1) + n = 5n + 2$?

b. i. Solve the equation $-2w - 2 = -2(w - 1)$ for w. If you are not sure of the solution go on to parts ii and iii.

   ii. Substitute $w = 0$ into the equation in part i. Is the statement true for $w = 0$? Substitute $w = 2$ and -2 into the equation in part i. Is the statement true for these values?

   iii. What is the solution to $-2w - 2 = -2(w - 1)$?

12. Identify each of the following equations as an identity or a contradiction.

   a. $x + 5 = x + 7$

   b. $10p - 5 = -5(1 - 2p)$

   c. $5m = 4(m + 1) + (m - 4)$

   d. $6(y + 5) = 6y + 5$

13. A Registered Letter Revisited. A registered letter mailed from San Francisco to Chicago costs $5.09 for the first ounce and $1.20 for each additional ounce. Earlier in the term, you came up with the model $C = 5.09 + 1.20(w - 1)$, where w is the weight of the package in ounces and C is the cost in dollars. Algebraically determine the weight of a letter that can be sent for $15.00.

14. The Car Rental Problem Revisited. Recall that the Cars-4-U car company rents a compact car for $25 a day and 23¢ a mile and the Beaters-R-Us car company rents a compact car for 35¢ a mile. Earlier you came up with the models $C = 25 + 0.23m$ for Cars-4-U and $C = 0.35m$ for Beaters-R-Us, where C represents the cost and m represents the number of miles driven. Algebraically determine the number of miles that result in the same cost for both companies.

15. Landscaping the Yard. The Yard and Garden Center will deliver barkdust. They charge $10 plus $3.25 per cubic yard of barkdust.

   a. Write an equation for the cost in terms of the number of cubic yards of barkdust.

   b. If the Sampson family has budgeted $25 for barkdust, how many cubic yards can they have delivered?

16. Mailing Costs. The director of a nonprofit organization is preparing to mail out a newsletter. The newsletters can be mailed bulk rate or first class. A bulk mailing permit costs $85 for the year and the cost per piece is then 9.5¢. First-class mailing costs 33¢ per piece. How many newsletters must the organization mail during the year to have it cost the same for either bulk rate or first class?

17. Reception Costs. Persimmon's Country Club rents their banquet hall for receptions. Their charges for a cake and beverage reception are $500 for up to 25 guests plus $2.95 per guest for additional guests.

   a. Assuming more than 25 guests attend an event, write an equation for the cost of the reception in terms of the number of guests.

   b. If the Wylies budgeted $1500 for a reception at Persimmon's, how many guests can attend?

18. Jazz Concert. Last spring, several of the Mt. Hood Community College (MHCC) jazz groups held a joint concert. Tickets sold for $5.50 for the general public and for $3.00 for students and staff.

a. Let G represent the number of general tickets and M represent the number of tickets sold to MHCC students and staff. Write an equation for the income I from the tickets sold in terms of G and M.

b. If it is known that the income from the concert was $1811 and that 193 MHCC students and staff attended, how many from the general public attended the jazz concert?

19. Investments. The Smiths invested $2500 over the last year. They invested the total amount for the entire year. Part of the $2500 was invested in a variable account that paid 12.5% interest last year. The remainder of the money was invested in an account that paid 5% interest. The Smiths earned a total of $200 from both of these investments last year. How much did they invest in each account?

3.4 Solving Linear Literal Equations

Activity Set 3.4

1. Consider the formula $K = NR + NT$.
   a. Substitute $K = 10$, $N = 2$, and $T = 3$ into the formula. Solve the formula for R when $K = 10$, $N = 2$, and $T = 3$.
   b. Substitute $K = 3$, $N = 4$, and $T = 5$ into the formula. Solve the formula for R when $K = 3$, $N = 4$, and $T = 5$.
   c. Substitute $K = 24$, $N = 4$, and $T = 7$ into the formula. Solve the formula for R when $K = 24$, $N = 4$, and $T = 7$.
   d. Using your results from parts a–c, complete the first three rows of the following table.
   e. Solve the formula for R in terms of K, N, and T.

K	N	T	Formula	R
10	2	3	10 = 2R + 6	2
3	4	5		
24	4	7		
K	N	T		

2. a. Solve $3t = 24$ for t. Check your solution.
   b. Solve $PT = K$ for T. Your solution will be a formula for T in terms of P and K.
   c. To verify your result, you need to obtain a value for T. Starting with your result from part b, choose values (other than 0, 1, or 2) for P and K. Substitute these values into your result and evaluate to obtain a value for T.

   Now that you have values for all three variables, substitute all three values into the *original* formula, $PT = K$. Does this substitution result in a true statement? If the statement is true, you have verified your solution using numerical substitution.

3. a. Solve $12 = 4 + 2x$ for x. Check your solution.
   b. Solve $c = a + bx$ for x. Your solution will be a formula for x in terms of a, b, and c.
   c. To verify your result, you need to obtain a value for x. Starting with your result, choose values (other than 0, 1, or 2) for a, b, and c. Substitute these values into your result, and evaluate to obtain a value for x.

   Now that you have values for all four variables, substitute all four values into the *original* formula, $c = a + bx$. Does this substitution result in a true statement?

4. Solve the following literal equations. Verify your solutions using methods similar to those in Activities 2c and 3c.
   a. $PD_p = D_c D_f$ for D_p b. $t_1 = 4 + vt_2$ for t_2

5. a. Solve $9 = 3(x - 2)$ for x. Check your solution.
   b. Solve $A = 2(P - R)$ for P. Your solution will be a formula for P in terms of A and R.
   c. To verify your result, you need to obtain a value for P. Starting with your result, choose values (other than 0, 1, or 2) for A and R. Substitute these values into your result, and evaluate to obtain a value for P.

   Now that you have values for all of the variables, substitute all values into the *original* formula. Does this substitution result in a true statement?

6. a. Solve $6 = 3(9 - x)$ for x. Check your solution.

b. Solve $A = A_0(P - R)$ for R. Your solution will be a formula for R in terms of A, A_0, and P.

c. Choose values (other than 0, 1, or 2) for A, A_0, and P. Substitute these values into your solution from part b to determine a value for R. Now that you have values for all of the variables, substitute all values into the *original* formula. Does this substitution result in a true statement?

7. Solve the following literal equations. Verify your solutions using methods similar to those in Activity 6c.

a. $Q = P(M - T)$ for M
b. $T_d = 3(T_2 - T_1)$ for T_1

8. a. Solve the equation for the indicated variable. Check your solution.

$$6 = \frac{3 + r}{3} \quad \text{for } r$$

b. Solve the equation for the indicated variable. Verify your solution using numerical substitution.

$$S = \frac{C + B}{C} \quad \text{for } B$$

9. Solve the following literal equations. Verify your solutions using numerical substitution.

a. $F = \dfrac{2E_1 + 2E_2}{5}$ for E_2
b. $D = \dfrac{F - G}{C}$ for G

10. a. Solve the equation for the indicated variable. Check your solution.

$$24 = \frac{13T}{5} \quad \text{for } T$$

b. Solve the equation for the indicated variable. Verify your solution using numerical substitution.

$$E = \frac{mv^2}{2} \quad \text{for } m$$

11. a. Solve the equation for the indicated variable. Check your solution.

$$36 = \frac{6(w - 3)}{15} \quad \text{for } w$$

b. Solve the equation for the indicated variable. Verify your solution using numerical substitution.

$$K = \frac{KR(T_2 - T_1)}{D} \quad \text{for } T_2$$

Discussion 3.4

Solving Literal Equations

In this section, we will be solving literal equations (formulas). A literal equation is an equation that contains more than one variable. When solving a literal equation, we solve for one variable in terms of the other variables.

Suppose we have a formula that converts degrees Celsius to degrees Fahrenheit, $F = \frac{9}{5}C + 32$, and we want to convert 90°F to Celsius. We substitute 90 for F and solve for C. This is an efficient way to convert 90°F to Celsius. Now suppose we have ten temperature readings in degrees Fahrenheit that we want to convert to Celsius. We could substitute each value in separately and solve for C, which means we would need to solve the equation ten different times. Alternatively, we can solve the original formula for C in terms of F and substitute the ten different temperatures into our new formula to find values for C.

Using the first method, we must solve the same formula ten times. Using the second method, we solve the formula once and then use substitution and arithmetic to find the desired values. The second method is much more efficient. You will have a chance to do this problem in the problem set.

We will see in the next couple of examples that solving a literal equation is similar to solving an equation in one variable. The difference comes when we need to verify our results. As you may have experienced in the activities, the process of verifying the solution to a literal equation requires more steps.

Example 1

Write an equation for the area A of the figure. The figure consists of a square and four congruent triangles. Solve your formula for y, and verify your results by numerical substitution.

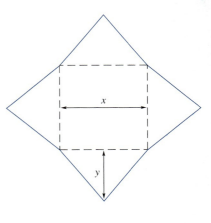

Solution

To find the area of the figure, we need to determine the area of the square and the area of the four triangles. All four triangles in the figure are congruent; therefore,

$$\text{area of figure} = \text{area of square} + 4(\text{area of triangle})$$

The area of the square is x^2, and the area of each triangle is $\frac{1}{2}xy$. Therefore, the area of the figure is given by

$$A = x^2 + 4\left(\frac{1}{2}xy\right)$$

When we solve an equation in one variable, we need to collect and isolate all terms that contain that variable. This formula has several variables, and we are solving for one of them. To make this process easier, once both sides are simplified, we identify the term or terms that contain the variable we are solving for. Then we proceed to collect and isolate these terms.

$$A = x^2 + 4\left(\frac{1}{2}xy\right)$$
$$A = x^2 + \underline{2xy} \qquad \text{Simplify the right-hand side, and identify the term containing } y.$$
$$A - x^2 = 2xy \qquad \text{Subtract } x^2 \text{ from both sides to isolate the term containing } y.$$
$$\frac{A - x^2}{2x} = \frac{2xy}{2x} \qquad \text{Divide both sides by the coefficient of } y, 2x.$$
$$\frac{A - x^2}{2x} = y$$

Now that we solved for y, we need to verify our result. Normally, when we solve an equation for y, we substitute the value for y into the original equation to check our result. We can see that, in solving the literal equation, y is equal to an expression, not a value. Therefore, to verify our result, we pick values for all of the variables except for y to obtain a value to use in our check.

> **Strategy** — *Verifying the Results of a Literal Equation*
>
> 1. Obtain a value for the variable you solved for.
>   a. Choose values for all of the variables *except* the one you solved for. Do not choose 0, 1, or 2.
>   b. Substitute your chosen values into your *result*, and calculate the value of the variable you solved for.
> 2. Substitute the values of all variables (this includes the calculated value) into the *original* equation, and check to see if the statement is true.

To verify our result, we need to obtain a value for *y*. We start with our solution and choose values for all of the variables except the one we solved for. We arbitrarily let $x = 3$ and $A = 12$. We substitute these values into our result to determine a value for *y*.

$\dfrac{A - x^2}{2x} = y$ Start with the result.

$\dfrac{12 - 3^2}{2 * 3} = y$ Substitute the chosen values into the result.

$0.5 = y$ Evaluate the expression to obtain a value for the variable we solved for.

Next, we substitute *all three values* into the *original equation* and check to see that the statement is true for these values.

$A = x^2 + 4\left(\dfrac{1}{2}xy\right)$ Return to the original equation.

$12 \stackrel{?}{=} 3^2 + 4\left(\dfrac{1}{2} * 3 * 0.5\right)$ Substitute all values into this equation.

$12 = 12$ ✔   Evaluate each side to determine if the statement is true.

We conclude that

$$y = \dfrac{A - x^2}{2x}$$

Example 2

Solve

$$R = \dfrac{C - S}{t} \quad \text{for } S$$

and verify your solution.

Solution

$R = \dfrac{C - S}{t}$

$t * R = \dfrac{C - S}{t} * t$ Multiply both sides by *t*.

$Rt = C - S$ Simplify both sides.

$Rt = C + {}^-1S$ Rewrite subtraction as addition of the opposite.

$Rt - C = {}^-1S$ Subtract *C* from both sides.

${}^-1 * (Rt + {}^-1C) = ({}^-1S) * {}^-1$ Multiply both sides by ${}^-1$.

${}^-Rt + C = S$

NOTE: In the second to last step in the preceding example, we multiplied both sides of the equation by $^-1$. Normally at this stage we would divide both sides of the equation by the coefficient of S, $^-1$; however, because multiplying an expression by negative one is equivalent to dividing an expression by negative one we can perform either operation. For example, $^-1 * 15 = {}^-15$ and $\frac{15}{-1} = {}^-15$. Multiplying an algebraic expression by negative one usually results in a less complex expression than dividing. If the coefficient of the variable we are solving for is $^-1$, we usually multiply both sides by $^-1$ to complete the solution.

Since we solved the formula for S, we can verify our results by choosing values for the other variables and then use our result to determine a value for S. Let $R = 3$, $t = 4$, and $C = 5$.

$^-Rt + C = S$ Start with the result.
$^-3 * 4 + 5 = S$ Substitute the chosen values into the result.
$^-7 = S$ Evaluate the expression to obtain a value for S.

Now substitute all four values into the original equation.

$R = \dfrac{C - S}{t}$ Return to the original equation.

$3 \stackrel{?}{=} \dfrac{5 - (^-7)}{4}$ Substitute all values into this equation.

$3 = 3$ ✔   Evaluate each side of the expression to determine if the statement is true.

We conclude that $S = {}^-Rt + C$.

Example 3

Solve $0.25(3w + M) = T - 0.30w$ for w, and verify your solution.

Solution

$0.25(3w + M) = T - 0.30w$

$0.25 * 3w + 0.25 * M = T - 0.30w$ Distribute 0.25 over the sum.

$\underline{0.75w} + 0.25M = T - \underline{0.30w}$ Simplify the left-hand side, and identify the terms containing w.

$0.75w + 0.25M + 0.30w = T - 0.30w + 0.30w$ Add $0.30w$ to both sides to collect terms containing w on the left.

$1.05w + 0.25M = T$ Simplify both sides.

$1.05w + 0.25M - 0.25M = T - 0.25M$ Subtract $0.25M$ from both sides to isolate the term containing w.

$1.05w = T - 0.25M$ Simplify both sides.

$\dfrac{1.05w}{1.05} = \dfrac{T - 0.25M}{1.05}$ Divide by the coefficient of w, 1.05.

$w = \dfrac{T - 0.25M}{1.05}$

To verify, we need to obtain a value for w. Let $T = 10$ and $M = 12$.

$w = \dfrac{T - 0.25M}{1.05}$ Start with the result.

$w = \dfrac{10 - 0.25 * 12}{1.05}$ Substitute the chosen values into the result.

$w \approx 6.67$ Evaluate to obtain a value for w.

Now substitute all of the values into the original equation.

$$0.25(3w + M) = T - 0.30w$$
$$0.25(3 * 6.67 + 12) \stackrel{?}{\approx} 10 - 0.30 * 6.67$$
$$8.0025 \approx 7.999 \quad ✓$$

We conclude that

$$w = \frac{T - 0.25M}{1.05}$$

Did you notice that our verification values were not exactly equal? Because we rounded the value that we obtained for w, this is expected. However, the values should be approximately equal. If you are not sure if your values are close enough, you can redo your verification using a more accurate rounded value.

As you can see from these examples, the strategy for solving literal equations is the same as for solving equations in one variable. First we simplify both sides of the equation or if the equation contains fractions we multiply by a common denominator. Second, we collect and isolate all terms containing the variable we are solving for. And finally, we divide by the coefficient of the variable we are solving for, which may include literal factors.

Verifying the results of literal equations is somewhat different from checking the solution to an equation in one variable. When checking the solution to an equation, we substitute the value of our solution into the original equation. When we solve a literal equation for a variable our result is not a value. Therefore, to verify the results to a literal equation, first obtain a value for the variable that was solved for. To do this, choose values for all other variables. Do not choose 0, 1, or 2. Substitute these values into the result to obtain a value for the solved for variable. Now, substitute the values of all variables, including the calculated value, into the original and check to see if the statement is true.

Problem Set 3.4

1. a. Identify the numerical coefficient of the term $-5WRT$.
    b. Identify the coefficient of T in the term $-5WRT$.
    c. Identify the coefficient of R in the term $-5WRT$.

2. a. Identify the numerical coefficient of the term $0.83x^2y$.
    b. Identify the coefficient of x^2 in the term $0.83x^2y$.

3. a. Identify the numerical coefficient of the term $MP(K + 5)$.
    b. Identify the coefficient of M in the term $MP(K + 5)$.
    c. Identify the coefficient of P in the term $MP(K + 5)$.

4. Solve each literal equation for the indicated variable. Verify your solutions.
    a. $A = P + Prt$ for r
    b. $F = mv^2$ for m
    c. $\frac{S + F}{S} = 4$ for S
    d. $b^2 - 4ac = d$ for c
    e. $A = \frac{1}{2}h(b_1 + b_2)$ for b_2
    f. $W - kr = (k - 1)R$ for r
    g. $a = \frac{5a + b}{3}$ for a
    h. $A = 2\pi r^2 + 2rh$ for h
    i. $G = \frac{t_1 + t_2 + t_3 + t_4}{4}$ for t_4

5. Solve each literal equation for the indicated variable. Verify your solutions.

   a. $E = I\left(R + \dfrac{r}{n}\right)$ for R

   b. $R = W + 0.20R$ for R

   c. $G = \dfrac{a}{1 - r}$ for r

   d. $P(Q + 6) = P - 12$ for Q

   e. $0.25q + 0.75Q = 2V$ for q

   f. $z = \dfrac{x - m}{s}$ for m

   g. $F = \dfrac{km_1 m_2}{d^2}$ for m_1

   h. $W = \dfrac{3T_1 + T_2}{4}$ for T_1

6. a. Solve the equation $3x - 4y = 9$ for y.
   b. Use your result from part a to complete the following table.
   c. Plot the points from your completed table.

x	-2	-1	0	1	2	3	4
y							

7. In addition to the Fahrenheit and Celsius temperature scales, there is a temperature scale called the Kelvin scale. The Kelvin scale is a temperature scale measured in degrees Celsius from a point called absolute zero, which is about $-273°C$. The formula

   $$\tfrac{5}{9}(F - 32) \approx K - 273$$

   relates F, the temperature in degrees Fahrenheit, to, K, the temperature in kelvins.

   a. Solve the formula for F in terms of K.
   b. Use your formula from part a to complete the following table.
   c. Plot the points from your completed table.

Temperature in kelvins	0	100	200	273	280	300
Temperature in °F						

8. The formula for converting temperature from degrees Celsius to degrees Fahrenheit is $F = \tfrac{9}{5}C + 32$.

   a. Solve this formula for C in terms of F.
   b. Use your result from part a to complete the following table.
   c. Explain why you would want to solve the formula for C before completing the table.

Degrees Fahrenheit	32	45	50	61	82	95	100
Degrees Celsius							

9. **Math Grade.** Ton is currently taking a mathematics course. So far, there have been two worksheets and one exam. She scored $\frac{24}{25}$ on the first worksheet, $\frac{18}{25}$ on the second worksheet, and $\frac{41}{50}$ on the first exam.

   a. What is Ton's current average in the course?

   b. The second exam is worth 75 points. The following is a formula to determine a student's percent, as a decimal, after the first two worksheets and first two exams.

   $$P = \frac{w_1 + w_2 + E_1 + E_2}{175}$$

   where P = percent written as a decimal
   w_1 = score on the first worksheet
   w_2 = score on the second worksheet
   E_1 = score on the first exam
   E_2 = score on the second exam

   Explain where the 175 comes from.

   c. Use the formula in part b to determine the score that Ton needs on the second exam to bring her average up to 88%.

   d. Ton is not the only student who has a question about the second exam. Three more students want to know what score they need on the second exam to achieve a certain percentage. Their current scores and desired percentage are recorded in the following table. Solve the formula in part b for E_2 in terms of the other variables (if you have not already done so), and complete the following table.

Name	Worksheet 1	Worksheet 2	Exam 1	Exam 2	Desired Percent (%)
Alma	20	15	42		80
Brown	15	18	38		75
Devon	22	23	42		92

10. **Budget.** A private agency increased its 1998 budget by 15% from the previous year. Its 1999 budget is now $7,475,000. What was the 1998 budget? What was the dollar increase in the budget from 1998 to 1999?

11. **Selling Price.** A local bookstore marks up all books by 22% of the *selling price*. If the bookstore buys a book for $15, what will the price of the book be to customers?

3.5 Solving Linear Literals Using Common Factoring

Activity Set 3.5

1. Complete the following table by writing the products as sums (or differences) using the distributive property. Then identify the factors of each of the terms of the resulting sums (or differences).

Original Expression	Factors of Original Expression	Expression Written as a Sum or Difference	Factors of Each Term		
			First Term	Second Term	Third Term
$x(2 + y)$	$x, 2 + y$	$2x + xy$	$2, x$	x, y	None
$y(x - 4y)$					
$3x(x + y)$					
$y(3 - x + y)$					
$xy(2x - y)$					

2. In the first expression in Activity 1, both the term $2x$ and the term xy had a factor of x. We refer to the x as a **common factor.** Identify the common factor(s) for the terms of each of the expressions. What relationship exists between the common factors and the factors of the original expressions?

3. a. Write the factors of each term in the expression $5m + mp$.
   b. What factor(s) do all of the terms have in common?
   c. Rewrite this expression as a product.
   d. Apply the distributive property to your result. After applying the distributive property, did you get the original expression?

4. In each of the following expressions, look for common factors in all of the terms, then rewrite the expressions as products. Apply the distributive property to verify your result.
   a. $3x^2 - 8x$
   b. $AB + AC$
   c. $3xy - 4x$
   d. $13x^2 + 5xy + x$

5. Consider the formula $K = NR + NT$.
   a. Substitute $K = 18$, $R = 4$, and $T = 5$ into the formula. Simplify the right-hand side. What is the coefficient of N? Solve the formula for N when $K = 18$, $R = 4$, and $T = 5$.
   b. Substitute $K = 6$, $R = 3$, and $T = 9$ into the formula. Simplify the right-hand side. What is the coefficient of N? Solve the formula for N when $K = 6$, $R = 3$, and $T = 9$.
   c. Substitute $K = 36$, $R = 1$, and $T = -5$ into the formula. Simplify the right-hand side. What is the coefficient of N? Solve the formula for N when $K = 36$, $R = 1$, and $T = -5$.

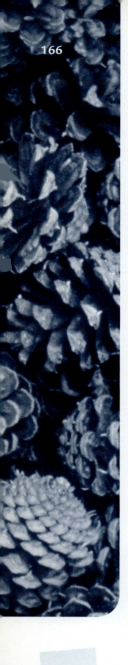

6. a. Using your results from Activity 5, complete *the second and third rows* of the following table.

b. In the formula $K = NR + NT$, what is the coefficient of N in terms of R and T? Solve the formula for N in terms of K, R, and T.

K	R	T	Formula After Substitution	Coefficient of N	Solution for N
18	4	5	$18 = 4N + 5N$	9	2
6	3	9			
36	1	−5			
K	R	T			

7. a. Solve $5t - 3t = 24$ for t. Check your solution.

b. Solve $4M - MT = K$ for M. Verify your solution by numerical substitution.

8. a. Solve $x = 2(5 - x)$ for x. Check your solution.

b. Solve $Q_1 = P(Q_2 - Q_1)$ for Q_1. Verify your solution by numerical substitution.

9. a. Solve $12x + 5 = 2(4x - 5)$ for x. Check your solution.

b. Solve $QC + T = 4(QB + D)$ for Q. Verify your solution by numerical substitution.

10. a. Solve the equation for the indicated variable. Check your solution.

$$6 = \frac{3 + r}{r} \quad \text{for } r$$

b. Solve the equation for the indicated variable. Verify your solution by numerical substitution.

$$P = \frac{A + B}{B} \quad \text{for } B$$

Discussion 3.5

Factoring Common Factors

In this section we will be solving linear literal equations that require factoring. In the activities, we found that if each term of an expression has a factor in common with all of the other terms, then the expression can be rewritten as a product by factoring out the common factor.

Example 1

Rewrite the expression $4m - 7mp + m^2$ as a product by factoring. Verify your results.

Solution The expression $4m - 7mp + m^2$ has three terms. The term $4m$ has two factors, 4 and m. The term $7mp$ has three factors, 7, m, and p. The term m^2 has m as a factor two times. Because all three terms have m as a factor in common, we can rewrite $4m - 7mp + m^2$ as a product of m and the sum or difference of the remaining factors.

$$4m - 7mp + m^2$$
$$= m(4 - 7p + m) \quad \text{The common factor } m \text{ is factored out of the expression.}$$

To verify, apply the distributive property to your result.

$$m(4 - 7p + m)$$
$$= 4m - 7mp + m^2 \quad \checkmark$$

We conclude that $4m - 7mp + m^2 = m(4 - 7p + m)$.

We can see from this example that factoring out common factors and using the distributive property are processes that are the reverse of each other. Factoring out common factors changes an expression with two or more terms into an equivalent expression that is a product. Applying the distributive property changes an expression with two or more factors into an equivalent expression written as the sum or difference of terms. By applying the distributive property to the previous result, we are able to get back to the original expression.

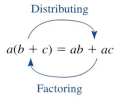

More Linear Literals

As you solve more complex linear literal equations, you may find the following strategy helpful. You will find many of these steps to be the same as the steps for solving linear equations in one variable.

Strategy

Solving a Linear Literal Equation

1. Simplify both sides of the equation by applying the distributive property or by combining like terms (or both). If the equation contains fractions, you may want to multiply both sides of the equation by a common denominator to eliminate the fractions.
2. Identify the terms containing the variable for which you are solving.
3. Add or subtract terms from both sides of the equation to collect all terms containing the variable for which you are solving on one side and all other terms on the opposite side. Again combine like terms if possible.
4. If the variable for which you are solving occurs in more than one term, factor out the common variable.
5. Divide both sides by the coefficient of the variable for which you are solving.
6. Verify your solution (see Section 3.4).

Example 2

Pictured here is a Norman window, which consists of a rectangle with a semicircle on top. Write a formula for the perimeter of the window in terms of L and d. Solve your formula for d, and verify your results by numerical substitution.

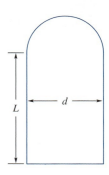

Solution

The window consists of two shapes: a rectangular portion with three sides and a semicircle.

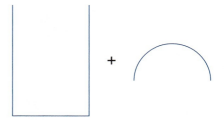

The perimeter of the rectangular portion is $2L + d$. The perimeter of the semicircle is $\frac{\pi d}{2}$, half the circumference of a circle. If we let P represent the perimeter of the window, then a formula for the perimeter of the window is

$$P = 2L + d + \frac{\pi d}{2}$$

Next, we need to solve this formula for d.

$P = 2L + d + \dfrac{\pi d}{2}$	
$2 * P = 2 * \left(2L + d + \dfrac{\pi d}{2}\right)$	Multiply both sides by 2 to eliminate the fraction.
$2P = 4L + 2d + \pi d$	Distribute to simplify the left-hand side.
$2P = 4L + \underline{2d + \pi d}$	Identify the terms that contain the variable d.
$2P - 4L = 2d + \pi d$	Subtract $4L$ from both sides to isolate the terms containing the variable d.
$2P - 4L = d(2 + \pi)$	Factor out d on the right-hand side.
$\dfrac{2P - 4L}{2 + \pi} = \dfrac{d(2 + \pi)}{2 + \pi}$	Divide both sides by the coefficient of d, $(2 + \pi)$.
$\dfrac{2P - 4L}{2 + \pi} = d$	

To verify our result, we start with our solution and choose values for all of the variables except the one we solved for. We arbitrarily let $P = 20$ and $L = 3$. We substitute these values into our result to determine a value for d.

$$\frac{2 * 20 - 4 * 3}{2 + \pi} = d$$

$$d \approx 5.4$$

Next, we substitute *all three values* into the *original equation* and check to see that the statement is true for these values.

$$P = 2L + d + \frac{\pi d}{2}$$

$$20 \stackrel{?}{\approx} 2*3 + 5.4 + \frac{\pi * 5.4}{2}$$

$$20 \approx 19.9 \quad \checkmark$$

Notice that because our value for d was approximate, our verification is approximate.

We conclude that

$$d = \frac{2P - 4L}{2 + \pi}$$

In this section we learned how to factor an expression that has common factors. Factoring common factors allows us to rewrite expressions that are written as sums or differences as a product. We can use this technique to solve linear literal equations if the variable we are solving for occurs in more than one term.

Problem Set 3.5

1. Rewrite each expression as a product by factoring. Verify your results.
    a. $4x + xy$
    b. $6ab - 13a$
    c. $12w + 7w^2 - 3wy$
    d. $K - KC$
    e. $WT^2 - 5WT$
    f. $5x^2 + 3x^2y$
    g. $PT + RT$
    h. $-8QT + Q$
    i. $4GK - 9GK^2 - G$

2. a. Identify the coefficient of x in the term $x(4 + y)$.
   b. Identify the coefficient of K in the term $K(1 - C)$.
   c. Identify the coefficient of G in the term $G(4K - 9K^2 - 1)$.
   d. Identify the coefficient of w in the term $pw(1 - x)$.

3. Solve each literal equation for the indicated variable. Verify your solution.
    a. $E = IR_1 + IR_2 + IR_3$ for I
    b. $y = mx + b$ for x
    c. $A = P + Prt$ for P
    d. $R = W + rR$ for R
    e. $P(Q + 6) = P - 12$ for P
    f. $P = 2\pi r + 2L + 2r$ for L
    g. $P = 2\pi r + 2L + 2r$ for r
    h. $(y - y_1) = d_f(x - x_1)$ for d_f
    i. $D = ad - bc$ for b

4. Solve each literal equation for the indicated variable. Verify your solution.

   a. $r = \dfrac{d}{t}$ for t

   b. $S = L - 0.07L$ for L

   c. $\dfrac{s_1 + s_2 + s_3 + s_4 + M}{T} = 0.70$ for M

   d. $x + 2ay = 5ay$ for a

   e. $P = \dfrac{S}{S + F}$ for F

   f. $P = \dfrac{S}{S + F}$ for S

   g. $F = 2.05(V_2 - V_1) + 3.75(V_2 - V_1)$ for V_2

5. a. Write a formula for the surface area (the sum of the area of each of the faces) of the *rectangular solid* pictured here.

   b. Solve your formula from part a for the height (H) of the solid.

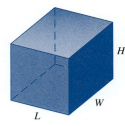

6. **Automotive Horsepower.** In automotive engineering a formula for an engine's horsepower in terms of the average pressure per piston, piston stroke length, piston area, crankshaft revolutions per minute, and number of pistons is modeled by the formula

   $$H_p = \dfrac{1}{396{,}000} P_p P_\ell A r n$$

   where H_p = horsepower

   P_p = average piston pressure in pounds per square inch

   P_ℓ = piston stroke length in inches

   A = area of the piston in square inches

   r = crankshaft revolutions per minute

   n = number of pistons

   a. Suppose it is your job to determine the piston stroke length for several cases. Solve the formula for P_ℓ.

   b. Use your formula from part a to complete the following data table. Round your answers to the nearest hundredth of an inch.

H_p (horsepower)	P_p (lb/in.^{2})	A (in.^{2})	r (rev/min)	n	P_ℓ (in.)
150	120	9	2400	6	
175	120	9	2400	6	
180	120	9	2500	6	
180	140	10	2500	6	
180	140	10	2500	8	

7. **Mark-Up on Books.** A local bookstore marks books up 22% of the retail cost of the books. The formula for this can be modeled as

   retail cost = wholesale cost + 0.22 ∗ retail cost

   You have a shipment that needs to be priced before it is placed on the shelves. The invoice for each book lists the wholesale cost.

   a. Solve the formula for the *retail cost* in terms of the *wholesale cost*.
   b. Use the formula to complete the following table.
   c. Explain why you would want to solve the formula for the retail cost before substituting in the values in the table and determining each of the retail costs.

Wholesale Cost of Book ($)	Retail Cost of Book
3.50	
5.95	
12.00	
25.80	
45.90	

8. List three solutions to the equation $y = x + 2$.

9. Graph each of the following equations by plotting points.
   a. $y = x + 2$
   b. $y = 5x + 3$
   c. $y = x^2 - 1$
   d. $y = \frac{1}{2}x^3$
   e. $y = 4(1 - 2x) + 6x$

10. Suppose you are asked to solve each of the following equations for the indicated variable. Without actually solving them, rank them in order from easiest to hardest. (Label the easiest #1, second easiest #2, and so on.)
    a. $10P = 4AP + 1$ for A _____
    b. $10P = 4AP + 1$ for P _____
    c. $\dfrac{20 + 34 + W}{3} = 25$ for W _____

11. **Raise.** Margaret just received her contract for the new work year. On the contract, it says that her annual salary for the new work year will be $40,248. She knows that she received a 7.5% raise over the previous year, but she does not remember her salary from the previous year. What is the dollar increase of her raise?

12. **Video Arcade.** Joel owns a video arcade. In 1999, he found that the video games were used for a total of 100,037 hours. This was the first year that he kept track of the usage. Joel estimates that this was a 40% increase over the previous year.
    a. If his assumption is correct, how many hours were the video games used the previous year?
    b. If the increase remains the same, how many hours will the games be used in 2000?

Chapter Three Summary

In Chapter 3 we learned how to simplify algebraic expressions and solve algebraic equations. In the process, we learned some new vocabulary that helped us discuss this process.

In a single term, the **coefficient** of a specified factor consists of the product of all the other factors. The **numerical coefficient** of a term is the numerical factor of that term. Terms that contain the same variables to the same powers are called **like terms.**

To simplify an algebraic expression means to perform all of the indicated operations.

- To add or subtract algebraic expressions we can combine like terms. To combine like terms, we add or subtract the numerical coefficients of the terms and keep the same literal factors. It is often helpful to rewrite subtraction as addition of the opposite.
- To multiply algebraic expressions, we multiply the numerical coefficients and rewrite the factors of the variables using exponents if necessary.
- To simplify the product of a single term and a sum, $a(b + c)$, we apply the distributive property.

> **The Distributive Property**
>
> For any real numbers a, b, and c, $a(b + c) = ab + ac$
>
> $a(b + c) = ab + ac$

To verify a simplified result, we choose values for all of the variables and substitute these values into the original expression and into the result. If the two values agree, we have confidence that the expressions are equivalent. If the values do not agree, we know that we made a mistake. It is important to note that this verification process is not a guarantee. It only gives us confidence in our result. In addition, this process cannot determine whether our result is completely simplified.

In this chapter we solved linear equations in one variable and linear literal equations. To solve an equation means to find the value of the variable that makes the equation true. To solve a linear equation in one variable we can use the following strategy.

Strategy

Solving Linear Equations in One Variable

1. Simplify both sides of the equation by applying the distributive property or combining like terms (or both). If the equation contains fractions, you may want to multiply both sides of the equation by the least common denominator to eliminate the fractions.
2. Add or subtract terms from both sides of the equation to collect all terms containing the variable on one side and all other terms on the opposite sides. Again combine like terms if possible.
3. Divide both sides by the coefficient of the variable.
4. Check your solution in the original equation.

To solve a linear literal equation we follow a similar process. The only difference is that the equation contains more than one variable and the variable we are solving for may be in two terms that cannot be combined. Therefore, we slightly modify the process.

Strategy

Solving a Linear Literal Equation

1. Simplify both sides of the equation by applying the distributive property or combining like terms (or both). If the equation contains fractions, you may want to multiply both sides of the equation by the least common denominator to eliminate the fractions.
2. Identify the terms containing the variable you are solving for.
3. Add or subtract terms from both sides of the equation to collect all terms containing the variable on one side and all other terms on the opposite sides. Again combine like terms if possible.
4. If the variable you are solving for occurs in more than one term, factor out the common variable.
5. Divide both sides by the coefficient of the variable.
6. Verify your solution.

To check a solution to a linear equation in one variable, substitute the value of the solution into the original equation. If this value makes the equation true, we know we have a correct solution.

When we solve a literal equation, our result is not a value; therefore, to verify the result to a literal equation we must first find a value for the variable that we solved for. We can use the following strategy.

Strategy

Verifying the Results to a Literal Equation

1. Obtain a value for the variable that you solved for.
   a. Choose values for all of the variables *except* the one that you solved for. Do not choose 0, 1, or 2.
   b. Substitute your chosen values into your *result*, and calculate the value of the variable that you solved for.
2. Substitute the values of all variables (this includes the calculated value) into the *original* equation, and check to see if the statement is true.

Chapter Review 1-3

1. For each of the numerical expressions do the following.

   i. Determine the number of terms in each expression.
   ii. Underline the terms.
   iii. Evaluate the expression showing all steps.
   iv. Evaluate the expression *in one step* on your calculator. Compare your results to those from part iii. If your results are different, find where you have made an error.

   a. $\dfrac{5 + 3}{6}$
   b. $5 + 3 \div 6$
   c. $(-5)^2 - 10 + 15$
   d. $-5^2 - 10 + 15$
   e. $-5\sqrt{4 + 12}$
   f. $-5\sqrt{4} + 12$
   g. $5 * 2^3 + \dfrac{8}{-4}$
   h. $\dfrac{(5 * 2)^3 + 8}{-4}$

2. For each expression do the following.

   i. Estimate the expression *without using a calculator.* Show your steps.
   ii. Evaluate the expression in one step using a calculator. Write the expression as it is displayed on your calculator. Round the result to the nearest thousandth.

   a. $49.3 - 11.2 * 7.65$
   b. $\dfrac{37.5 + 2 * 28.4 + 3 * 31.8}{6}$
   c. $\dfrac{\sqrt{6.2 * 9.8} - 12.6}{0.23 * 18}$

3. a. Identify the coefficient of r in the term $5tr$.
   b. Identify the coefficient of t in the term $5tr$.
   c. Identify the numerical coefficient in the term $5tr$.

4. Use the expression $-3xy^2 + xy(a + b) - ab + 7\sqrt{x + y}$ to answer the following questions.
   a. How many terms are in the expression?
   b. What is the numerical coefficient of the first term? The third term?
   c. What is the coefficient of x in the first term? The second term?
   d. What are the factors of the first term? The last term?

5. Write the following numbers in scientific notation.
   a. 47,350,000
   b. 0.00056
   c. 2700.
   d. 0.008500
   e. $200 * 10^{-5}$
   f. $0.05 * 10^{-5}$

6. Measure the following figure to determine its perimeter and area. On a sketch of the figure, indicate your measurements including units. Include units in all computations.

7. Explain the difference between the expressions 25 m² and (25 m)².

8. a. A team of students was asked to convert 100 cm² to square feet. Some of the students got a result of about 3 ft², but the other students obtained a result of about 0.1 ft². Without converting the measurements, decide which of these answers is reasonable and explain how you arrived at your decision.

   b. Another team of students was asked to determine the number of gallons in one cubic yard. The following is their solution process. Is this process correct? If so, explain each step. If not, explain what the mistakes are.

$$1 \text{ yd}^3$$
$$= 1 \text{ yd}^3 * \frac{3 \text{ ft}^3}{1 \text{ yd}^3} * \frac{7.481 \text{ gal}}{1 \text{ ft}^3}$$
$$= 3 * 7.481 \text{ gal}$$
$$\approx 22 \text{ gal}$$

9. Use unit fractions to answer the following questions.
   a. How many gallons are in 12.5 liters?
   b. How many square inches are in one square meter?
   c. How many feet per second are in 80. kilometers per hour?
   d. How many gallons per minute are equivalent to 125 cubic meters per hour?

10. In each of the following tables, observe the pattern and use it to complete the table. For the last row in each table, write an expression in terms of the variable that generalizes the pattern.

a.
C	P
2	$-2^2 + 2*5 + 1$
4	$-4^2 + 4*5 + 1$
6	$-6^2 + 6*5 + 1$
8	
C	

b.
t	R
5	45(10)
10	45(15)
15	45(20)
20	
t	

c.
x	Y
1	$15*0 + 0$
2	$15*1 + 1$
3	$15*2 + 2$
4	
x	

d.
a	B
1	$3*1 + \dfrac{2\pi}{2}$
2	$3*2 + \dfrac{4\pi}{2}$
3	$3*3 + \dfrac{6\pi}{2}$
4	
a	

e.
m	N
1	2
2	$2*3$
3	$2*3*3$
4	$2*3*3*3$
5	
m	

f.
x	y
1	2
2	$3*3$
3	$4*4*4$
4	
5	
x	

11. For each expression, determine the numerical value and units of the result.

   a. 2.0 ft (825 ft + 140 ft)

   b. $\dfrac{\pi(4.3 \text{ cm})^2 * 24 \text{ cm}}{155 \text{ sec}}$

   c. $1.8 \dfrac{\text{km}}{\text{min}} * \dfrac{1 \text{ mile}}{1.609 \text{ km}} * \dfrac{60 \text{ min}}{1 \text{ h}}$

   d. $\dfrac{1600 \text{ m}}{50 \dfrac{\text{m}}{\text{min}}}$

12. A pattern of structures is pictured in the following figures.

   a. Determine the number of blocks needed to construct the 50th structure.

   b. Using n as the structure number, write an equation for the number of blocks needed to build the structure.

Structure 1

Structure 2

Structure 3

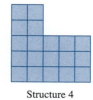
Structure 4

13. In each of the following situations, decide which variable is the dependent variable and which is the independent variable.
    a. You plan to replace the roof on your house. Let C represent the cost of the roofing material and F represent the area of your roof in square feet.
    b. You plan to fence your yard. Let L represent the distance that you plan to fence in meters and H represent the number of hours it will take to put up the fence.
    c. You fill your ice cube trays with water and place them in your freezer. Let M represent the length of time since the water was placed in the freezer and T represent the temperature of the water.
    d. You want to refinance your home. Let P represent your monthly mortgage payment and R represent your annual interest rate.

14. **Force on Flag.** A flag blowing in the wind on a flagpole exerts a force on the pole. To determine the strength of that force we can use the formula
    $$F = 3.0 * 10^{-4} A v^{1.9}$$
    where F = force on the flagpole in pounds,
    A = area of the flag in square feet, and
    v = velocity of the wind in miles per hour
    a. Determine the force on the flagpole with a flag that is 48 ft^2 when the wind is blowing at 15 mph.
    b. Determine the force on a flagpole with a flag that is 71 in. by 52 in. when the wind is blowing at 21 mph.

15. **Annual Milk Consumption.** The following graph shows the annual consumption of milk per person by Americans from 1970 to 1994. *Use the graph* to answer the following questions.
    a. How much milk was consumed by an American in 1982? 1990?
    b. In what year did milk consumption fall below 27 gallons per person?
    c. How much does the annual milk consumption decrease each year after 1970?

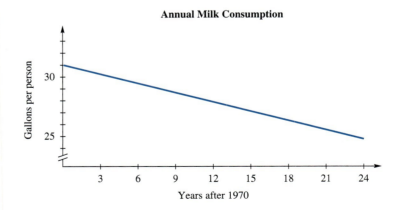

16. a. Write a single expression that represents the area of the following figure.
    b. Write a single expression that represents the perimeter of the figure.

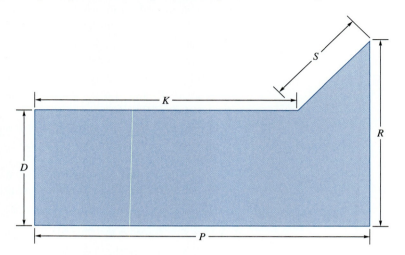

17. For each part a–f, do the following.
    i. Identify if the number is in scientific notation, ordinary notation, or neither.
    ii. Rewrite the number in the other notation(s).

    a. In 1992, the United States sent $2,240,000 in aid to countries in the Western Hemisphere.
    b. The population of the United States is about $2.56 * 10^8$.
    c. Lake Mead, a reservoir for the Hoover Dam in Nevada, is $28.250 * 10^6$ acres.
    d. The diameter of a particular atom is 0.00000001 inches.
    e. The population of China is 1,169,000,000 people.
    f. A dusty salamander's DNA weighs $3.5 * 10^{-12}$ grams.

18. **A Cellular Call.** Western-net Company charges its cellular phone customers $27.95 a month for the first 30 minutes of air time plus $0.39 for each additional minute beyond 30 minutes.
    a. What is the monthly cost for 30 minutes of air time? 35 minutes? 50 minutes?
    b. Make a table of at least five values for the monthly cost of a cellular phone. Include air times between 30 and 60 minutes in your table.
    c. Draw a graph of this situation. (*Hint:* First decide which variable is most likely the dependent variable and which is most likely the independent variable. Use this to decide how you should label the horizontal and vertical axes.)
    d. Write an equation for the monthly cost of a cellular phone in terms of the number of minutes of air time. (*Note:* Your model will only be valid if you talk for more than 30 minutes in a given month.)
    e. Find the monthly cost to talk for 37 minutes.
    f. How long could you talk in one month for $35.00?

19. Simplify each expression. You may find it helpful to rewrite subtraction as addition of the opposite. Verify your results by numerical substitution.
    a. $-5P + M + 3P - 4M$
    b. $3.2R - 3.8RQ + 6.7RQ - 12.3$
    c. $2(3b + 4a + 5)$
    d. $-5(2x^2 - 3xy)$
    e. $4M - 3(2MN + M)$

20. Simplify each expression. You may find it helpful to rewrite subtraction as addition of the opposite. Verify your results by numerical substitution.
    a. $7b - 2b^2 + b^2 - b$
    b. $3T(15P + 41)$
    c. $18Q(3R + 5 - 7T)$
    d. $-M(3M + 14 + MN)$
    e. $8(x^2 - 2xy) - 2x(3y - x)$

21. Rewrite each expression as a product by factoring. Verify your results.
    a. $5m + 2mn$
    b. $12x - 5xy + x^2$
    c. $pk + p$
    d. $10w^2p - 3wp$
    e. $18RT + PT$
    f. $\pi r^2 + 2\pi r - \pi$

22. a. Identify the numerical coefficient of the expression $\frac{x}{5}$.
    b. Identify the numerical coefficient of the expression
    $$\frac{3(x - 5)}{2}$$

23. **Medical Insurance.** A medical insurance policy uses the following formula to determine how much the patient should pay.
    $$E = [(T - D) * (1.00 - P)] + D$$
    where E = total expense to the patient,
    T = total of the hospitalization bill,
    D = deductible that the patient must pay first, and
    P = decimal percentage that the insurance company pays after the patient meets the deductible

    Suppose after a short hospital stay, your total hospitalization bill was $953.73. You are required to pay the first $100 of your hospital expenses (this is known as the deductible).
    a. Substitute the known values into the preceding formula.
    b. After substituting in the known values, simplify the formula from part a.
    c. If the total expense to you is $270.75 for this hospital stay, algebraically determine the percentage that the insurance company pays.

24. **Styrofoam.** One cubic foot of Styrofoam weighs about 6 ounces. Use unit fractions to determine the density of Styrofoam in grams per cubic centimeter.

25. **Folding Box.** A formula for the volume of an open box made from a rectangular sheet of cardboard by cutting four squares from each corner and folding up the sides is given as
    $$V = x(L - 2x)(W - 2x)$$
    where V = volume of the box
    L = overall length of the sheet of cardboard
    W = overall width of the sheet of cardboard, and
    x = height of the box after cutting squares of side x from each corner

    a. Is this a general formula or a unit-specific formula?
    b. Determine the volume of a box made from a 2 ft by 3 ft rectangular sheet of cardboard if a square, 3 in. on a side, is cut from each corner and the sheet is then folded into the shape of an open box.

26. U.S. Public Debt. The following graph closely approximates the public debt of the United States for the years between 1975 to 1990. Use the graph to answer the questions.

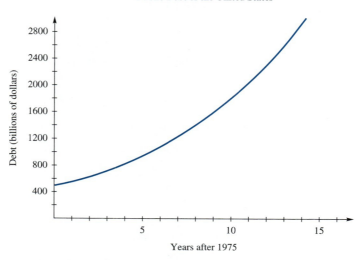

Public Debt of the United States

a. What was the public debt of the United States in 1975? 1980?
b. In what year did the public debt reach $1000 billion? In what year did the public debt reach $2,500,000,000,000?
c. Where does the graph touch the vertical axis? What does it mean in the context of this problem?

27. Graph each of the following equations.

a. $y = 2x - 7$
b. $y = 1 - x^2$
c. $y = (x - 5)^2$
d. $y = 3(x - 4) + 2x$

28. Suppose you are asked to solve each of the following equations. Without actually solving them, rank them in order from easiest to hardest. (Label the easiest #1, second easiest #2, and so on.)

a. $2(x - 5) + 4 = 8x$ _____
b. $4w - 18 = 32$ _____
c. $\dfrac{k}{3} = 7$ _____
d. $38.50 = 2.50 - 3.25p$ _____
e. $\dfrac{R}{3} - 5R = 10$ _____

29. Following is a student's work, showing how he solved the equation $24(x - 2) + 5 = 29 + 4x$. Explain each step in the process.

$$24(x - 2) + 5 = 29 + 4x$$
$$24x - 48 + 5 = 29 + 4x$$
$$24x - 43 = 29 + 4x$$
$$24x - 43 - 4x = 29 + 4x - 4x$$
$$20x - 43 = 29$$
$$20x - 43 + 43 = 29 + 43$$
$$20x = 72$$
$$\dfrac{20x}{20} = \dfrac{72}{20}$$
$$x = 3.6$$

30. Solve the following equations, and check your solutions.
   a. $-2x + 5 = 17$
   b. $2 = 4 + 3(1 + B)$
   c. $\dfrac{14}{5} = \dfrac{60 - 2x}{15}$
   d. $5w - 15w = 10w + 10$
   e. $3(Y + 4) - 5(2Y - 5) = 125$
   f. $\dfrac{w}{5} - \dfrac{3w}{2} = w + 1$

31. Solve the following equations, and check your solutions.
   a. $\dfrac{-2x}{3} - 5 = \dfrac{7}{10}$
   b. $0.2(3 - x) - 0.3(x - 2) = 0.1x$
   c. $8 - \dfrac{4x}{5} = 3x$
   d. $6(x - 1) = 3(x - 2) + 6$
   e. $\dfrac{9}{15} = \dfrac{3}{x}$

32. **Safe Dosage.** The dosage of a particular drug given to dogs is determined by the weight of the dog. For an 80-lb dog, a safe dosage range is 45 to 105 mg. What is the safe dosage range for a 50-lb dog?

33. **Vacation Rentals.** Island Rental Surfboards rents surfboards for $7.50 for the first hour and $5.00 each additional hour. The Surf Shack rents surfboards for $9.15 the first hour and $3.90 each additional hour.
   a. For each company, how much does it cost to rent a surfboard for 1 hour? 2 hours? 6 hours?
   b. For each company, make a table of at least five values for the cost of renting a surfboard for several lengths of time.
   c. For each company, draw a graph of the cost of renting a surfboard in terms of the length of time it is rented. Draw both graphs on the same coordinate axes.
   d. For each company, write an equation for the cost to rent a surfboard in terms of the number of hours rented.
   e. Determine how long can you rent a surfboard for $40 from each of the companies.
   f. At what point is the cost of renting a surfboard the same for both companies?
   g. Which of the two companies is trying to capture the portion of the market that uses surfboards for a long time each day?

34. A wedge from a circle is called a sector. Plots of land can be in the shape of a sector. To find the area of such a sector, for real estate purposes, we can use the following formula.

   $A = 2.0034 * 10^{-7} D r^2$

   where A = area of the sector in acres,
   D = measure of the angle of the wedge in degrees, and
   r = radius of the sector in feet

   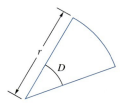

   Determine the area of a sector with an angle of 52° and a radius of 37.25 yards.

35. Regulations for a ramp for the handicapped specify that the slope be in a ratio of 1:12. This means that for every foot of vertical rise the ramp must cover 12 feet horizontally. You are building a ramp to reach the same height as a staircase that is 3.7 feet high. Determine the length of the ramp.

36. a. Identify the coefficient of r^2 in the term $pr^2(s + t)$.
  b. Identify the coefficient of $(s + t)$ in the term $pr^2(s + t)$.
  c. Identify the numerical coefficient in the term $pr^2(s + t)$.

37. Suppose you are asked to solve each of the following equations for the indicated variable. Without actually solving them, rank them in order from easiest to hardest. (Label the easiest #1, second easiest #2, and so on.)
  a. $PK = R - PT$ for T _____
  b. $PK = R - PT$ for K _____
  c. $PK = R - PT$ for R _____
  d. $PK = R - PT$ for P _____

38. The following is a student's work, showing how she solved the equation
$$\frac{KM + P}{T} = KR \quad \text{for } K$$
Explain each step in the process.

$$\frac{KM + P}{T} = KR$$

$$T * \frac{KM + P}{T} = KR * T$$

$$KM + P = KRT$$

$$KM + P - KM = KRT - KM$$

$$P = KRT - KM$$

$$P = K(RT - M)$$

$$\frac{P}{RT - M} = \frac{K(RT - M)}{RT - M}$$

$$\frac{P}{RT - M} = K$$

39. Solve each literal equation for the indicated variable. Verify your solution.
  a. $P = 2LW + 4LH$ for L
  b. $ax + by = c$ for y
  c. $A = P + Prt$ for P
  d. $R = W + 0.25R$ for R
  e. $R = \dfrac{CVL}{M}$ for L
  f. $A = \pi r^2 + LW$ for W

40. Solve each literal equation for the indicated variable. Verify your solution.

  a. $P = \pi d + 2L + W + 2\pi d$ for L

  b. $0.25x + 0.40(P - x) = 0.30P$ for P

  c. $\dfrac{T + P}{T} = 1.25$ for T

  d. $3K + 4 = \dfrac{T + P}{T}$ for T

  e. $D = 2R(C - P)$ for C

  f. $G = \dfrac{PT}{3P - T}$ for P

41. Solve the following equations. Express answers as specified. Check your solutions by numerical substitution into the original equations.

  a. $43 + y = 0.57(100 + y)$ nearest tenth

  b. $\dfrac{1.45P}{3.14} = 7.82 - 3.91P$ nearest hundredth

  c. $4.3(k + 3.2) = 8.1(k - 2.4)$ two significant digits

  d. $46.3(2.32 * 10^5 x + 5710) = 4.89 * 10^5$ three significant digits

Chapter Four

Linear Equations in Two Variables

In Chapter 2, we learned that we can graph any equation by making a table of values, plotting the points, and connecting the points with a smooth curve. However, we also discussed the fact that with this strategy we cannot be sure that our graph is entirely correct. We only know that the points we plotted are correct. In Chapter 4, we will look at a specific class of equations in two variables called linear equations. We will learn how to recognize an equation as linear. Once we know that an equation is linear, we only need to plot two points because two points determine a line. In addition, we will be able to write the equations of many different lines and use these to model problem situations.

4.1 Linear Relationships: Numerically and Graphically

1. For each of the following parts, complete the table, and graph the equation by plotting the points.

a.
Input	Output	Change
x	y = x − 3	in Output
−1	−4	
0	−3	+1
1		
2		
3		

b.
Input	Output	Change
x	y = −2x + 1	in Output
−1	3	
0		
1		
2		
3		

c.
Input	Output	Change
x	y = x² − 3	in Output
−1	−2	
0		
1		
2		
3		

d.
Input	Output	Change
x	y = 2^x	in Output
1		
2		
3		
4		
5		

2. a. Which equations from Activity 1 graph as straight lines?
   b. What features do the equations that graph as straight lines have in common that the other two do not?
   c. What features do the tables of values that graph as straight lines have in common that the other two do not?

3. Complete the following table, and graph the equation by plotting points. Does $y = 0.5x + 1$ graph as a straight line? Does it satisfy the features you described in Activity 2? Does this table or graph lead you to modify your response to Activity 2?

Input	Output	Change
x	y = 0.5x + 1	in Output
1		
2		
3		
4		
6		
8		

4. Based on your observations in Activities 1−3 answer the following questions.
   a. Describe how to determine from a table of values whether the data will lie in a straight line.
   b. Describe how to determine from an equation whether the equation will graph as a straight line.

Discussion 4.1

As you may have discovered in the activities, the graph of any equation of the form $y = mx + b$, where m and b are constants, is a line. For this reason the equation $y = mx + b$ is called a **linear equation.** In this section we focus on the numerical and graphical representations of a linear relationship. In the next section, we make connections between the numerical, graphical, and algebraic representations.

In the activities we found that the numerical view of a *linear equation shows a constant change in output for a constant change in input.*

> **Numerical Representations of Linear Equations**
> Any linear equation shows a constant change in output for a constant change in input.

Example 1

Using numerical methods, determine which of the following tables describe relationships that graph as straight lines. Explain how you decided.

a.

Input	Output
1	4
2	7
3	10
4	13

b.

Input	Output
1	30
2	28
3	25
4	21

c.

Input	Output
0	6.0
3	5.5
6	5.0
12	4.0

Solution We compute the change in output for each table.

a. In the table in part a, the output increases by 3 each time the input increases by 1. Because the change in output is constant for a constant change in input, the relationship described by the table graphs as a straight line.

Input	Output	Change in Output
1	4	
2	7	+3
3	10	+3
4	13	+3

Let's plot these points to visualize the relationship between the change in output and the change in input.

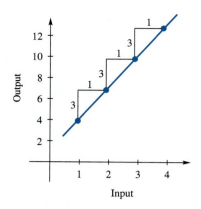

Notice that the change in output, 3, is represented as a vertical change on our graph. The input change of 1 is represented horizontally.

b. In the table in part b, the output decreases by a different amount when the input increases by 1. Therefore, the relationship described by the table does *not* graph as a straight line.

Input	Output	Change in Output
1	30	
		−2
2	28	
		−3
3	25	
		−4
4	21	

Let's plot these points to visualize the relationship between the change in output and the change in input.

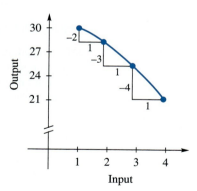

Again, the changes in output and input are represented on our graph by vertical and horizontal changes, respectively. We can visualize that the different changes in outputs for a constant change in input produce a curve rather than a straight line.

c. In the first three rows of the table in part c, the output decreases by 0.5 when the input increases by 3. In the third and fourth rows of the table, the output decreases by 1.0 when the input increases by 6, which is the same as a decrease of 0.5 for an increase of 3. This represents a constant change in output for a constant change in input. Therefore, the relationship described by the table graphs as a straight line.

4.1 Linear Relationships: Numerically and Graphically

Change in Input	Input	Output	Change in Output
	0	6.0	
3			−0.5
	3	5.5	
3			−0.5
	6	5.0	
6			−1.0
	12	4.0	

Let's plot these points to visualize the relationship between the change in output and the change in input.

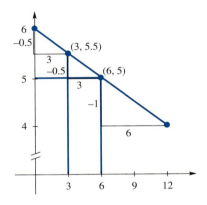

Between the points (0, 6.0) and (3, 5.5) the vertical change is −0.5, and the horizontal change is 3. Between the points (6, 5.0) and (12, 4.0) the vertical change is −1, and the horizontal change is 6. These changes are exactly twice the changes between the first two points. We can see this results in a straight line.

We now know that a linear relationship produces a constant change in output for a constant change in input. One way to see this relationship is to look at the *ratio of the change in output to the change in input*. For instance, consider Example 1c. In the first three rows of the table in part c, the output decreases by 0.5 when the input increases by 3. Written as a ratio, this is

$$\frac{-0.5}{3} = -\frac{1}{6}$$

In the third and fourth rows of the table, the output decreases by 1.0 when the input increases by 6. Written as a ratio, this is

$$\frac{-1.0}{6} = -\frac{1}{6}$$

By writing the relationship between the change in output and the change in input as a ratio, it is easier to see that the relationship is constant. For this reason we define this ratio to be the slope of the line.

Definition
The **slope** of a line is the *ratio of the vertical change to the horizontal change*. That is,

$$\text{slope} = \frac{\text{vertical change}}{\text{horizontal change}}$$

> The vertical change is often referred to as the **rise,** and the horizontal change is referred to as the **run.** Using this terminology, we have
>
> $$\text{slope} = \frac{\text{rise}}{\text{run}}$$

In any linear relationship, the ratio of the vertical change to the horizontal change is constant. This means that a horizontal increase of one unit produces a constant vertical change. In other words, a run of 1 unit produces a constant rise.

Example 2

Determine the slope of each of the following lines.

a.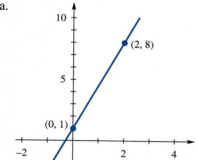

b.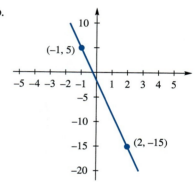

Solution

a. Two points are given on the graph in part a, the point (0, 1) and the point (2, 8). Moving from (0, 1) to (2, 8) the vertical change is 7 units up, and the horizontal change is 2 units to the right. Therefore, the

$$\text{slope} = \frac{\text{vertical change}}{\text{horizontal change}}$$

$$\text{slope} = \frac{7 \text{ units}}{2 \text{ units}} = \frac{7}{2}, \text{ or } 3.5$$

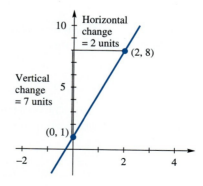

b. In graph b two points are given: (−1, 5) and (2, −15). In moving from (−1, 5) to (2, −15) the vertical change is −20 units (down 20 units), and the horizontal change is 3 units to the right.

Therefore, the

$$\text{slope} = \frac{\text{vertical change}}{\text{horizontal change}}$$

$$\text{slope} = \frac{-20 \text{ units}}{3 \text{ units}} = -\frac{20}{3}$$

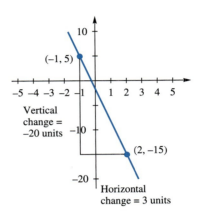

What is the slope of a vertical line? What is the slope of a horizontal line? For more information of slopes of vertical and horizontal lines, refer to Problem 7 in Problem Set 4.1.

Many graphs have horizontal intercepts and vertical intercepts.

Definition

A **horizontal intercept** is a point where the graph intersects the horizontal axis. A **vertical intercept** is a point where the graph crosses the vertical axis.

Because any point on the horizontal axis has a second coordinate of zero, any horizontal intercept has the form (#, 0). Similarly, because any point on the vertical axis has a first coordinate of zero, any vertical intercept has the form (0, #).

Example 3

What are the intercepts of the following graphs?

a.

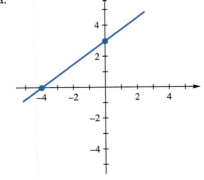

b.

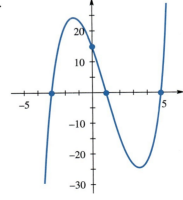

Solution a. The straight line in graph a crosses the horizontal axis once and the vertical axis once. Therefore, it has one horizontal intercept and one vertical intercept. It crosses the horizontal axis at ⁻4. Because we know that any point on the horizontal axis has a second coordinate of zero, the horizontal intercept is the point (⁻4, 0). Because any point on the vertical axis has a first coordinate of zero, the vertical intercept is the point (0, 3).

b. The graph in part b crosses the horizontal axis three times. Therefore, it has three horizontal intercepts, (⁻3, 0), (1, 0), and (5, 0). It crosses the vertical axis only once, and the vertical intercept is (0, 15).

Example 4

a. Draw the graph of the line passing through the point (2, 3) with a slope of $\frac{1}{4}$.
b. Draw the graph of the line passing through the point (⁻3, 2) with a slope of ⁻2.
c. Draw the graph of the line with a horizontal intercept of (⁻1, 0) and a slope of $\frac{2}{5}$.

Solution a. Because we must start at a known point, we first plot the point (2, 3). Then, because the slope is $\frac{1}{4}$, we know that

$$\frac{\text{vertical change}}{\text{horizontal change}} = \frac{1}{4}$$

This can be represented with a vertical change of 1 for a horizontal change of 4. Therefore, starting at the point (2, 3), we can find another point on the line by moving *up* 1 unit and *right* 4 units. The coordinates of this new point are (6, 4). We could continue this process to find more points.

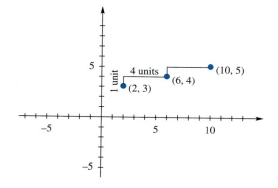

Once we have at least two points we can use a straight edge to draw a line through them. Following is the graph of the line passing through the point (2, 3) with a slope of $\frac{1}{4}$.

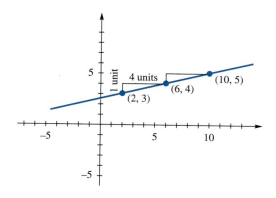

b. We want a line passing through the point (⁻3, 2) with a slope of ⁻2. First let's plot the point (⁻3, 2). Then, because the slope is ⁻2 and $-2 = \frac{-2}{1}$, we know that

$$\frac{\text{vertical change}}{\text{horizontal change}} = \frac{-2}{1}$$

4.1 Linear Relationships: Numerically and Graphically   193

This can be represented with a vertical change of −2 for a horizontal change of 1. Therefore, starting at the point (−3, 2), we can find another point on the line by moving *down* 2 units and *right* 1 unit. The coordinates of this new point are (−2, 0). We can then draw a line through these two points.

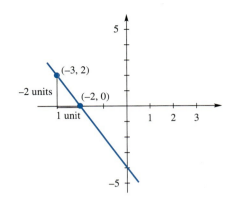

c. We want a line with a horizontal intercept of (−1, 0) and a slope of $\frac{2}{5}$. First we plot the intercept (−1, 0). From there we want a vertical change of 2 and a horizontal change of 5. This means we move up 2 units and to the right 5 units to find another point on the line. This puts us at the point (4, 2). Once we have two points, we can draw the straight line through them.

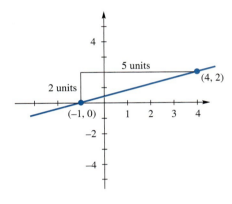

Example 5

Lap Swimming. A recreational lap swimmer has determined that she can swim about 10 laps in 15 minutes and 30 laps in 45 minutes. Assume that the relationship between the number of laps and the number of minutes is linear.

a. Graph the data. Assume time is the independent variable.
b. Using your graph, predict the number of laps that this person can swim in half an hour.
c. Using your graph, predict the amount of time it will take for this person to swim 40 laps.
d. Determine the slope of your line. What are the units associated with the slope? What does the slope tell you about the problem situation?
e. If one lap is 50 meters, determine the speed of this swimmer in miles per hour.

Solution a. Because we are told to use time as the independent variable, time goes on the horizontal axis. The number of laps is then the dependent variable and goes on the vertical axis.

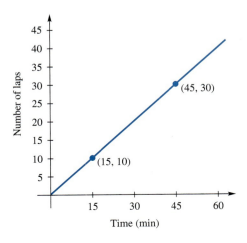

b and c.

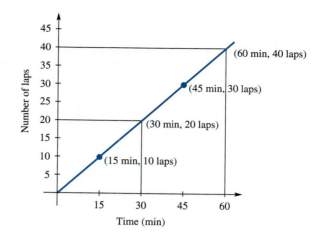

From the graph, we determine that this person can swim approximately 20 laps in half an hour and 40 laps will take about an hour.

d. Next, we want to determine the slope of the line. If we include the units as we find the slope, it will be more meaningful.

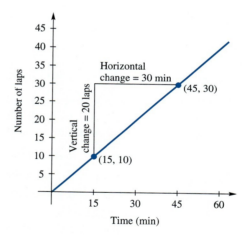

Between the two marked points on our graph the rise is 20 laps and the run is 30 minutes. Therefore, the slope is

$$\frac{\text{rise}}{\text{run}} = \frac{20 \text{ laps}}{30 \text{ min}} = \frac{2}{3} \text{ laps per minute.}$$

This means that the swimmer can swim 20 laps every 30 minutes or equivalently $\frac{2}{3}$ lap every minute.

e. Because we know that the swimmer can swim 20 laps every 30 min we can use unit fractions to convert to miles per hour.

$$\frac{20 \text{ laps}}{30 \text{ min}} * \frac{50 \text{ m}}{1 \text{ lap}} * \frac{1 \text{ km}}{1000 \text{ m}} * \frac{1 \text{ mile}}{1.609 \text{ km}} * \frac{60 \text{ min}}{1 \text{ h}} = \frac{1.25 \text{ miles}}{\text{h}} = 1.24 \text{ mph}$$

This swimmer swims at a rate of $1\frac{1}{4}$ miles per hour.

4.1 Linear Relationships: Numerically and Graphically

In this section we began to look in detail at the graphs of **lines.** At this point, we can recognize a linear model when we are given the data in numerical or graphical form. We know that a constant vertical change occurs for a constant horizontal change, that is, the slope is constant for a linear model. From a scaled graph, we can determine, at least approximately, the **slope, horizontal intercept,** and **vertical intercept.** We looked briefly at the algebraic form of a linear model and will look at this in much more depth in Section 4.2.

Problem Set 4.1

1. Without plotting the points, identify which of the following tables graph as straight lines. Explain how you decided.

a.
Input	Output
1	11
2	15
3	19
4	23

b.
Input	Output
1	45
2	47
3	51
4	57

c.
Input	Output
0	10
3	8
6	6
9	4

d.
Input	Output
1	−5
2	0
4	10
6	20
8	30

e.
Input	Output
0	23
1	19
2	14
3	8
4	1

2. Determine the slope of each of the following lines. In this problem, assume that dots are on integer coordinates.

a.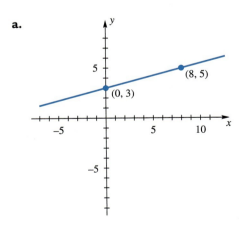

b.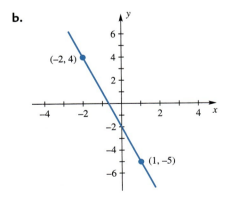

3. Determine the slope of each of the following lines. Assume that dots are on integer coordinates.

a.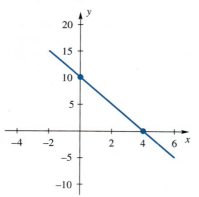
b.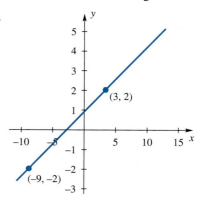

4. The following is the graph from Problem 2b. Use the points $(^-2, 4)$ and $(^-1, 1)$ to determine the slope. Do you get the same result as before?

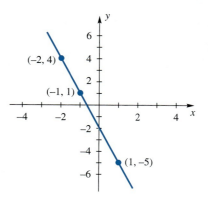

5. Determine the slope of each of the following lines. In this problem, assume that dots are on integer coordinates.

a.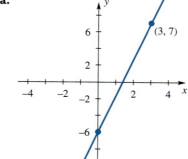
b.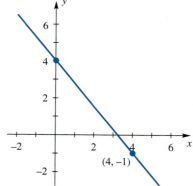
c.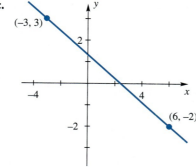
d.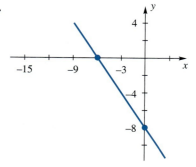

6. a. Line K has a slope of 2 and passes through the origin. On the axes provided, plot three points that are on line K, and draw the line through them.

   b. Line L has a slope of 1 and passes through the origin. On the same axes provided, plot three points that are on line L, and draw the line through them in a different color.

   c. Line M has a slope of $\frac{1}{2}$ and passes through the origin. On the same axes provided, plot three points that are on line M, and draw the line through them in a third color.

   d. Write down any observations that you have about the three lines.

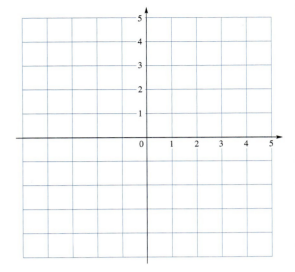

7. On the graph, which line has a larger slope? Explain.

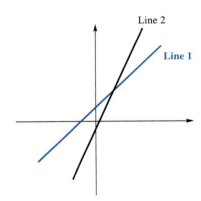

8. Determine the slope of each line.

   a.

   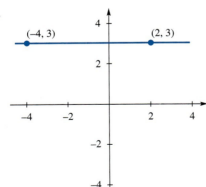

   b.

   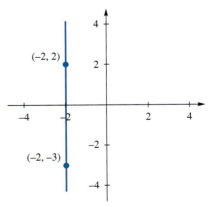

   c. Draw any horizontal line. Label two points on your line. Determine the slope of your line. What is the slope of any horizontal line?

   d. Draw any vertical line. Label two points on your line. Determine the slope of your line. What is the slope of any vertical line?

9. For each part, draw the graph of the line that satisfies the given conditions. On each graph, label two points on the line.
   a. Passing through the point (−3, 4), with a slope of $\frac{1}{2}$
   b. Passing through the point (0, 0), with a slope of $\frac{-3}{5}$
   c. Passing through the point (−5, 7), with a slope of −3
   d. Passing through the point (0, 150), with a slope of 25
   e. Passing through the point ($\frac{-1}{2}, \frac{3}{4}$), with a slope of $\frac{1}{4}$

10. a. On the same axes, draw the graphs of the following:

   The line passing through the point (2, −5), with a slope of 4
   The line passing through the point (−1, 3), with a slope of 4
   The line passing through the point (0, 0), with a slope of 4

   b. What is the relationship between the three lines you drew in part a?

11. **Manufacturing Costs.** A manufacturing company has found that the cost to produce any one of its items is related to the time required to produce the item. The following graph shows a plot of the cost of an item in terms of the time needed to produce the item.
   a. Estimate the slope of the line. What are the units associated with the slope?
   b. What does the slope tell you about the problem?

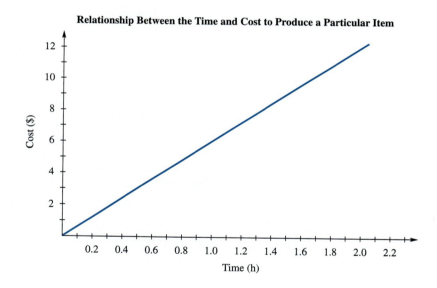

Relationship Between the Time and Cost to Produce a Particular Item

12. **Population.** The following graph represents the population of a midwestern town in terms of the number of years after 1980.
    a. Does the graph show a linear relationship between the number of years after 1980 and the population of the town?
    b. Draw a line between the data points representing the population in 1980 and 1990. Determine the slope of this line, including units.
    c. Draw a line between the data points representing the population in 1980 and 1995. Determine the slope of this line, including units.
    d. If we assume that this town's growth will continue to follow the pattern observed in the graph, would the slope of the line between the population in 1980 and 2000 be greater or less than the slope you computed in part c?
    e. What is the vertical intercept, including units? Interpret the meaning of the vertical intercept in the context of the problem.

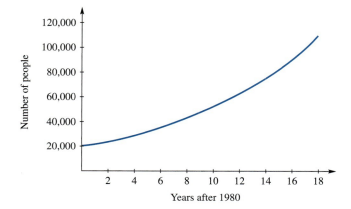

13. **The Car Rental Revisited.** The following graph shows the cost of renting a car for one day in terms of the number of miles driven for two different rental companies. Use the graph to answer the following questions.
    a. Estimate the slope, including units, of the line labeled Cars-4-U.
    b. Estimate the slope, including units, of the line labeled Beaters-R-Us.
    c. Interpret the meaning of the slope of each line in the context of this problem.
    d. Determine the vertical intercept, including units, of each line. Interpret the meaning of the vertical intercept of each line in the context of this problem.

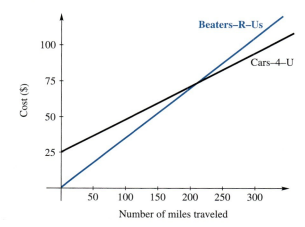

4.2 The Graphs of Linear Equations

1. a. Complete the following table. Show your substitution in the first row of the table.
   b. Draw a graph on the axes provided by plotting the points from your table.
   c. Determine the slope and identify the vertical intercept. Complete the sentences beneath the graph.

x	$y = 2x + 4$
-2	
-1	
0	
1	
2	

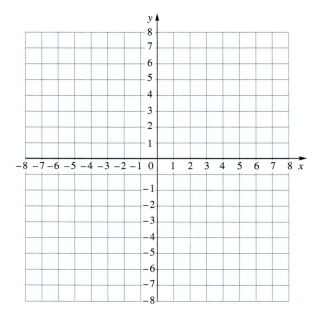

The slope of line $y = 2x + 4$ is _____. The vertical intercept is (___, ___).

2. a. Complete the following table. Show your substitution in the first row of the table.
   b. Draw a graph on the axes provided by plotting the points from your table.
   c. Determine the slope, and identify the vertical intercept. Complete the sentences beneath the graph.

x	$y = \frac{1}{3}x - 2$
−6	
−3	
0	
3	
6	

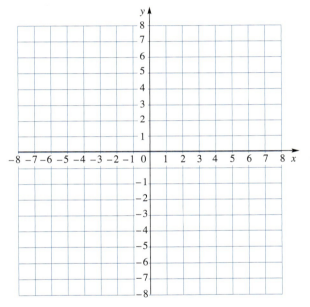

The slope of line $y = \frac{1}{3}x - 2$ is _____. The vertical intercept is (____, ____).

3. **a.** Complete the following tables. Show your substitution in the first row of each table.
   b. For each equation, draw a graph on the axes provided by plotting the points from your table. You may want to use different colors to distinguish the graphs.
   c. For each equation, determine the slope and identify the vertical intercept. Complete the sentences beneath the graph.

x	y = ⁻4x
-2	
-1	
0	
1	
2	

x	y = ⁻4x − 3
-2	
-1	
0	
1	
2	

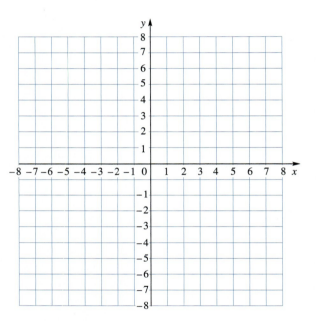

The slope of line $y = {}^-4x$ is _____. The vertical intercept is (____, ____).

The slope of line $y = {}^-4x - 3$ is _____. The vertical intercept is (____, ____).

4. Based on your observations from Activities 1–3, describe the relationship between the slope of a line and its equation.

5. Based on your observations from Activities 1–3, describe the relationship between the vertical intercept of a line and its equation.

6. If two lines are parallel, what can you say about their equations?

7. **a.** Answer the following questions *without graphing* the equation.
   What is the slope of the line $y = 10x + 24$?
   What is the vertical intercept of the line $y = 10x + 24$?

   b. Graph the equation $y = 10x + 24$ on your calculator. Check your answers from part a using your graph. Explain how you did this.

   c. What is the slope of the line $y = mx + b$?
   What is the vertical intercept of the line $y = mx + b$?

8. **a.** Answer the following questions without graphing the equation.
   What is the slope of the line $y = 10 - 3x$?
   What is the vertical intercept of the line $y = 10 - 3x$?

   b. Graph the equation $y = 10 - 3x$ on your calculator. Check your answers from part a using your graph.

9. Write the equation of the line shown in the following graph. Check your equation.

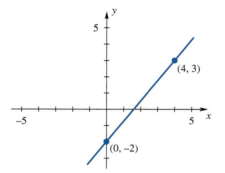

Discussion 4.2

The Graphs of Linear Equations

Frequently, the variable x is used to represent the independent variable on the horizontal axis, and y is used to represent the dependent variable on the vertical axis. For this reason, the horizontal intercept is often referred to as the *x*-intercept and the vertical intercept as the *y*-intercept. When the variables x and y are used in an equation, assume that x is the independent variable and y is the dependent variable unless otherwise stated.

Now, let's go back to our linear equation $y = mx + b$. As you discovered in the activities, the coefficient of x represents the slope and the constant term represents the vertical coordinate of the vertical intercept. For this reason the equation $y = mx + b$ is called the slope–intercept equation of a line. We could also write this as $y = slope * x + vertical\ intercept$.

> **The Slope-Intercept Equation of a Line**
> The **slope–intercept form** of a linear equation is
> $$y = slope * x + vertical\ intercept$$
> or
> $$y = mx + b$$
> where $m = $ slope and $(0, b) = $ vertical intercept

Using this information, if we are given a linear equation we can draw the graph or if we are given the slope and vertical intercept of the graph of a linear equation we can write the equation.

In the equation $y = mx + b$ we see that the variable y is isolated on one side of the equation. A linear equation may also be written in the form $Ax + By = C$, where A, B, and C are constants. This equation graphs as a straight line, provided A and B are not both equal to zero. The equation $Ax + By = C$ is called the **general equation of a line.** Most linear equations can also be written in slope–intercept form, that is, $y = mx + b$, where m and b are constants. Equations in slope–intercept form are usually easier to interpret and graph.

Because $y = mx + b$ and $Ax + By = C$ are two simplified forms of a linear equation, we know that any equation that does not fit these two forms does not graph as a line. This means that if the variables in an equation are raised to any power other than 1, the equation is not linear. If the variable occurs in the denominator or under a radical sign, the equation is not linear. However, the numerical coefficients may include powers, roots or fractions.

Example 1

Graph the equation $y = 3 - 2x$.

Solution First we need to decide if the equation graphs as a line. If it does, then we only need to find two points to draw the graph because two points determine a line. If the equation does not graph as a line, then we need to plot several points to get an idea of the shape.

We see that the equation $y = 3 - 2x$ can be written as $y = -2x + 3$, which is slope–intercept form for a line. Therefore, this equation graphs as a line. There are three options we can use to graph a line.

Option 1 We can find two points by making a table.

x	$y = 3 - 2x$
0	$3 - 2 * 0 = 3$
1	$3 - 2 * 1 = 1$
2	$3 - 2 * 2 = -1$

We can draw a graph by plotting these points. You may have noticed that the table shows three points when only two are necessary. Using three can help us spot simple errors, and it never hurts to have a spare.

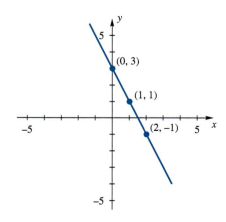

Option 2 We could also find two points on the line by using the relationship between the equation and the graph. To do this we must identify the vertical intercept and the slope of the line from the equation.

$$y = 3 - 2x$$
$$y = -2x + 3$$

In this form we see that the vertical intercept is (0, 3) and the slope is -2. Therefore, we can plot two points by first plotting the vertical intercept and then using the slope to find a second point. The slope is -2. This is equivalent to the fraction $\frac{-2}{1}$. So, from the point (0, 3), we need to move down 2 units and to the right 1 unit. This takes us to the point (1, 1). If we continue this process using the slope, the next point would be (2, -1).

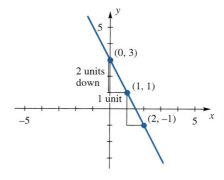

Option 3 A third option is to determine the two intercepts. Earlier we wrote the equation in slope–intercept form.

$$y = -2x + 3$$

We saw that the vertical intercept is (0, 3).

Next we want to find the horizontal intercept. All of the points on the *x*-axis have one thing in common; the *y*-coordinate of an ordered pair on the *x*-axis is always zero. That is, any point on the *x* axis can be written as (#, 0). So, to find the horizontal intercept of a graph, we can let $y = 0$ in the equation and then solve for *x*.

$$y = -2x + 3 \qquad \text{To find the horizontal intercept, set } y = 0.$$
$$0 = -2x + 3$$
$$0 - 3 = -2x + 3 - 3$$
$$-3 = -2x$$
$$\frac{-3}{-2} = \frac{-2x}{-2}$$
$$1.5 = x$$

The horizontal intercept is (1.5, 0).

Now we can plot the two intercepts, (0, 3) and (1.5, 0), and use them to draw the graph of the equation $y = 3 - 2x$.

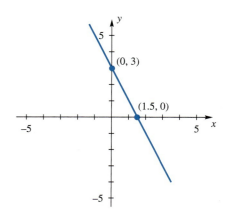

In Example 1, we graphed a linear equation using three different methods. In the future we will choose only one of the methods to create a graph.

Example 2

Graph the equation $y = 3x - 5$.

Solution First we determine if the equation graphs as a line. We can see that the equation $y = 3x - 5$ is in slope–intercept form. Therefore, we know it graphs as a line. From Example 1, we know that three options are possible for graphing lines. Next we discuss these three options and pick the one that seems the most efficient.

The first option was to make a table of values. To do this it is helpful if the equation is solved for one of the variables. The equation $y = 3x - 5$ is already solved for y. This makes it easier to create a table.

The second option is to use the slope and vertical intercept to determine two points. Again, to do this it is easier if the equation is in slope–intercept form, and $y = 3x - 5$ is.

The third option is to find the vertical and horizontal intercepts. We find the vertical intercept by letting $x = 0$ and solving for y. Similarly, we find the horizontal intercept by letting $y = 0$ and solving for x.

Because the equation $y = 3x - 5$ is already in slope–intercept form and the constants are reasonably small numbers, we use the slope and the vertical intercept to graph this line. We know that the point $(0, -5)$ is the vertical intercept and the slope is 3. The slope of 3 is equivalent to $\frac{3}{1}$. This can be thought of as a vertical change of 3 and a horizontal change of 1. With this we can graph the equation.

We start at the point $(0, -5)$. We can then determine a second point by using the slope. From the point $(0, -5)$, we move up 3 units and to the right 1 unit to the point $(1, -2)$. This gives us two points through which we can draw a line.

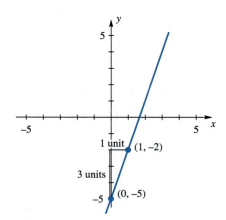

Example 3

Without graphing, determine if each equation graphs as a line. Then graph the equation.

a. $y = \frac{3}{2}x + 1$ b. $y = 2x^2 + 1$ c. $1.05x + y = 5.75$ d. $y = \frac{1}{x}$

Solution

a. The equation $y = \frac{3}{2}x + 1$ graphs as a line because it is in the form of $y = slope * x + vertical\ intercept$. The slope is $\frac{3}{2}$ and the vertical intercept is the point (0, 1). To graph a line we only need two points. We can start at the vertical intercept, the point (0, 1). From the point (0, 1) we move up 3 units and to the right 2 units to the point (2, 4). This gives us two points through which we can draw a line.

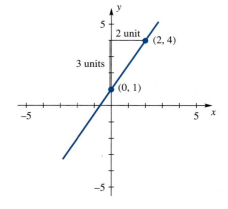

b. The equation $y = 2x^2 + 1$ does not graph as a line because it does not fit the general form of a line or the slope–intercept form. To match either form the power on x must be 1.

Because the equation $y = 2x^2 + 1$ is not linear, we need to plot several points to draw this graph. It is a good idea to select both positive and negative values to substitute for x. We make a table of values from -3 to 3, then use these values to plot points.

x	$y = 2x^2 + 1$
-3	$2 * (-3)^2 + 1 = 19$
-2	$2 * (-2)^2 + 1 = 9$
-1	3
0	1
1	3
2	9
3	19

Notice that when we substituted a negative number for x in the table, we had to insert parentheses. Parentheses are necessary to denote the square of a negative number.

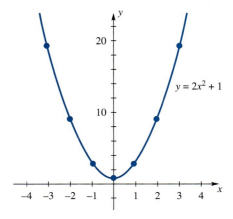

It is important to note that because we have not studied graphs in which x is squared, we don't *really* know if this graph is entirely correct. We only know for sure that the seven points plotted from the table are correct. We make an *assumption* that a smooth curve connecting the points is what the graph will look like.

We could also have graphed the equation using a graphing calculator. But this graph may not give an accurate or complete picture either, because graphing technology has some limitations and you will only see the graph that appears in the window that you select. Later, you will study more types of nonlinear graphs so that you will know what the complete graph should be.

c. The equation $1.05x + y = 5.75$ is written in the form $Ax + By = C$; therefore, this is a linear equation. Because we know it graphs as a line, we only need to plot two points to draw the graph. We use the method of finding the two intercepts to graph this line. We could also use the vertical intercept and the slope to graph this line, but because of the decimal values this would be more difficult.

To find the horizontal intercept, we substitute $y = 0$ and solve for x.

$$1.05x + y = 5.75$$
$$1.05x + 0 = 5.75$$
$$1.05x = 5.75$$
$$\frac{1.05x}{1.05} = \frac{5.75}{1.05}$$
$$x \approx 5.48$$

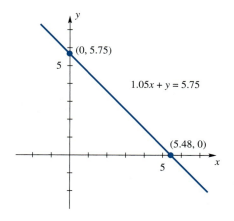

The horizontal intercept is approximately $(5.48, 0)$.

Similarly, to find the vertical intercept we substitute $x = 0$ and solve for y.

$$1.05x + y = 5.75$$
$$1.05(0) + y = 5.75$$
$$0 + y = 5.75$$
$$y = 5.75$$

The vertical intercept is $(0, 5.75)$.

d. The equation $y = \frac{1}{x}$ does not graph as a line because a linear equation is written as the slope *times x,* not *divided* by x. Because the equation $y = \frac{1}{x}$ is not linear, we need to plot several points to draw this graph. Again, it is a good idea to select both positive and negative values to substitute for x. We make a table of values from $^-3$ to 3, then use these values to plot points.

x	$y = \frac{1}{x}$
$^-3$	$-\frac{1}{3}$
$^-2$	$-\frac{1}{2}$
$^-1$	$^-1$
0	Undefined
1	1
2	$\frac{1}{2}$
3	$\frac{1}{3}$

Notice that when $x = 0$, there is no value for y because division by zero is undefined. We need to add some more points closer to zero to determine what the graph should look like near $x = 0$.

x	$y = \dfrac{1}{x}$
-3	$-\dfrac{1}{3}$
-2	$-\dfrac{1}{2}$
-1	-1
$-\dfrac{1}{2}$	-2
$-\dfrac{1}{3}$	-3
0	Undefined
$\dfrac{1}{3}$	3
$\dfrac{1}{2}$	2
1	1
2	$\dfrac{1}{2}$
3	$\dfrac{1}{3}$

Now we plot these points. We again assume that the graph is a smooth curve. So we connect the points *on each side* of the y-axis. Remember that we do not want to connect the graph across the y-axis because at $x = 0$, y is undefined.

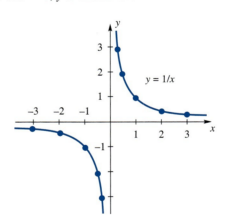

Example 4

Write an equation for the line shown in the following graph. Check your equation.

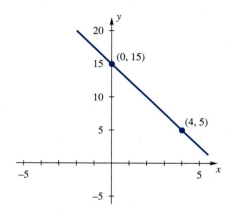

Solution We know the equation of a line can be written in the form $y = mx + b$. Therefore, we need to determine the slope and vertical intercept to write the equation. From the graph we can see that the vertical intercept is the point $(0, 15)$. We can also find the slope from the graph by determining the ratio $\frac{\text{rise}}{\text{run}}$. Going from the point $(0, 15)$ to the point $(4, 5)$, the slope is $\frac{-10}{4}$, which can also be written as $\frac{-5}{2}$, or -2.5.

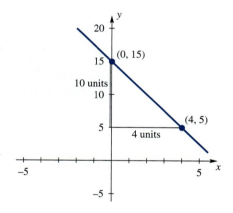

Now that we know the slope and vertical intercept, we can write the equation.

$$y = -2.5x + 15$$

To check the equation, we need to see if the two labeled points work in the equation. First we try $(0, 15)$.

$$y = -2.5x + 15$$
$$15 \stackrel{?}{=} -2.5 * 0 + 15$$
$$15 = 15 \quad ✔$$

Many lines pass through the point $(0, 15)$; to be certain our equation is correct, we must check both points. So, next we try $(4, 5)$.

$$5 \stackrel{?}{=} -2.5 * 4 + 15$$
$$5 = 5 \quad ✔$$

We conclude that the equation of the line is $y = -2.5x + 15$.

Example 5

Telephone Bill. Chris and Cameron recently changed long distance phone companies. On their first bill, they were charged $44.15 for 372 minutes of long distance calls. This included a monthly service charge of $6.95.

a. Assuming that the relationship is linear, write an equation for the cost of their long distance calls in terms of the number of minutes.

b. Use your equation to determine the cost for Chris and Cameron if they have 550 minutes worth of long distance calls in a month.

c. What are the units associated with the slope of the line? What does the slope tell you about the problem situation?

d. What is the relationship between the vertical intercept of the line and the problem situation?

Solution

a. From the problem statement, we know that 0 minutes cost $6.95 and 372 minutes cost $44.15. Because the cost depends on the number of minutes, we know that the number of minutes is the independent variable and the cost is the dependent variable. To start, let's plot these two points. Then, because we are assuming that this relationship is linear, we draw a line through our two points.

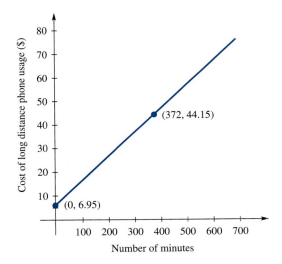

Because we know the equation of a line can be written in the form $y = mx + b$, we need to determine the slope and vertical intercept. The vertical intercept we can see from the graph. It is the point (0, 6.95). We can also find the slope from the graph by determining the ratio $\frac{rise}{run}$. Going from the point (0, 6.95) to the point (372, 44.15), the graph rises 37.2 and runs 372. Therefore, the slope is

$$\text{slope} = \frac{37.2}{372}$$

$$= 0.10$$

Now that we know the slope and vertical intercept, we can write the equation.

Let n = number of minutes, and
C = cost of long distance phone usage in dollars.

Then, the equation for this line is

$C = 0.10n + 6.95$

Before we use our equation to answer the rest of the questions, we should check to see that it is correct. Because we used the vertical intercept to write the equation, we can check the equation using the other point, (372, 44.15).

$$44.15 \stackrel{?}{=} 0.10 * 372 + 6.95$$
$$44.15 = 44.15 \quad \checkmark$$

b. To determine the cost for 550 minutes of long distance calls, we need to substitute $n = 550$ into our equation.

$$C = 0.10 * 550 + 6.95$$
$$C = 61.95$$

Therefore, 550 minutes of long distance costs Chris and Cameron $61.95. We can also see that this answer agrees with our graph.

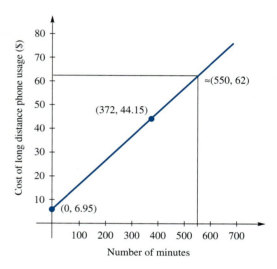

c. In part a, we found the slope of the line to be 0.10. To determine the units associated with the slope, we must look at the units on each axis. Slope is rise over run. The units on the vertical axis (rise) are dollars, and the units on the horizontal axis (run) are minutes. Therefore, the units associated with the slope are dollars per minute, ($\frac{\text{dollar}}{\text{minute}}$). If we put the units together with the numerical value, the slope is 0.10 dollars per minute or 10 cents per minute.

The slope tells us that Chris and Cameron are charged 10 cents for each minute of long distance phone usage.

d. The vertical intercept is (0 min, $6.95). This represents the monthly service charge.

Putting this information together with part c, we know that Chris and Cameron are charged $6.95 per month plus 10 cents per minute.

Example 6

How Long Can We Talk? Chris and Cameron have budgeted $75 per month for their long distance phone calls.

    a. Write a linear equation for a constant cost of $75. Determine the slope and vertical intercept of this line.

    b. Graph this equation along with the linear model from the previous example.

    c. How many minutes of long distance usage can they have and stay within their budget?

Solution

a. A constant cost of $75 can be represented by the equation $C = 75$, where C is the cost in dollars. What are the vertical intercept and slope of this equation? If we rewrite $C = 75$ as $C = 0 * n + 75$, we can see that the slope of the equation is 0 and the vertical intercept is the point (0, 75).

b. Because the vertical intercept of the graph of $C = 75$ is the point (0, 75) and the slope is 0, we know that the graph of $C = 75$ is a horizontal line that passes through the point (0, 75). When we add this to our previous graph, we obtain the following. We may need to extend our previous graph to see where the two lines intersect.

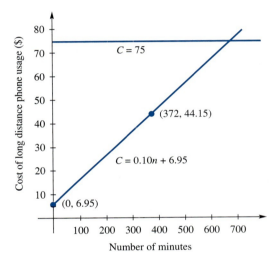

c. The intersection of the two graphs tells us how long Chris and Cameron can use the phone. The point of intersection appears to be approximately (670, 75).

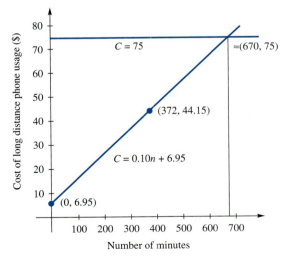

Using our approximation from the graph, we conclude that they can use about 670 minutes of long distance calls and still stay within their budget.

In this situation, we could also have solved the equation $75 = 0.10n + 6.95$ algebraically, in which case we would have found that they can actually have 680 minutes of long distance usage and still stay within their budget.

In this section, we learned that any equations in the form $Ax + By = C$ and $y = mx + b$ graph as straight lines. We saw that if a linear equation is in **slope-intercept form,** $y = slope * x + vertical\ intercept,$ we can use the intercept and the slope to find two points to graph the line. We also saw how to determine **horizontal** and **vertical intercepts** from the equation. Linear models were used to describe problem situations and to answer questions about the situation.

Additionally, we discussed how to differentiate between equations that graph as lines and those that do not. Only two points are necessary to graph a line; however, we need several points to graph nonlinear equations. To graph nonlinear equations we used a table with several values (preferably positive and negative values) to sketch the graphs. Using a table of values to graph a nonlinear equation does not provide enough information to ensure that our graph is entirely correct.

Finally, we saw that we can determine the equation for a line if we are given the vertical intercept and an additional point on the line. Many times two points are given to us in a problem situation like the "Telephone Bill" example in this section.

Problem Set 4.2

1. Without using a graphing calculator, identify all of the following equations that graph as straight lines. Explain how you decided.

   a. $y = 3x + 2$

   b. $y = x$

   c. $y = 3x^2 + 2$

   d. $y = \frac{1}{3}x - 5$

   e. $y = \frac{5}{x}$

   f. $y = \frac{x}{5}$

   g. $y = 4.8 - 2.2x$

   h. $5x + 12y = 60$

2. Determine the vertical intercept and slope of each line. Use these to write the equation for each line. Check that your equations are correct. In this problem, assume that dots are on integer coordinates.

   a.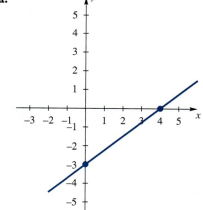

   b.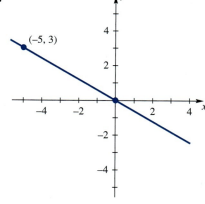

c.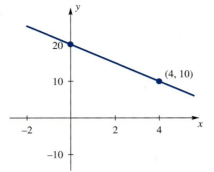

d.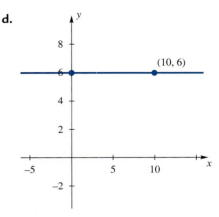

3. Write the equations of the lines graphed here. Check that your equations are correct. In these problems, assume that dots are on integer coordinates.

a.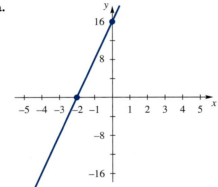

b.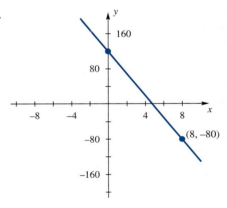

c.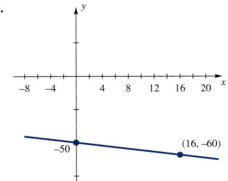

d.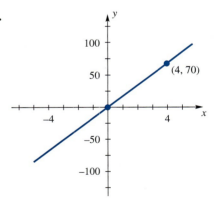

4. Rewrite each equation in slope–intercept form (if needed). For each equation, determine the slope and the vertical intercept. Use these to draw a graph.

   a. $y = \frac{1}{2}x - 5$

   b. $y = \frac{-2}{3}x + 4$

   c. $y = 3x + 22$

   d. $y = x$

   e. $y = {}^-35 + 55 * x$

   f. $y = 4(x + 1) - 3(5 - x)$

   g. $3x + y = 5$

5. Algebraically determine the horizontal and vertical intercepts of each of the following equations. Use the intercepts to draw the graph of each line.

   a. $3x - 5y = 45$ b. $6y + 10x = 12$ c. $2.5x + 0.5y = 3.2$

6. Determine the horizontal and vertical intercepts of the line $Ax + By = C$, in terms of A, B, and C.

7. Graph the following equations. (You may want to look at your results from Problem 1.)

   a. $y = 3x + 2$

   b. $y = x$

   c. $y = 3x^2 + 2$

   d. $y = \frac{1}{3}x - 5$

   e. $y = \frac{5}{x}$

   f. $y = \frac{x}{5}$

   g. $y = 4.8 - 2.2x$

   h. $5x + 12y = 60$

8. Graph the linear equation $y = x$ in each of the following windows using your graphing tool. (*Note:* The window settings are given in the form [xmin, xmax] by [ymin, ymax].)

   a. [$^-$10, 10] by [$^-$10, 10]

   b. [$^-$40, 40] by [$^-$10, 10]

   c. [$^-$10, 10] by [$^-$40, 40]

   d. Do the slopes of the lines in parts a–c *appear* to be similar or different? Explain your response.

9. Sketch the graphs described.

   a. A line whose horizontal intercept is the origin and whose slope is $^-1$

   b. A vertical line through the point ($^-$3, 0)

   c. The line that passes through the point (0, 5) and has the same slope as the line $y = {}^-2x + 3$

   d. A line whose vertical intercept is (0, $^-$5) and whose slope is 0

10. **Picking a Plumber.**   For service calls, Pat's Plumbing charges $28 per half hour on the job. Flow Right Plumbers charges $25 for the service call plus $22 per half hour on the job.

    a. For each company, write an equation for the cost of a plumber's service call in terms of the number of hours on the job.

    b. Determine graphically and algebraically the number of hours for which the cost of a service call is the same for both companies.

    c. For each graph, what is the slope, including units? What does the slope mean?

    d. For each graph, what is the vertical intercept, including units? What does the vertical intercept mean?

    e. If you need to hire a plumber, how would you decide which company to call?

11. **Landscape.** George hired some people to work on his yard one day per week for a month. At the end of the month, George received a bill. On the bill he was charged $230 for 12 hours worth of work, including the monthly service fee of $50.

    a. Assuming that the relationship between the amount of the bill and the number of hours worked is linear, write an equation for this relationship.

    b. What units are associated with the slope of the line? What does the slope tell you about the problem situation?

    c. What is the relationship between the vertical intercept of the line and the problem situation?

12. **Newspaper Circulation.** In Section 2.2, you answered some questions from the following graph. The graph shows the circulation of morning and evening newspapers over a period of time. We can now interpret the graph more completely.

    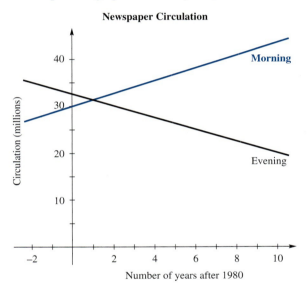

    a. For each line, determine the slope, including units. Interpret the meaning of the slopes in the context of this problem.

    b. For each line, determine the vertical intercept, including units. Interpret the meaning of the intercepts in the context of this problem.

13. **Perimeter and Area of a Quarter Circle.** The figure shown here is enclosed by a quarter circle and two straight sides.

    a. Complete the following table.

Radius of the quarter circle (in.)	3.00	4.00	5.00	6.00	7.00
Perimeter of the figure					
Area of the figure					

    b. Plot the graph of the perimeter in terms of the radius. Plot the graph of the area in terms of the radius. Which graph appears to be linear?

    c. Write an equation for the perimeter of the figure in terms of the radius of the quarter circle. Write a formula for the area of the figure in terms of the radius of the quarter circle. Which equation is linear?

    d. Determine the slope of the linear relationship. Explain how you determined the slope.

14. **Charles' Law.** In chemistry, Charles' law states that for a given volume of gas, as the temperature increases, the volume increases linearly. If a balloon is inflated with helium to a volume of 4.1 L at 0°C and then brought inside where the room temperature is 25°C, the volume will increase to 4.5 L. Assuming the relationship is linear, write an equation for the volume of helium in the balloon in terms of the temperature.

15. **The Leaking Aquarium.** The base of a rectangular aquarium tank measures 4.30 meters by 5.70 meters and the tank is 9.40 meters deep. When filled to a height 1 foot below the top, it holds approximately 58,800 gallons. Matt notices that, one week after the tank was filled, the water level has dropped about 8 inches. He calculates that approximately 57,500 gallons of water are now in the tank.

   a. Assuming that the tank continues leaking at the same rate, write an equation for the amount of water in the tank in terms of the number of days after it is filled.

   b. What units are associated with the slope? What does the slope tell you about the problem situation?

   c. What is the vertical intercept? What does the vertical intercept tell you about the problem situation?

   d. What should Matt report to the aquarium managers about the leak in the tank?

Chapter Four Summary

In this chapter we studied a specific class of equations in two variables called **linear equations.** We learned how to identify a linear relationship from a table of values and from an equation, we learned how to graph a linear equation by plotting *only two* points, and we learned how to write equations for specific linear relationships.

From a table of values, we can determine whether a relationship graphs as a line by looking at the change in output and change in input. If the change in output is constant for a constant change in input, then the data graphs as a line. Another way to look at this relationship is to look at the ratio

$$\frac{\text{change in output}}{\text{change in input}}$$

> ### Linear Relationships: Numerically
> A table of data graphs as a line if and only if the ratio
>
> $$\frac{\text{change in output}}{\text{change in input}}$$
>
> is constant for all pairs of data in the table.

Because this ratio is constant for any line, we call this ratio the **slope** of the line. On a graph, we refer to this ratio as

$$\frac{\text{vertical change}}{\text{horizontal change}} \quad \text{or} \quad \frac{\text{rise}}{\text{run}}$$

The slope, together with its units, can tell us important information about a problem situation.

> ### Linear Relationships: Algebraically
> Algebraicallly, we found that linear equations are seen in two forms:
>
> - slope–intercept form of a linear equation
>
> $y = mx + b$
>
>   where m = slope
>
> $(0, b)$ = vertical intercept
>
> - general form of a linear equation
>
> $Ax + By = C$
>
>   where A and B are not both zero

Because these are the only two simplified forms of a linear equation, we know that any equation that does not fit these two forms does not graph as a line. This means that if the variables in an equation are raised to any power other than one, then the equation is not linear. If the variable occurs in the denominator or under a radical sign, the equation is not linear. However, the numerical coefficients may include powers, roots, or fractions.

With this information, we are able to inspect an equation to determine whether the equation graphs as a line. When an equation graphs as a line, we only need to plot two points, because two points determine a line.

We learned three options for determining two possible points through which to draw our line.

Option 1
Make a table of values. Again we only need two points if we know the graph is a line.

Option 2
Use the slope and intercept. Start by plotting the vertical intercept. Then, determine a second point by using the slope to move up or down and over the appropriate amount.

Option 3
Algebraically determine both the vertical intercept and the horizontal intercept. The vertical intercept can be found by substituting $x = 0$ into the equation and solving for y. Similarly, the horizontal intercept can be found by substituting $y = 0$ into the equation and solving for x.

The method we choose for graphing a linear equation depends on the form of the equation, the magnitude of the numbers in the equation, and our preference on methods.

If the equation we are trying to graph is not linear, then we must use the techniques we learned in Chapter 2 to get an idea of how the graph will look. As you proceed in your mathematics courses, you will learn more about different classes of equations and their graphs.

Finally, if we know or can determine the vertical intercept and slope of a line, we can write its equation by substituting these values into the slope–intercept equation. The information we use to determine the vertical intercept and slope may be given in graphical form or in an application. If data are given in an application, it is sometimes helpful to graph the data or put the data in a table before trying to write an equation.

Chapter Five

Positive Integer Exponents

Exponents are used frequently in mathematics. In this chapter, we will investigate positive integer exponents and how to appropriately use them with signed number bases.

In Chapter 3, we learned how to add, subtract, and multiply expressions involving exponents. In this chapter, we will create a set of properties that summarize the techniques we used in Chapter 3.

5.1 Language and Evaluation

If n is a positive integer and x is a real number, then x^n (read "the nth power of x" or "x to the nth power") is the product of n factors of x. For example $x^5 = x * x * x * x * x$. Assuming that x is a nonzero real number, x^0 is defined as being equal to 1, and 0^0 is undefined. In the expression x^n, x is called the **base** and n is called the **exponent.**

Activity Set 5.1

1. How is each expression read? Evaluate each expression without a calculator.
   a. 3^2 b. $(-3)^2$ c. -3^2

2. a. Evaluate each expression when $x = 5$. Show your substitutions. Do you get the same result for both expressions? Should you?

   $x * x$ and x^2

   b. Repeat part a with $x = {}^-5$.

   $x * x$ and x^2

3. a. Evaluate each expression when $x = 5$. Show your substitutions. Do you get the same result for both expressions? Should you?

   ${}^-x * {}^-x$ and $({}^-x)^2$

   b. Repeat part a with $x = {}^-5$.

   ${}^-x * {}^-x$ and $({}^-x)^2$

4. a. Evaluate each expression when $x = 5$. Show your substitutions. Do you get the same result for both expressions? Should you?

   ${}^-x * x$ and ${}^-x^2$

   b. Repeat part a with $x = {}^-5$.

   ${}^-x * x$ and ${}^-x^2$

Discussion 5.1

Positive Integer Exponents: Language and Evaluation

In this section, we will learn what expressions with positive integer exponents mean and how to evaluate them correctly. Learning how to read these expressions will help us to precisely interpret what they mean. The following examples give us some practice reading expressions that include positive integer exponents. We will see how careful reading can help in correctly evaluating expressions.

Example 1

For each expression, determine how the expression is read. Then use the reading to help evaluate the expression.

a. 5^2 b. -5^2 c. $(-5)^2$

Solution

a. The expression 5^2 can be read as *five squared* or *the square of five*. The square of 5 is 25. Therefore,

$$5^2 = 25$$

b. Because an exponent always applies only to what is on its immediate left, the square in the expression -5^2 applies only to the 5. The expression -5^2 can therefore be read as *the opposite of the square of five*. To evaluate this expression, first we must square 5 and then take the opposite of the result. The square of 5 is 25 and the opposite of this is -25. Therefore,

$$-5^2 = -25$$

c. In the expression $(-5)^2$, the parentheses are to the immediate left of the exponent. Therefore, the exponent applies to the quantity inside the parentheses, -5. The expression $(-5)^2$ can be read as *the square of negative five*. The square of negative 5 is $-5 * -5 = 25$. Therefore,

$$(-5)^2 = 25$$

Example 2

Evaluate each expression when $x = 3$.

a. x^2 b. $(-x)^2$ c. $-x^2$

Solution

a. The expression x^2 is *the square of x*. Because x is 3, the expression becomes the square of 3, or 3^2.

$$3^2 = 9$$

b. The expression $(-x)^2$ is *the square of the opposite of* x. Because x is 3, the expression becomes the square of the opposite of 3, or $(-3)^2$. We have substituted 3 into the expression for x.

$$(-3)^2 = 9$$

c. The expression $-x^2$ is *the opposite of the square of* x. Because x is 3, the expression becomes the opposite of the square of 3, or -3^2. In this case, we do not need parentheses. The expression could also be written as $-(3)^2$ to show very clearly that only 3 is to be squared and not -3.

$$-3^2 = -9 \quad \text{or} \quad -(3)^2 = -9$$

Example 3 Evaluate each expression when $x = {}^-3$.

a. x^2 b. $({}^-x)^2$ c. ${}^-x^2$

Solution

a. The expression x^2 is *the square of* x. Because x is ${}^-3$ we must square ${}^-3$. The only way to square a negative number is to enclose it in parentheses. The expression x^2 becomes $({}^-3)^2$ when $x = {}^-3$. Even though the original expression x^2 has no parentheses, when we substitute a negative number for x we must include the parentheses.

$$({}^-3)^2 = 9$$

b. The expression $({}^-x)^2$ is *the square of the opposite of* x. Because x is ${}^-3$, the opposite of x is 3. The expression $({}^-x)^2$ becomes $(3)^2$ when $x = {}^-3$. If we directly substitute $x = {}^-3$ into $({}^-x)^2$, the expression could also be written as $({}^{--}3)^2$. We must be careful here to square the opposite of x and not just the value of x.

$$(3)^2 = 9 \quad \text{or} \quad ({}^{--}3)^2 = 9$$

c. The expression ${}^-x^2$ is *the opposite of the square of* x. Because x is ${}^-3$, the square of x is the square ${}^-3$ and then we must take the opposite of this result. To square ${}^-3$, we must enclose it in parentheses. The opposite, which is not being squared, must be outside of these parentheses. Therefore, the expression ${}^-x^2$ becomes ${}^-({}^-3)^2$ when $x = {}^-3$. In the following expression, the square applies to ${}^-3$. Then the opposite of that result is taken.

$${}^-({}^-3)^2 = {}^-9$$

From these examples, we can see that the only way to raise a negative number to a power is to include parentheses around the negative number and place the exponent just outside the parentheses. This means that to substitute a negative number for a variable that is raised to a power, we must include parentheses around the negative number, and the exponent needs to be outside the parentheses. This applies even when the original expression does not contain parentheses.

> **Substituting Numbers in Power Expressions**
>
> To substitute a negative number for a variable that is raised to a power, we must include parentheses around the negative number, and the exponent needs to be outside the parentheses.

The three examples in this section have demonstrated that we need to be aware of how an exponent is applied. This knowledge requires us to carefully read an expression with positive integer exponents. This may help us in evaluating the expressions. We also saw that whenever a negative number is substituted for a variable in an expression involving exponents, that negative number must be placed in parentheses to ensure that we apply the exponent to the correct value.

Problem Set 5.1

1. Translate each of the following English phrases into correct symbolic mathematics. Do not simplify.
   a. x used as a factor six times
   b. the fifth power of the opposite of x
   c. the square of negative four
   d. the opposite of the square of four
   e. the square of the opposite of x
   f. the opposite of the square of x

Evaluate the following numerical expressions. Write all noninteger results as fractions.

2. a. -8^3 b. $(-8)^3$ c. -8^2 d. $(-8)^2$ e. -8^0

3. a. $(-3)^3$ b. -3^3 c. $(-3)^4$ d. -3^4 e. $(-3)^0$

4. a. $\left(\frac{2}{3}\right)^2$ b. $\left(\frac{2}{3}\right)^3$ c. $\left(\frac{4}{7}\right)^2$

5. Complete the following table. Show your substitutions.

x	x^2	x^3	x^4	x^5
-3	$(-3)^2 = 9$			
0				
3				

6. If x is a positive number, determine whether the following expressions are positive or negative. Compare your results with your table from Problem 5.
   a. x^2 b. x^3 c. x^4 d. x^5

7. If x is a negative number, determine whether the following expressions are positive or negative. Compare your results with your table from Problem 5.
   a. x^2 b. x^3 c. x^4 d. x^5

8. Evaluate each expression when $x = 2$. Show your substitutions.
   a. x^4 b. x^5 c. $-x^4$ d. $-x^5$ e. $(-x)^4$ f. $(-x)^5$

9. Evaluate each expression when $x = -2$. Show your substitutions.
   a. x^4 b. x^5 c. $-x^4$ d. $-x^5$ e. $(-x)^4$ f. $(-x)^5$

10. a. Show your substitution of $x = -3$ into the expression x^2.
    b. Without evaluating the expression, decide whether the expression is positive or negative for the given substitution.
    c. Evaluate the expression.

11. a. Show your substitution of $x = -3$ into the expression x^3.
    b. Without evaluating the expression, decide whether the expression is positive or negative for the given substitution.
    c. Evaluate the expression.

12. a. Show your substitution of $x = -3$ into the expression $-x^2$.
    b. Without evaluating the expression, decide whether the expression is positive or negative for the given substitution.
    c. Evaluate the expression.

13. a. Show your substitution of $x = {^-}3$ into the expression $({^-}x)^2$.
    b. Without evaluating the expression, decide whether the expression is positive or negative for the given substitution.
    c. Evaluate the expression.

14. Evaluate the following expressions for the given values of the variables. Express any approximate results to the nearest hundredth.
    a. x^2y^3 for $x = {^-}2$ and $y = {^-}3$
    b. $\dfrac{7(r^2t)^3}{14t}$ for $r = 2.41$ and $t = {^-}2.01$
    c. $2x^3 + y^2$ for $x = 6.52$ and $y = {^-}3.20$
    d. $3m^2(m - n^2)$ for $m = {^-}5$ and $n = 7$
    e. $\dfrac{a^2 + 2b^3}{2}$ for $a = {^-}16$ and $b = {^-}4$
    f. $a^2 + b^3$ for $a = {^-}16$ and $b = {^-}4$

15. a. Do you think
    $$\dfrac{3x^2 + y^3}{3} \quad \text{simplifies to} \quad x^2 + y^3?$$
    b. Evaluate the expression
    $$\dfrac{3x^2 + y^3}{3} \quad \text{when } x = 5 \text{ and } y = 4$$
    c. Evaluate the expression $x^2 + y^3$ when $x = 5$ and $y = 4$.
    d. Did you get the same value in parts b and c? What does this say about your response in part a?

16. Graph the following equations.
    a. $y = \tfrac{1}{2}x + 4$ b. $y = x^2 - 5$ c. $y + 2x = 3$

5.2 Properties and Simplification

1. Simplify the following expressions. Verify your results with numerical substitution.
   a. $R^2 R^3$ b. $3x^2 * 5x$

2. Simplify the following expressions. Verify your results with numerical substitution.
   a. $m(m^3 + m^2)$ b. $5y^3(3y^2 - 2) + y^5$

3. Simplify each of the following expressions. Verify your results by numerical substitution.
   a. $x^2 x^3$ b. $w^5 w^2$ c. $y^3 y y^4$

4. Based on your results from Activity 3, when you multiply powers of x, what do you do to the exponents?

5. Write out the meaning of each expression (for example, x^3 means $x * x * x$), then simplify the expression. Assume that all variables are nonzero. Verify your results by numerical substitution.
   a. $\dfrac{x^3}{x^2}$ b. $\dfrac{m^6}{m^2}$ c. $\dfrac{p^2}{p^5}$ d. $\dfrac{y^4}{y^9}$

6. Based on your results from parts a and b in Activity 5, when you divide powers of x and the power in the numerator is greater than the power in the denominator, what do you do to the exponents?

7. Based on your results from parts c and d in Activity 5, when you divide powers of x and the power in the numerator is less than the power in the denominator, what do you do to the exponents?

8. Write out the meaning of each expression, then simplify the expression. Verify your results by numerical substitution.
   a. $(x^3)^2$ b. $(A^5)^3$ c. $(rate^2)^5$

9. Based on your results from Activity 8, when you simplify a power of a power of x, what do you do to the exponents?

10. Write out the meaning of each expression, then simplify the expression. Assume all variables are nonzero. Verify your results by numerical substitution.
    a. $(xy)^3$ b. $\left(\dfrac{a}{b}\right)^4$ c. $\left(\dfrac{2m^4}{n^3}\right)^3$

11. Based on your results from Activity 10, when you raise a product to a power, what do you do with the exponent?

12. Based on your results from Activity 10, when you raise a quotient to a power, what do you do with the exponent? Why must we say that the variable in the denominator is not equal to zero?

Discussion 5.2

Properties of Exponents

In the activities you were asked to simplify expressions that involve exponents. You probably noticed that the first three activities involved simplifying expressions that you already knew how to simplify. This means that much of what you will be asked to do in this section is not new to you. However, we will be extending these ideas so that, in later sections, we can simplify expressions involving negative exponents. To do this it is helpful and necessary to have some *rules* (properties) for simplifying exponents. Practicing the properties of exponents on positive integer exponents now will make it easier to use them later as we work with more complex expressions.

We list the rules or properties below, but realize that there is no need to memorize these rules. You can rediscover them at any time by looking at simple examples. Suppose you need to rewrite $x^{23}x^{12}$ and cannot remember the rule for multiplying expressions involving powers of x. It is not reasonable to write out what $x^{23}x^{12}$ means, but we can reason that this expression is the product of 23 factors of x and 12 factors of x. This means that there is a total of 23 + 12, or 35, factors of x. Therefore, $x^{23}x^{12} = x^{23+12} = x^{35}$. We could also rediscover the rule by looking at a simpler expression, for example, x^2x^3 is similar to $x^{23}x^{12}$.

The expression x^2x^3 means

$$x^2x^3 \qquad\qquad x^2x^3$$
$$= (xx)(xxx) \qquad = x^{2+3}$$
$$= xxxxx \qquad\qquad = x^5$$
$$= x^5$$

Therefore, we add exponents when we are multiplying expressions involving powers of x. So $x^{23}x^{12} = x^{23+12} = x^{35}$.

Similarly, if we are asked to simplify $(x^4)^{10}$, we can write out what this expression means (which could get tedious) or we can rediscover the rule by looking at a simpler example such as $(x^2)^3$.

$$(x^2)^3 \qquad\qquad (x^2)^3$$
$$= x^2x^2x^2 \qquad = x^{2*3}$$
$$= (xx)(xx)(xx) \qquad = x^6$$
$$= xxxxxx$$
$$= x^6$$

Therefore, we multiply exponents when simplifying a power of a power, so $(x^4)^{10} = x^{4*10} = x^{40}$.

Some people have difficulty remembering when they need to multiply exponents and when they need to add exponents. Some of these people create their own *incorrect* rule that says if there are parentheses then they should multiply exponents. Consider the following four expressions.

$$x^3 * x^4$$
$$x^3x^4$$
$$x^3 \cdot x^4$$
$$x^3(x^4)$$

All four of these expressions represent the product of x^3 and x^4. When we multiply powers of x we need to add the exponents. Therefore, all four of the expressions simplify to $x^{3+4} = x^7$. Notice that the last expression contains parentheses, yet we still add the exponents.

> ### Properties of Positive Integer Exponents
> Let x and y be nonzero real numbers. Let n and m be positive integers. Then the following are true.
>
> 1. $x^m x^n = x^{m+n}$ When *multiplying powers of* x, add the exponents.
>
> 2. $\dfrac{x^m}{x^n} = x^{m-n}$ if $m > n$ When *dividing powers of* x, subtract the exponents.
>
> 3. $\dfrac{x^m}{x^n} = \dfrac{1}{x^{n-m}}$ if $n > m$ When *dividing powers of* x, subtract the exponents.
>
> 4. $(xy)^m = x^m y^m$ The *power of a product* can be simplified by applying the power to each factor.
>
> 5. $\left(\dfrac{x}{y}\right)^m = \dfrac{x^m}{y^m}$ The *power of a quotient* can be simplified by applying the power to each factor in the numerator and each factor in the denominator.
>
> 6. $(x^m)^n = x^{m*n}$ When simplifying a *power of a power of* x, multiply the exponents.

We only multiply the exponents when we are simplifying a power of a power. A power of a power like $(x^3)^4$ always contains parentheses and simplifies to $x^{3*4} = x^{12}$. However, other expressions that are not powers of powers may also contain parentheses.

Example 1

Simplify the following expressions, assuming that w is a nonzero real number. Verify your results numerically.

     a. $3m^2 * 5m^4$ b. $\dfrac{w^5}{w^7}$ c. $5x^2 - (4x)^2$

Solution   a. The expression $3m^2 * 5m^4$ has one term. To simplify this expression we need to perform the multiplication. We see that the product of the numerical factors is 15. Because m^2 means that m is used as a factor two times, and m^4 means that m is used as a factor four times, in this term m is used as a factor a total of six times. This can be simplified as m^6. Therefore,

$$3m^2 * 5m^4$$
$$= 15m^6$$

You may have noticed that we did not use anything new in simplifying this expression. However, we could have used the *multiplying powers of* x property of exponents to simplify the product of m^2 and m^4. This property tells us that when we are multiplying powers of a variable (provided the variables are the same), we can add the exponents.

$$3m^2 * 5m^4$$
$$= 3 * 5 * m^2 * m^4 \quad \text{Rearrange the factors.}$$
$$= 15 * m^{2+4} \quad \text{When multiplying powers of } m \text{, add the exponents.}$$
$$= 15m^6$$

VERIFY Let $m = 8$.
$$3m^2 * 5m^4 \stackrel{?}{=} 15m^6$$
$$3 * 8^2 * 5 * 8^4 \stackrel{?}{=} 15 * 8^6$$
$$3{,}932{,}160 = 3{,}932{,}160 \quad ✔$$

We conclude that $3m^2 * 5m^4$ simplifies to $15m^6$.

b. The expression $\dfrac{w^5}{w^7}$ has one term. To simplify this expression we need to perform the division. We see that the numerator has w used as a factor five times and that the denominator has w used as a factor seven times. We can cancel five factors of w from the numerator and the denominator. This leaves two factors of w in the denominator.

$$\dfrac{w^5}{w^7}$$
$$= \dfrac{\cancel{wwwww}}{\cancel{wwwww}ww} \qquad \text{Cancel five factors of } w \text{ from the numerator and the denominator.}$$
$$= \dfrac{1}{w^2}$$

Again we did not use anything new in simplifying this expression. However, we could have used the *dividing powers of* x property of exponents to simplify the quotient of w^5 and w^7. This property tells us that when we divide variables with exponents (provided the variables are the same), we can subtract the exponents.

$$\dfrac{w^5}{w^7}$$
$$= \dfrac{1}{w^{7-5}} \qquad \text{Because the power in the denominator is larger we use property 3, and subtract 5 from 7 to avoid negative exponents.}$$
$$= \dfrac{1}{w^2}$$

VERIFY Let $w = 4$.
$$\dfrac{w^5}{w^7} \stackrel{?}{=} \dfrac{1}{w^2}$$
$$\dfrac{4^5}{4^7} \stackrel{?}{=} \dfrac{1}{4^2}$$
$$0.0625 = 0.0625 \quad ✔$$

We conclude that $\dfrac{w^5}{w^7}$ simplifies to $\dfrac{1}{w^2}$.

c. The expression $5x^2 - (4x)^2$ contains two terms. The first step is to simplify each of the terms. The first term is already simplified. To simplify the square of the product of 4 and x, we write out what that expression means and then simplify.

$$5x^2 - (4x)^2$$
$$= 5x^2 - 4x * 4x \qquad \text{Write out the meaning of } (4x)^2.$$
$$= 5x^2 - 16x^2$$
$$= {-}11x^2 \qquad \text{Combine like terms.}$$

This expression can also be simplified using the properties of exponents.

$$5x^2 - (4x)^2$$
$$= 5x^2 - 4^2 x^2 \qquad \text{Apply the second power to both factors in the second term.}$$
$$= 5x^2 - 16x^2$$
$$= {-}11x^2 \qquad \text{Combine like terms.}$$

VERIFY Let $x = 7$.
$$5x^2 - (4x)^2 \stackrel{?}{=} {-}11x^2$$
$$5 * 7^2 - (4 * 7)^2 \stackrel{?}{=} {-}11 * 7^2$$
$$-539 = -539 \quad ✓$$

We conclude that $5x^2 - (4x)^2$ simplifies to $-11x^2$.

NOTE: Numerical verification helps us decide whether our final result is *equivalent* to the original expression. It does not help us determine whether we have completed the simplifying process. In part c if we had wanted to verify whether $5x^2 - (4x)^2 \stackrel{?}{=} 5x^2 - 16x^2$, it would have verified. The expression $5x^2 - 16x^2$ is equivalent to $5x^2 - (4x)^2$, however, it is not simplified because we can still combine like terms. Numerical verification lets us know if we made an error in the simplifying process.

In the previous example, we simplified each of the expressions both with and without using the properties of exponents. You may be asking, Why do we bother with a set of six new properties of exponents if we can simplify expressions without using them? This is a good question. You should find that with practice, knowing the rules and how they can be applied to simplifying expressions with exponents makes the process quicker and easier. For example, if a product contains y^{34} and y^{54}, it can be easily simplified by adding the exponents. More importantly, remember that the properties of exponents become necessary when you are introduced to negative and fractional exponents. Practicing the properties of exponents on positive integer exponents now will make them easier to use later as we work with more complex exponents. But remember, if you get confused simplifying an expression with exponents, you can always think about what the expression means and logically think it through.

As we begin to practice applying the properties of exponents to simplify expressions, we may find that applying the rules correctly can become confusing. We will look at several statements and decide which are always true. This will help us distinguish and clarify the properties of exponents. It is important to know when we can apply the properties of exponents, but it is just as important to know when the properties are not applicable. The following examples begin to help us make this distinction.

Example 2 For each of the following equations, first look at the properties of exponents and decide if you think that the statement is true for all values of the variables. Then substitute numerical values to verify your conjecture. Decide why your conclusion from substituting numerically is correct. If the statement is not true for all values, rewrite the right-hand side to make the statement true.

a. $(2m^3)^4 = 8m^{12}$ b. $(x + y)^2 = x^2 + y^2$ c. $m^3 + m^3 = 2m^3$

Solution a. Many simplified results "look correct." In this first example, it *appears* that the fourth power has been applied to each of the factors in the parentheses, so we might assume that this statement is true. However, it is always a good idea to substitute numerical values into an expression to determine if your guess is correct.

Let $m = 5$.
$$(2m^3)^4 \stackrel{?}{=} 8m^{12}$$
$$(2 * 5^3)^4 \stackrel{?}{=} 8 * 5^{12}$$
$$3{,}906{,}250{,}000 \neq 1{,}953{,}125{,}000 \quad ✗$$

When we substitute $m = 5$ into the statement, the statement is *false*. So our guess was not correct! Let's see if we can determine why this statement is false. From the properties of exponents we know that in the expression $(2m^3)^4$ we can apply the power to each factor in the parentheses.

$$(2m^3)^4$$
$$= 2^4(m^3)^4 \quad \text{Apply the power to each factor.}$$
$$= 16m^{3*4} \quad \text{Simplify } 2^4, \text{ and multiply exponents.}$$
$$= 16m^{12}$$

Our first guess was not correct because $2^4 \neq 8$. In the second step, notice that we take the fourth power of 2. We do *not* multiply the 2 and 4. We *never* multiply an exponent by its base. We conclude that the statement $(2m^3)^4 = 8m^{12}$ is *false*, and a correct statement is $(2m^3)^4 = 16m^{12}$.

b. In this example, each of the terms is raised to the second power. Looking through the properties for positive exponents, we see that there is no rule for raising a sum to a power. The expression *looks* correct, so we might guess that this statement is always true. However, without a rule to support our guess, we should definitely substitute numerical values into the equation to determine if we are correct.

Let $x = 3$ and $y = 4$.

$$(x + y)^2 \stackrel{?}{=} x^2 + y^2$$
$$(3 + 4)^2 \stackrel{?}{=} 3^2 + 4^2$$
$$49 \neq 25 \quad \text{✗}$$

The statement is *false* when we substitute values for x and y. Here it appeared that the exponent is being applied to each *term* in the sum. In the properties of exponents it is stated that the *power of a product* (not sum) can be simplified by applying the power to each factor (not term). That is, $(xy)^m = x^m y^m$. The expression on the left-hand side of our statement is a power of a sum and the properties of exponents do not include powers of sums of differences. To write this expression correctly, we must write out its meaning.

$$(x + y)^2$$
$$= (x + y)(x + y)$$

Later we will learn how to determine the product of two sums. We conclude that the statement $(x + y)^2 = x^2 + y^2$ is *false*, and that a correct statement is $(x + y)^2 = (x + y)(x + y)$.

c. Should the sum of m^3 and m^3 be m^6, or possibly $2m^6$? Is this just the sum of like terms? Again, after you guess whether the statement is always true, you should verify it numerically.

VERIFY Let $m = 4$.
$$m^3 + m^3 \stackrel{?}{=} 2m^3$$
$$4^3 + 4^3 \stackrel{?}{=} 2 * 4^3$$
$$128 = 128 \quad \text{✔}$$

The statement is *true* when $m = 4$.

By looking at the statement, can we see why the statement is always be true? Remember that m^3 and m^3 are like terms, and when we combine like terms we add their coefficients (exponents do not change). Therefore, we conclude that the statement $m^3 + m^3 = 2m^3$ is a *true* statement for all values of m.

> Because exponents are a shorthand notation for repeated multiplication, the exponents only change when we simplify products, quotients, and powers. Exponents do not change when we simplify sums or differences.

Example 3

Simplify the following expressions. Assume that all variables are nonzero real numbers. Round any approximate numerical coefficients to two significant digits. Verify your result numerically.

a. $\dfrac{k^5}{4k^5}$

b. $\dfrac{k^5}{(4k)^5}$

c. $\dfrac{(2m^4)^3}{m^5}$

d. $\left(\dfrac{2m^4}{m^5}\right)^3$

Solution

a. To simplify this expression we cancel k^5, which appears in the numerator and the denominator.

$$\dfrac{\cancel{k^5}}{4\cancel{k^5}}$$

$$= \dfrac{1}{4}, \text{ or } 0.25 \qquad \text{Cancel the factor } k^5.$$

VERIFY Let $k = 3$.

$$\dfrac{k^5}{4k^5} \stackrel{?}{=} 0.25$$

$$\dfrac{3^5}{4 * 3^5} \stackrel{?}{=} 0.25$$

$$0.25 = 0.25 \quad \checkmark$$

We conclude that $\dfrac{k^5}{4k^5}$ simplifies to 0.25.

b. We see in the denominator that the product is raised to the fifth power. We must first simplify the power of the product in the denominator before canceling.

$$\dfrac{k^5}{(4k)^5}$$

$$= \dfrac{k^5}{4^5 k^5} \qquad \text{Apply the exponent to each factor in the denominator.}$$

$$= \dfrac{1}{4^5} \qquad \text{Cancel the factor } k^5.$$

$$= \dfrac{1}{1024} \qquad \text{We could stop with this fraction or find the approximate decimal.}$$

$$\approx 0.00098$$

VERIFY Let $k = 7$.

$$\dfrac{k^5}{(4k)^5} \stackrel{?}{\approx} 0.00098$$

$$\dfrac{(7)^5}{(4*7)^5} \stackrel{?}{\approx} 0.00098$$

$$0.00098 \approx 0.00098 \quad \checkmark$$

We conclude that $\dfrac{k^5}{(4k)^5}$ simplifies to $\dfrac{1}{1024}$, or approximately 0.00098.

c. In the numerator we see that the cube applies to the product of 2 and the fourth power of *m*. We must first simplify the cube of the product in the numerator before simplifying the quotient.

$$= \frac{(2m^4)^3}{m^5}$$

$$= \frac{2^3(m^4)^3}{m^5} \quad \text{Apply the exponent to each factor in the numerator.}$$

$$= \frac{2^3 m^{4*3}}{m^5} \quad \text{Multiply exponents.}$$

$$= \frac{2^3 m^{12}}{m^5}$$

$$= 2^3 m^{12-5} \quad \text{Subtract exponents.}$$

$$= 8m^7$$

VERIFY Let $m = 7$.

$$\frac{(2m^4)^3}{m^5} \stackrel{?}{=} 8m^7$$

$$\frac{(2*7^4)^3}{7^5} \stackrel{?}{=} 8*7^7$$

$$6{,}588{,}344 = 6{,}588{,}344 \quad \checkmark$$

We conclude that $\frac{(2m^4)^3}{m^5}$ simplifies to $8m^7$.

d. In this expression, the cube applies to the entire quotient. Therefore, we can simplify the quotient within the parentheses first and then apply the cube to the quotient.

$$\left(\frac{2m^4}{m^5}\right)^3$$

$$= \left(\frac{2}{m^{5-4}}\right)^3 \quad \text{Subtract the exponents.}$$

$$= \left(\frac{2}{m}\right)^3$$

$$= \frac{2^3}{m^3} \quad \text{Apply the exponent to the factor in the numerator and the factor in the denominator.}$$

$$= \frac{8}{m^3}$$

VERIFY Let $m = 5$.

$$\left(\frac{2m^4}{m^5}\right)^3 \stackrel{?}{=} \frac{8}{m^3}$$

$$\left(\frac{2*5^4}{5^5}\right)^3 \stackrel{?}{=} \frac{8}{5^3}$$

$$0.064 = 0.064 \quad \checkmark$$

We conclude that $\left(\frac{2m^4}{m^5}\right)^3$ simplifies to $\frac{8}{m^3}$.

NOTE: In the preceding example, the cube could have been applied to each of the factors in the numerator and the denominator before simplifying the quotient. The results would be the same.

5.2 Properties and Simplification

Compare the expressions in parts a and b in the previous example. How are they alike? How are they different? What about parts c and d?

Example 4

Simplify each expression, and verify your results.

a. $3p(5p^2 - mp)$ b. $x^8(x^4 - 1) + (2x^3)^4$

Solution   a.
$3p(5p^2 - mp)$
$= 3p * 5p^2 - 3p * mp$ Distribute $3p$ over the difference.
$= 15p^3 - 3mp^2$ Simplify each term. The first term has factors of 3, p, 5, p, and p so it simplifies to $15p^3$. Similarly for the second term.

VERIFY   Let $m = 3$ and $p = 4$.
$3p(5p^2 - mp) \stackrel{?}{=} 15p^3 - 3mp^2$
$3 * 4(5 * 4^2 - 3 * 4) \stackrel{?}{=} 15 * 4^3 - 3 * 3 * 4^2$
$816 = 816$ ✓

We conclude that $3p(5p^2 - mp)$ simplifies to $15p^3 - 3mp^2$.

b.
$x^8(x^4 - 1) + (2x^3)^4$
$= x^8 * x^4 - x^8 * 1 + 2^4(x^3)^4$ Distribute x^8 over the difference in the first term, and apply the exponent to each factor in the last term. There are now three terms in the expression.

$= x^{8+4} - x^8 + 16x^{3*4}$ The first term is a product of powers, so we add the exponents. The third term contains a power of a power, so we multiply the exponents.

$= x^{12} - x^8 + 16x^{12}$ Simplify the first and third terms.
$= 17x^{12} - x^8$ Combine like terms by adding their coefficients.

VERIFY   Let $x = 5$.
$x^8(x^4 - 1) + (2x^3)^4 \stackrel{?}{=} 17x^{12} - x^8$
$5^8(5^4 - 1) + (2 * 5^3)^4 \stackrel{?}{=} 17 * 5^{12} - 5^8$
$4{,}150{,}000{,}000 = 4{,}150{,}000{,}000$ ✓

We conclude that $x^8(x^4 - 1) + (2x^3)^4$ simplifies to $17x^{12} - x^8$.

Exponents Applied to Units

We will now revisit the process of evaluating formulas. When we substitute values with units into formulas, the rules for exponents often need to be applied.

Example 5

Find the area of a circle whose radius is 5.2 cm.

Solution
$A = \pi r^2$
$A = \pi(5.2 \text{ cm})^2$ To write the square of 5.2 cm, we must write 5.2 cm in parentheses.
$A = \pi * 5.2^2 * \text{cm}^2$ Apply the exponent to both factors in the product.
$A \approx 85 \text{ cm}^2$

The area of the circle is approximately 85 cm².

Example 6

How many cubic feet are in a cubic yard?

Solution Previously we did this unit fraction conversion as

1 yd^3
$= 1 \text{ yd} * \text{ yd} * \text{ yd} * \dfrac{3 \text{ ft}}{1 \text{ yd}} * \dfrac{3 \text{ ft}}{1 \text{ yd}} * \dfrac{3 \text{ ft}}{1 \text{ yd}}$
$= 1 * 3 * 3 * 3 \text{ ft}^3$
$= 27 \text{ ft}^3$

With the rules of exponents, we can use a shorthand notation to do this same conversion.

1 yd^3
$= 1 \text{ yd}^3 * \left(\dfrac{3 \text{ ft}}{1 \text{ yd}}\right)^3 \quad$ Recall, $\dfrac{3 \text{ ft}}{1 \text{ yd}} * \dfrac{3 \text{ ft}}{1 \text{ yd}} * \dfrac{3 \text{ ft}}{1 \text{ yd}} = \left(\dfrac{3 \text{ ft}}{1 \text{ yd}}\right)^3.$
$= 1 \text{ yd}^3 * \dfrac{3^3 \text{ ft}^3}{1^3 \text{ yd}^3} \quad$ Apply the exponent to each factor in the quotient.
$= 1 * 3^3 \text{ ft}^3$
$= 27 \text{ ft}^3$

Either of these methods is correct. Use the one that makes the process clearest to you.

We have looked at and applied the properties of **positive integer exponents.** We saw that we could simplify expressions using these properties as well as using methods we learned previously. Because exponents are a shorthand for multiplication, the properties of exponents apply to expressions involving products, quotients, and powers. Exponents do not change when simplifying sums or differences.

Problem Set 5.2

1. Explain the difference between $(-2)^4$ and -2^4. How is each expression read?

2. Which of the following terms are like terms?
   a. $3m^2 p$ b. $5mp$ c. $3m^2$ d. $12pm^2$ e. $m(4mp)$

3. Simplify each of the following expressions. Assume that all variables are nonzero real numbers. Verify your result.
   a. $b^5 * b^2$
   b. $\dfrac{x^3}{x^7}$
   c. $\dfrac{x^3 * x^4}{x^2}$
   d. $3k^4 * 7k^5$
   e. $\dfrac{20p^6}{4p^3}$
   f. $3x(4p^2 x^3)$

4. **Always True?** Determine if each of the following statements is true for all values of the variables. If the statement is not true for all values, rewrite the right-hand side to make it true (if possible).
   a. $(AB)^2 = A^2 B^2$
   b. $B^2 + B^2 = B^4$
   c. $(3AB)^2 = 3AB^2$
   d. $(5x^3)^2 = 10x^6$
   e. $3 * 3^2 = 9^2$
   f. $2^3 * 3^4 = 6^7$

5. Simplify each of the following expressions. Assume that all variables are nonzero real numbers. Verify your result.

   a. $m^3 + (2m)^3$

   b. $w^5(w^2 + 1)$

   c. $k^2k^4 - 4(k^3)^2$

   d. $\dfrac{ab^3}{b} + (ab)^2 - ab^2$

   e. $a(-2b)^3$

   f. $MP^2(MP^4 + 1)$

   g. $2\pi\left(\dfrac{L^2}{g^3}\right)^2$

   h. $(5.2m^2n^3)^2$

6. Simplify each of the following expressions. Assume that all variables are nonzero real numbers. Verify your result.

   a. $\left(\dfrac{a}{b^2}\right)^3$

   b. $\dfrac{6m^{10}}{3m(m^2)}$

   c. $\left(\dfrac{R^2}{2R^4}\right)^3$

   d. $\dfrac{mp^4}{p^2} + \dfrac{(3mp)^2}{m}$

   e. $5a - 2a(b - 3)$

   f. $12x^6 - 4x^2 * x^3 + (-2x^2)^3$

   g. $-k(5k^3 - 1) + 7k$

7. **Always True?** Determine if each of the following statements is true for all values of the variables. If the statement is not true for all values, rewrite the right-hand side to make it true (if possible).

   a. $(A + B)^3 = A^3 + B^3$

   b. $\left(\dfrac{A}{B}\right)^3 = \dfrac{A^3}{B^3}$

   c. $(3A^2B)^3 = 9A^6B^3$

   d. $A^{2m} * A^m = A^{3m}$

   e. $B^2 + B^2 = 2B^2$

   f. $B^5 + B^4 = B^9$

8. Determine half of 2^{40}, and write the exact answer in exponential form with a base of 2.

9. a. Choose a number for the side length of a square. Determine the area of your square. Now double the side length of your square. Determine the new area. Double the side length again, and compute the area. Compute the ratio of consecutive areas. What happens to the area of a square when you double the side length?

   b. Use a strategy similar to the one used in part a to determine what happens to the area of a square when the side length is tripled.

10. What happens to the volume of a cube when you double the side length? What happens to the volume of a cube when you triple the side length?

11. **Allowance.** Jerry's daughter Marty has asked him to consider a change in her allowance. She has proposed that he pay her 2¢ this week and that he double the amount each week. It sounds good for this week, but Jerry is suspicious!

    a. How much would Jerry pay in week 2? week 3? week 10?
    b. Make a table of values for the amount of allowance Jerry pays Marty in terms of the week.
    c. Graph the amount of allowance Jerry pays Marty in terms of the week for the first eight weeks.
    d. Write an equation for the amount of allowance Jerry pays Marty in terms of the week.
    e. How much would Jerry pay in week 50?
    f. Should Jerry accept this proposal? Explain your response.

12. The surface area S of a sphere is given by the formula $S = 4\pi r^2$, where r is the radius of the sphere. Assuming the earth is a sphere, find the surface area of the earth given that the diameter is approximately 6,370,000 meters. Express your result in scientific notation.

13. Find the volume, in gallons, of a rectangular storage tank with a length of 5.6 meters, a width of 3.2 meters, and a height of 7.5 meters.

14. The height of an object thrown into the air is given by the formula

    $$S = \tfrac{1}{2}gt^2 + v_0 t$$

    where S = the object's height,

    g = acceleration due to gravity,

    v_0 = initial speed of the object, and

    t = length of time since the object was thrown

    Find the value of S in the preceding formula, given $g = -9.8$ m/sec^2, $t = 10$ sec, and $v_0 = 50$ m/sec.

15. **Silo Volume.** The volume of a silo formed by a cylinder with a hemisphere on the top is given in the formula

    $$V = \pi r^2 h + \frac{2\pi r^3}{3}$$

    where V = volume,

    r = radius of the silo, and

    h = height of the silo

    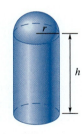

    Find the volume if the radius is 6.2 feet and the height is 24 feet.

Chapter Five Summary

Throughout mathematics we use positive integer exponents to represent repeated multiplication. However, to use them correctly, we must remember how they are applied in numerical expressions.

> **Exponents in Numerical Expression**
>
> Exponents apply to the item that is to the exponent's immediate left.

For example, in the expression $8 * 3^2$, the exponent applies only to the 3. In the expression -5^4, the exponent applies only to the 5. Therefore, -5^4 is read as *the opposite of the fourth power of 5* and simplifies to -625. In the expression $(-5)^4$, the exponent applies to -5 and simplifies to 625.

In this chapter, we learned that the only way to raise a negative number to a power is to include parentheses around the negative number and place the exponent just outside of the parentheses. This means that to substitute a negative number for a variable that is raised to a power, we must include parentheses around the negative number. This applies even when the original expression does not contain parentheses.

In Chapter 3 we added, subtracted, and multiplied expressions involving exponents. Most of the expressions that we simplified in this chapter could have been simplified using the techniques we learned in Chapter 3. However, as we continue in this course we will expand our definition of exponents to include negative integer exponents. Then, we will need the **properties of exponents** that we learned in this chapter. Practicing with these properties now will be helpful for later sections in the textbook.

The two properties of exponents that are often confused are $x^m x^n$ and $(x^m)^n$. Using the language that accompanies these properties can help us keep them straight.

- $x^m x^n$ When *multiplying powers of x,* add the exponents.
- $(x^m)^n$ When simplifying a *power of a power of x,* multiply the exponents.

Finally, remember that there are *no* properties where an exponent is applied to a sum or difference. The only way to rewrite a power of a sum or difference is to write out the meaning of the exponent. For example, $(a + b)^2$ can be rewritten as $(a + b)(a + b)$.

Chapter Six

Working with Algebra

In Chapter 2, we created formulas to describe block patterns. We found several different ways to express the same relationship. In Chapter 3, we were able to see algebraically why some of these expressions were equivalent, but others we still could not simplify. In this chapter, we will extend our understanding of the distributive property to see how it applies to the product or power of multiple-term expressions and division.

Then, we will learn how to solve two new types of equations. Simple quadratic equations arise in many geometry applications, including the Pythagorean theorem. Rational equations contain algebraic fractions. They occur in applications from many disciplines.

6.1 Multiplication of Multiterm Expressions

Activity Set 6.1

1. For each collection of structures, follow directions i–iv.
   i. Draw the fourth and fifth structures that would extend the pattern.
   ii. Describe, in words, how you would construct the 50th structure.
   iii. How many blocks would be in the 50th structure?
   iv. If n is the structure number, write an algebraic expression for the number of blocks that are needed to build the structure.

Structure Number n	1	2	3	4	5	50
Number of Blocks B						

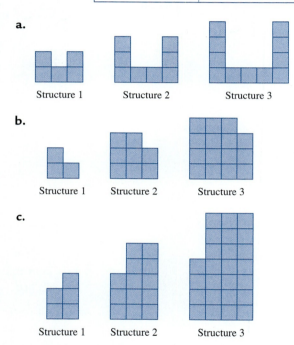

   a. Structure 1, Structure 2, Structure 3
   b. Structure 1, Structure 2, Structure 3
   c. Structure 1, Structure 2, Structure 3

2. Find the area of the following figure in two different ways according to the directions in parts a and b.

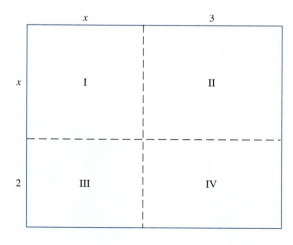

a. Write the lengths of each side of the figure as a sum. Then write the area of the figure as a product of these lengths.

b. Find the area of each of the four regions. Write the area of the figure as a sum of these four regions. Combine like terms if possible.

c. Because the expressions you wrote in parts a and b both represent the area of the whole rectangle, they must be equal. Write an equation expressing this equality.

3. Follow the directions in Activity 2 to find the area of the following figure in two different ways.

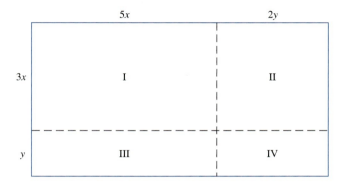

4. Follow the directions in Activity 2 to find the area of the following figure in two different ways.

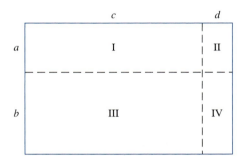

5. a. From your observations in Activities 2–4, what do you expect the product of $(2m + 3)$ and $(m + 5)$ to be? Verify your conclusion with numerical substitution.

b. From your observations in Activities 2–4, what do you expect the product of $(k + 4)$ and $(3p + 2k + 1)$ to be? Verify your conclusion with numerical substitution.

Discussion 6.1

Multiplying Multiterm Expressions

In the activities, we found that products of multiterm expressions can be illustrated using a divided-rectangle model. This product can then be rewritten as a sum by writing the area of the rectangle as a sum of the areas of its regions.

Example 1

Draw a divided-rectangle model to illustrate the product of the sum of m and 3 and the sum of $3m$ and 4. Then, rewrite the product as a sum.

Solution To illustrate $(m + 3)(3m + 4)$ with a divided-rectangle model, we need to draw a rectangle whose dimensions are $m + 3$ and $3m + 4$.

	$3m$	4
m	I	II
3	III	IV

Then, the area of region I is $m * 3m = 3m^2$.
The area of region II is $m * 4 = 4m$.
The area of region III is $3 * 3m = 9m$.
The area of region IV is $3 * 4 = 12$.

Therefore, the area of the whole rectangle is $3m^2 + 4m + 9m + 12$. Combining like terms, this becomes $3m^2 + 13m + 12$. We see that the product $(m + 3)(3m + 4)$ is equivalent to the sum $3m^2 + 13m + 12$. That is,

$$(m + 3)(3m + 4) = 3m^2 + 13m + 12$$

We saw in the activities and in Example 1 that when multiplying multiterm expressions, each term of the second factor is multiplied by each term of the first factor and the results are added. This pattern can be used to multiply any two multiterm expressions. It does not matter how many terms are in each factor. It can, however, get cumbersome if we have too many terms.

When multiplying two factors, in which each factor consists of two terms, we can write the product symbolically as

$$(a + b)(c + d) = ac + ad + bc + bd$$

Many people refer to this process as FOIL. FOIL stands for **first**, **outer**, **inner**, and **last**.

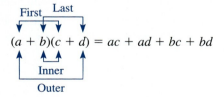

Remember: This mnemonic only works for a product of two factors consisting of two-term expressions.

Example 2

Use the previous pattern to rewrite each product as a sum. Simplify by combining like terms when possible. Verify your results by numerical substitution.

   a. $(2x + 7)(x - 4)$ b. $(3k - 2)(k^2 - 3k - 8)$ c. $(2m + 5)^2$

Solution a. It can be helpful to rewrite subtraction as addition of the opposite.

$$(2x + 7)(x - 4)$$
$$= (2x + 7)(x + {}^-4)$$ Rewrite subtraction as addition of the opposite.

$$= (2x + 7)(x + {}^-4)$$ Distribute the $2x$ over the sum, and distribute the 7 over the sum.

$$= 2x * x + 2x * {}^-4 + 7 * x + 7 * {}^-4$$
$$= 2x^2 + {}^-8x + 7x + {}^-28$$ Simplify each term.
$$= 2x^2 - x - 28$$ Combine like terms, and rewrite addition of the opposite as subtraction.

VERIFY Let $x = 3$.

$$(2x + 7)(x - 4) \stackrel{?}{=} 2x^2 - x - 28$$
$$(2 * 3 + 7)(3 - 4) \stackrel{?}{=} 2 * 3^2 - 3 - 28$$
$${}^-13 = {}^-13 \quad \checkmark$$

We conclude that $(2x + 7)(x - 4)$ simplifies to $2x^2 - x - 28$.

b.
$$(3k - 2)(k^2 + 3k - 8)$$
$$= (3k + {}^-2)(k^2 + 3k + {}^-8)$$ Rewrite subtraction as addition of the opposite.

$$= (3k + {}^-2)(k^2 + 3k + {}^-8)$$ Distribute the $3k$ over the sum, and distribute the ${}^-2$ over the sum.

$$= 3k * k^2 + 3k * 3k + 3k * {}^-8 + {}^-2 * k^2 + {}^-2 * 3k + {}^-2 * {}^-8$$
$$= 3k^3 + 9k^2 + {}^-24k + {}^-2k^2 + {}^-6k + 16$$ Simplify each term.
$$= 3k^3 + 7k^2 - 30k + 16$$ Combine like terms, and rewrite addition of the opposite as subtraction.

VERIFY Let $k = 5$.

$$(3k - 2)(k^2 + 3k - 8) \stackrel{?}{=} 3k^3 + 7k^2 - 30k + 16$$
$$(3 * 5 - 2)(5^2 + 3 * 5 - 8) \stackrel{?}{=} 3 * 5^3 + 7 * 5^2 - 30 * 5 + 16$$
$$416 = 416 \quad \checkmark$$

We conclude that $(3k - 2)(k^2 + 3k - 8)$ simplifies to

$$3k^3 + 7k^2 - 30k + 16$$

c. This problem is clearer if we rewrite the square as the product of two factors.

$$(2m + 5)^2$$
$$= (2m + 5)(2m + 5)$$
$$= 2m * 2m + 2m * 5 + 5 * 2m + 5 * 5 \quad \text{Distribute the } 2m \text{ over the sum, and distribute the 5 over the sum.}$$
$$= 4m^2 + 10m + 10m + 25 \quad \text{Simplify each term.}$$
$$= 4m^2 + 20m + 25 \quad \text{Combine like terms.}$$

VERIFY Let $m = 4$.
$$(2m + 5)^2 \stackrel{?}{=} 4m^2 + 20m + 25$$
$$(2 * 4 + 5)^2 \stackrel{?}{=} 4 * 4^2 + 20 * 4 + 25$$
$$169 = 169 \quad \checkmark$$

We conclude that $(2m + 5)^2$ simplifies to $4m^2 + 20m + 25$.

Notice that when we square the sum of $2m$ and 5, we do not get the sum of the square of $2m$ and the square of 5. That is, $(2m + 5)^2 \neq (2m)^2 + 5^2$. The rules of exponents state that the *power of a product* can be simplified by applying the power to each factor. That is, $(xy)^n = x^n y^n$. The exponent applies to each factor of a product or quotient. This same property does *not* hold true when the grouped quantity is a sum or difference. That is, $(x + y)^n \neq x^n + y^n$.

In the activities, we worked with block patterns and wrote algebraic expressions to describe the number of blocks in a given structure in terms of the structure number. In most problems we had more than one expression that worked. One way to verify that two different expressions are equivalent is to substitute values into both expressions. We can now show that these expressions are equivalent by algebraic methods.

Example 3

For the following block pattern, two of the expressions that can describe the number of blocks in the nth structure are: $(n + 1)^2 + n^2$ and $(n + 1)(2n + 1) - n$. Show algebraically that these two expressions are equivalent.

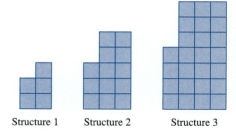

Structure 1   Structure 2   Structure 3

Solution

To show algebraically that $(n + 1)^2 + n^2$ and $(n + 1)(2n + 1) - n$ are equivalent we simplify each expression.

$$(n + 1)^2 + n^2 \qquad\qquad (n + 1)(2n + 1) - n$$
$$= (n + 1)(n + 1) + n^2 \qquad = 2n^2 + 1n + 2n + 1 - n$$
$$= n^2 + 1n + 1n + 1 + n^2$$
$$= 2n^2 + 2n + 1 \qquad\qquad = 2n^2 + 2n + 1$$

Because both expressions are equivalent to $2n^2 + 2n + 1$, we conclude that

$$(n + 1)^2 + n^2 = (n + 1)(2n + 1) - n$$

In the next example, we will look at products involving three factors.

Example 4

Multiply to write each product as a sum or difference. Verify your results.

a. $4(x + 3)(7 - x)$ b. $10 * (2m + 1) * 5$

Solution

a. When multiplying multiple factors, we multiply two factors at a time. In this example we could multiply the 4 and $(x - 1)$ first, we could multiply $(x - 3)$ and $(7 - x)$ first, or we could multiply 4 and $(7 - x)$ first.

If we multiply the 4 and the $(x - 1)$ first, the process looks like the following.

$4(x + 3)(7 - x)$
$= (4 * x + 4 * 3)(7 - x)$ Multiply the first two factors by distributing the 4 over the sum.
$= (4x + 12)(7 + {}^-1x)$ Simplify, and rewrite subtraction as addition of the opposite.
$= 4x * 7 + 4x * {}^-1x + 12 * 7 + 12 * {}^-1x$ Multiply the two remaining factors.
$= 28x + {}^-4x^2 + 84 + {}^-12x$ Simplify the expression.
$= {}^-4x^2 + 16x + 84$ Combine like terms.

VERIFY   Let $x = 10$.

$4(x + 3)(7 - x) \stackrel{?}{=} {}^-4x^2 + 16x + 84$
$4(10 + 3)(7 - 10) \stackrel{?}{=} {}^-4 * 10^2 + 16 * 10 + 84$
$-156 = -156$ ✔

Alternatively, we could multiply the second two factors first, which results in the following simplification process.

$4(x + 3)(7 - x)$
$= 4(x + 3)(7 + {}^-1x)$ Rewrite subtraction as addition of the opposite.
$= 4(x * 7 + x * {}^-1x + 3 * 7 + 3 * {}^-1x)$ Multiply the last two factors.
$= 4(7x + {}^-1x^2 + 21 + {}^-3x)$ Simplify the second factor.
$= 4({}^-1x^2 + 4x + 21)$ Combine like terms in the second factor.
$= {}^-4x^2 + 16x + 84$ Distribute the 4 over the sum.

Using either process, we can see that $4(x + 3)(7 - x)$ simplifies to ${}^-4x^2 + 16x + 84$.

b. In the expression $10 * (2m + 1) * 5$, we start by multiplying any two of the factors. Because the product of 10 and 5 is easy, let's multiply these factors first.

$10 * (2m + 1) * 5$
$= 50 * (2m + 1)$ Multiply the 10 and 5.
$= 100m + 50$ Distribute the 50 over the sum.

VERIFY   Let $m = 3$.

$10 * (2m + 1) * 5 \stackrel{?}{=} 100m + 50$
$10 * (2 * 3 + 1) * 5 \stackrel{?}{=} 100 * 3 + 50$
$350 = 350$ ✔

We conclude that $10 * (2m + 1) * 5$ simplifies to $100m + 50$.

Solving Literal Equations

Being able to multiply multiterm expressions allows us to solve some literal equations that we could not solve before.

Solve the equation $n = (n - 1)(R - r)$ for n.

Solution

$n = (n - 1)(R - r)$

$n = (n + {}^-1)(R + {}^-r)$ Rewrite subtraction as addition of the opposite.

$n = nR + {}^-rn + {}^-R + r$ Multiply the two factors on the right.

$\underline{n} = \underline{nR} + \underline{{}^-rn} + {}^-R + r$ Identify the terms containing n.

$n - nR + rn = {}^-R + r$ Subtract nR and add rn to both sides to get all terms containing n on one side.

$n(1 - R + r) = {}^-R + r$ Factor out the common factor of n.

$n = \dfrac{{}^-R + r}{1 - R + r}$ Divide by the coefficient of n.

VERIFY   Let $R = 3$ and $r = 5$. Substitute these values into our result to obtain a value for n.

$n = \dfrac{{}^-R + r}{1 - R + r}$

$n = \dfrac{-3 + 5}{1 - 3 + 5}$

$n = \dfrac{2}{3}$

Substituting all three values, $R = 3$, $r = 5$, and $n = \frac{2}{3}$, into our original equation gives

$n = (n - 1)(R - r)$

$\dfrac{2}{3} \stackrel{?}{=} \left(\dfrac{2}{3} - 1\right)(3 - 5)$

$\dfrac{2}{3} = \dfrac{2}{3}$ ✓

We conclude that

$n = \dfrac{{}^-R + r}{1 - R + r}$

In this section, we learned that to multiply two multiterm expressions, each term of the second factor is multiplied by each term of the first factor and the results are added. We used this pattern to rewrite algebraic expressions. This technique was also applied to solving literal equations.

Problem Set 6.1

1. Multiply to write each product as a sum or difference. Verify your results numerically.
   a. $(3x + 7)(x + 3)$
   b. $(2M - 5)(4M + 1)$
   c. $(3c + a)(4c - 5a)$
   d. $(2k + 11)(7 - 4m)$
   e. $(x + 5)(x^2 + 5x - 3)$
   f. $(p + 3)^2$
   g. $(p + 3)^2 * (2p + 1)$
   h. $(3xy - 5)(5x + y)$

2. Simplify the following algebraic expressions by doing the indicated operations. Verify your results numerically.
   a. $3r(2r + 7) - (r - 4)$
   b. $(2n + 3)(n - 4) + (3n)^2$
   c. $8(m + 4)(2m - 1)$
   d. $3 * 4 * (2x - 10)$
   e. $2(5 - 3K)8$
   f. $(x + y)^2 - (x - y)^2$
   g. $4ab^2 + (2ab - 3)(b + 5)$

3. Solve each literal equation for the indicated variable. Verify your results numerically.
   a. $A = \frac{1}{2}h(b_1 + b_2)$ for b_2
   b. $4k - m = (m + 3)(3k - 2)$ for m
   c. $A = 2\pi r^2 + 2\pi r h$ for h
   d. $\frac{2t}{q} = (t + 2)(q - 3)$ for t

4. a. Square each of the following expressions as indicated. Verify your results numerically.
   i. $(x + 5)^2$
   ii. $(m + 4)^2$
   iii. $(y - 3)^2$
   iv. $(6 - p)^2$
   v. $(3R + 2Q)^2$
   vi. $(5x + 3y)^2$

   b. From your results in part a, what pattern do you see for the generalized square of a sum, $(a + b)^2$? a difference $(a - b)^2$?

5. In Problem 4, you found the general pattern for the square of a sum, $(a + b)^2 = a^2 + 2ab + b^2$. Sometimes we want to go from the sum to the factored form. For example, we can factor $x^2 + 8x + 16$:

   $$x^2 + 8x + 16$$
   $$= x^2 + 2 * x * 4 + 4^2$$
   $$= (x + 4)^2$$

   Express each sum or difference as the square of a sum or difference.
   a. $n^2 + 12n + 36$
   b. $k^2 - 6k + 9$
   c. $4y^2 + 12y + 9$
   d. $q^2 - 10q + 25$
   e. $25m^2 + 10m + 1$

6. a. Each of the following expressions is the product of a sum and a difference. Rewrite each product as a sum or difference.
   i. $(x + 3)(x - 3)$
   ii. $(m + 5)(m - 5)$
   iii. $(p - 8)(p + 8)$
   iv. $(2p + 3)(2p - 3)$
   v. $(3x + 7y)(3x - 7y)$

   b. From your results in part a, what pattern do you see for the generalized product of the sum of a and b and the difference of a and b: $(a + b)(a - b)$?

7. In Problem 6, you found the general pattern for the difference of two squares by multiplying the sum of a and b by the difference of a and b. That is, $a^2 - b^2 = (a + b)(a - b)$. Express each of the following differences as an equivalent product.

   a. $x^2 - 64$
   b. $16 - k^2$
   c. $4m^2 - 25$
   d. $p^2 - 81$
   e. $r^2 - 1$
   f. $25 - 9x^2$

8. a. For the following pattern, determine the number of blocks needed to build the 50th structure.
   b. If n is the structure number, write an algebraic expression for the number of blocks needed to build the structure.

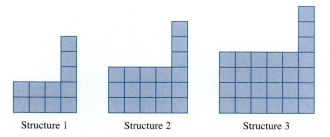

9. a. For the following pattern, determine the number of blocks needed to build the 50th structure.
   b. If n is the structure number, write an algebraic expression for the number of blocks that are needed to build the structure.

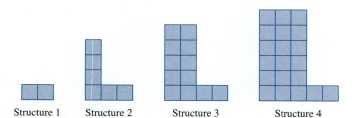

10. The three figures shown here form the beginning of a geometric pattern that can be continued.
    a. How many sticks are in the 100th structure?
    b. Write an algebraic equation for the number of sticks in the structure in terms of the structure number.

Structure 1   Structure 2   Structure 3

11. The following picture shows a square inside another square. Assume that the triangles are all the same size.

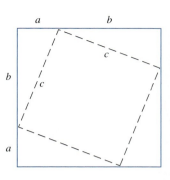

   a. What are the dimensions of the large square in terms of *a* and *b*? Write an expression for the area of the large square using these dimensions.

   b. Write an expression for the area of the large square by adding the area of the smaller square to the area of the four triangles.

   c. Write an equation showing the equality of the expressions in parts a and b. Simplify each side of your equation. Solve for c^2. What is your final conclusion? Some of you may recognize this result as the **Pythagorean theorem,** which states that in any right triangle, the square of the hypotenuse is equal to the sum of the squares of the legs.

6.2 Division

Discussion 6.2

We know that the product of a term and a sum of terms can be rewritten as a sum of terms by applying the distributive property of multiplication over addition. That is,

$$a(b + c) = ab + ac$$

This same property can be used to rewrite the quotient of a sum and a term.

Example 1

Rewrite the following as a sum. Verify your results.

$$\frac{4x^2 + 10}{2}$$

Solution

Dividing by 2 is equivalent to multiplying by $\frac{1}{2}$, the reciprocal of 2. Therefore,

$$\frac{4x^2 + 10}{2}$$

can be rewritten as $\frac{1}{2} * (4x^2 + 10)$.

$$\frac{4x^2 + 10}{2}$$

$= \frac{1}{2} * (4x^2 + 10)$ Rewrite division as multiplication by the reciprocal.

$= \frac{1}{2} * 4x^2 + \frac{1}{2} * 10$ Distribute the $\frac{1}{2}$ over the sum.

$= \frac{4x^2}{2} + \frac{10}{2}$ Rewrite multiplication by $\frac{1}{2}$ as division by 2.

$= 2x^2 + 5$

VERIFY   Let $x = 3$.

$$\frac{4x^2 + 10}{2} \stackrel{?}{=} 2x^2 + 5$$

$$\frac{4 * 3^2 + 10}{2} \stackrel{?}{=} 2 * 3^2 + 5$$

$$23 = 23 \quad \checkmark$$

We conclude that the equation

$$\frac{4x^2 + 10}{2}$$

simplifies to $2x^2 + 5$.

Notice, in Example 1 the quotient $\frac{4x^2 + 10}{2}$ was rewritten as $\frac{4x^2}{2} + \frac{10}{2}$ by applying the distributive property. In the first expression, $\frac{4x^2 + 10}{2}$, the sum is divided by 2. In the second expression, $\dfrac{4x^2}{2} + \dfrac{10}{2}$, each term in the sum is divided by 2. Because dividing by a term is equivalent to multiplying by its reciprocal, this simplification always works.

> **Distributing Division over Addition**
> Stated more generally, the quotient of the sum of a and b and c is equal to the sum of a divided by c and b divided by c. That is,
> $$\dfrac{a+b}{c} = \dfrac{a}{c} + \dfrac{b}{c}$$

Example 2

Rewrite each expression as a sum or difference. Verify your results.

a. $\dfrac{5m^3 - k^2 m^2 + 2m^2}{m^2}$

b. $\dfrac{3x^5 - x^2 y + xy}{xy}$

Solution

a.
$$\dfrac{5m^3 - k^2 m^2 + 2m^2}{m^2}$$
$$= \dfrac{5m^3}{m^2} - \dfrac{k^2 m^2}{m^2} + \dfrac{2m^2}{m^2} \quad \text{Divide each term in the numerator by } m^2.$$
$$= 5m^{3-2} - k^2 + 2 \quad \text{Apply properties of exponents, and cancel common factors.}$$
$$= 5m - k^2 + 2$$

VERIFY Let $m = 3$ and $k = 5$.
$$\dfrac{5m^3 - k^2 m^2 + 2m^2}{m^2} \stackrel{?}{=} 5m - k^2 + 2$$
$$\dfrac{5*3^3 - 5^2*3^2 + 2*3^2}{3^2} \stackrel{?}{=} 5*3 - 5^2 + 2$$
$$-8 = -8 \quad ✓$$

We conclude that $\dfrac{5m^3 - k^2 m^2 + 2m^2}{m^2}$ simplifies to $5m - k^2 + 2$.

b.
$$\dfrac{3x^5 - x^2 y + xy}{xy}$$
$$= \dfrac{3x^5}{xy} - \dfrac{x^2 y}{xy} + \dfrac{xy}{xy} \quad \text{Divide each term in the numerator by } xy.$$
$$= \dfrac{3x^{5-1}}{y} - x^{2-1} + 1 \quad \text{Apply the properties of exponents, and cancel common factors.}$$
$$= \dfrac{3x^4}{y} - x + 1$$

VERIFY Let $x = 7$ and $y = 11$.

$$\frac{3x^5 - x^2y + xy}{xy} \stackrel{?}{=} \frac{3x^4}{y} - x + 1$$

$$\frac{3*7^5 - 7^2*11 + 7*11}{7*11} \stackrel{?}{=} \frac{3*7^4}{11} - 7 + 1$$

$$648.82 = 648.82 \quad ✔$$

We conclude that $\frac{3x^5 - x^2y + xy}{xy}$ simplifies to $\frac{3x^4}{y} - x + 1$.

A common question arises as we begin to divide sums and differences. When can we cancel? The answer is, *if the numerator consists of more than one term then you cannot cancel,* unless the numerator is identical to the denominator. We can only cancel if the expression we are canceling is both a factor of the numerator and a factor of the denominator. We illustrate with the following example. Consider the expression $\frac{6x + 3}{3}$. The 3 in the numerator is a term and therefore *does not* cancel with the 3 in the denominator.

$$\frac{6x + 3}{3} \neq \frac{6x + \cancel{3}}{\cancel{3}}$$

Instead, we must divide each term in the numerator by 3.

$$\frac{6x}{3} + \frac{3}{3}$$

In the first term, 3 is a factor of the numerator and 3 is a factor of the denominator. These factors of 3 cancel, leaving $2x$. In the second term, 3 divided by 3 is equivalent to 1.

$$\frac{6x + 3}{3}$$
$$= \frac{6x}{3} + \frac{3}{3}$$
$$= 2x + 1$$

Example 3

Rewrite the following expression as a sum. Verify your results.

$$\frac{2pk^2 + (3k^2)^3}{4k^2}$$

Solution

$$\frac{2pk^2 + (3k^2)^3}{4k^2}$$

$$= \frac{2pk^2}{4k^2} + \frac{(3k^2)^3}{4k^2} \qquad \text{Divide each term by } 4k^2.$$

$$= \frac{p}{2} + \frac{3^3(k^2)^3}{4k^2} \qquad \text{Cancel common factors in the first term. Apply the power of 3 to each factor in the numerator of the second term.}$$

$$= \frac{p}{2} + \frac{27k^{2*3}}{4k^2} \qquad \text{Simplify the power of a power in the numerator of the second term by multiplying exponents.}$$

$$= \frac{p}{2} + \frac{27k^6}{4k^2}$$

$$= \frac{p}{2} + \frac{27k^4}{4}$$ Subtract exponents in the second term.

$$= \frac{1}{2}p + \frac{27}{4}k^4$$ Write the numerical coefficients of the terms as fractions.

or

$$= 0.5p + 6.75k^4$$ Alternatively, we can write the numerical coefficients as decimals.

VERIFY Let $p = 3$ and $k = 5$.

$$\frac{2pk^2 + (3k^2)^3}{4k^2} \stackrel{?}{=} 0.5p + 6.75k^4$$

$$\frac{2pk^2 + (3k^2)^3}{4k^2} \stackrel{?}{=} 0.5 * 3 + 6.75 * 5^4$$

$$4220.25 = 4220.25 \quad \checkmark$$

We conclude that $\dfrac{2pk^2 + (3k^2)^3}{4k^2}$ simplifies to $0.5p + 6.75k^4$.

In this section, we saw how the distributive property can be used to rewrite a quotient as a sum or difference. The quotient of the sum of a and b and c is equal to the sum of a divided by c and b divided by c. That is,

$$\frac{a + b}{c} = \frac{a}{c} + \frac{b}{c}$$

In using this property, it is important to know when it is appropriate to cancel. We can only cancel if the expression we are canceling is both a factor of the numerator and a factor of the denominator in a given term.

Problem Set 6.2

1. Rewrite each quotient as a sum or difference, and simplify each term. Verify your results.

   a. $\dfrac{21 - 6x}{3}$

   b. $\dfrac{13d^2t + 8d^3t^2 - 4dt}{d^2}$

   c. $\dfrac{4m^2y - 2m}{2m}$

   d. $\dfrac{(-3x^3y)^2 + 3x^3y^2}{x^2y^2}$

   e. $\dfrac{p^2 + 1}{p^2}$

   f. $\dfrac{-8w^4t^3 + 9w^2t}{-4w^2t}$

2. Rewrite each equation to put it in slope–intercept form, $y = mx + b$. For each equation, determine the slope and vertical intercept. Use this information to sketch a graph.

   a. $3x + 4y = 12$

   b. $x - 2y = {}^-8$

   c. $3x = 4y - 8$

   d. $4(x + 2) - y = 13$

   e. $5x = 3(y - 2) + 6$

   f. $2(3 - x) + 5(y - 7) = 15$

3. Rewrite the equation $Ax + By = C$ to put it in slope–intercept form, $y = mx + b$. Identify the coefficient of x in terms of A and B.

4. **Always True?** First, try to decide whether each of the following equations is true for all values of the variables. Then, substitute numerical values into the equation to verify your conjecture. If the statement is not true for all values, rewrite the right-hand side of the equation to make it true.

   a. $\dfrac{15xy^2}{-3x} = -5y^2$

   b. $\dfrac{5b - 3}{5} = b - 3$

   c. $\dfrac{5(b - 3)}{5} = b - 3$

   d. $\dfrac{3p(p + r)}{p(p - r)} = 3$

   e. $\dfrac{rt + 4}{dt + 4} = \dfrac{r}{d}$

   f. $\dfrac{r(t + 4)}{d(t + 4)} = \dfrac{r}{d}$

   g. $\dfrac{mn}{m + n} = \dfrac{mn}{m} + \dfrac{mn}{n}$

   h. $\dfrac{mn}{m^2 + n} = \dfrac{1}{m}$

5. Marty earns a weekly salary plus a commission on what he sells. Choose which of the following graphs most reasonably models this situation. Explain how you made your decision.

   a.
   b.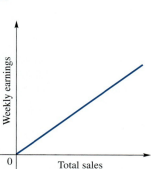
   c.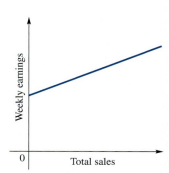

6. Tom placed the fish aquarium under the running faucet then ran to catch the telephone. While Tom was talking on the phone, the tank filled to the top and began to overflow. Choose the following graph below that most reasonably models this situation. Explain how you made your decision.

   a.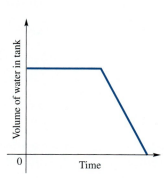
   b.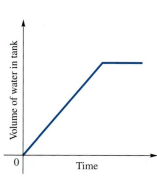
   c.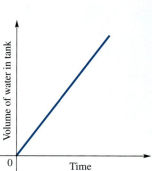

7. Each square in the following figures has dimensions 1 unit by 1 unit.
   a. Determine the perimeter of each figure.

   Figure 1   Figure 2   Figure 3

   b. Using your results and continuing the pattern, complete a table similar to the following.

Figure Number	1	2	3	4	5	n
Perimeter of Figure						

   c. Graph the perimeter of the figures in terms of the figure number.

8. For each of the following expressions, determine the numerical value and units of the result.

   a. $\dfrac{24 \text{ in.}^2 + 13 \text{ in.}^2}{5 \text{ in.}}$

   b. $\dfrac{123.5 \text{ miles} - 47.2 \text{ miles}}{1.75 \text{ h}}$

   c. $\dfrac{36 \, \frac{\text{ft}}{\text{sec}} + 12 \, \frac{\text{ft}}{\text{sec}}}{4.7 \text{ sec}}$

   d. $\dfrac{4.0 \text{ cm} \, (13.2 \text{ cm}^2 + 12.6 \text{ cm}^2)}{3.7 \text{ h}}$

9. Juliette is supposed to give her son 44 cc (cubic centimeters) of a liquid medication. How many tablespoons of medication should she give her son?

10. If Miccah runs 5 miles in 45 minutes and Abdull runs 10 kilometers in 1 hour and 5 minutes, who runs faster? What is the speed of each runner?

11. The Irvings just bought a piece of land. The land is rectangular and the dimensions are roughly 200 ft by 400 ft. What is the area of the Irvings' land in acres?

Activity Set 6.3

Simple Quadratics

1. **a.** What *two* numbers solve the equation $x^2 = 9$?
   b. What *two* numbers solve the equation $x^2 = 36$?
   c. In general, what can you say about the solution to the equation $x^2 = N$, where N is any positive number?

2. The area is given for each of the following squares. Find the length of the sides of each square.

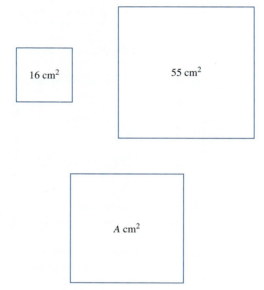

Discussion 6.3

In this section, we will learn how to solve equations of the form $ax^2 = b$, and $ax^2 + c = b$, where a, b, and c are constants and x is the variable. We call these equations **simple quadratic equations.** We saw in the activities that 3 and $^-3$ were both solutions to the equation $x^2 = 9$ because $3^2 = 9$ and $(^-3)^2 = 9$. Because 3 and $^-3$ are both square roots of 9, we might be tempted to say that the solution to $x^2 = 9$ is $x = \sqrt{9}$. But $\sqrt{9}$ means the principal (or positive) square root of 9 only, and our solution must include *all values* of x that make the equation a true statement. Therefore, to be correct we say that the solution to $x^2 = 9$ is $x = \sqrt{9}$ or $x = ^-\sqrt{9}$, that is, $x = 3$ or $x = ^-3$.

> The solution to $x^2 = N$ (where N is any positive number) is
>
> $x = \sqrt{N}$ or $x = ^-\sqrt{N}$
>
> This is sometimes written as $x = \pm\sqrt{N}$ and is read as "x is equal to *plus or minus* the square root of N."

Example 1

What is the solution to the equation $x^2 = 729$?

Solution

$x^2 = 729$

$x = \pm\sqrt{729}$ The solution is the positive or negative square root of 729.

$x = \pm 27$

We can check this numerically.

CHECK $x = 27$ and $x = -27$ in our original equation.

$27^2 \stackrel{?}{=} 729$ $(-27)^2 \stackrel{?}{=} 729$

$729 = 729$ ✔ $729 = 729$ ✔

Our solution is $x = 27$ or $x = -27$. This is sometimes written as $x = \pm 27$.

Example 2

What is the solution to the equation $5x^2 = 210$? Round answers to the nearest thousandth.

Solution

$5x^2 = 210$

$x^2 = 42$ Divide both sides by 5 to isolate the square of x.

$x = \pm\sqrt{42}$ The solution is the positive or negative square root of 42.

$x \approx \pm 6.481$

CHECK $x \approx 6.481$ and $x \approx -6.481$ in our original equation.

$5x^2 = 210$ $5x^2 = 210$

$5 * 6.481^2 \stackrel{?}{=} 210$ $5 * (-6.481)^2 \stackrel{?}{=} 210$

$210.017 \approx 210$ ✔ $210.017 \approx 210$ ✔

Our solution is $x \approx \pm 6.481$.

Example 3

Energy States. Niels Bohr introduced the idea that a single electron of hydrogen could occupy only certain energy states, called stationary states. Each stationary state is associated with a color on the visible spectrum. He found that the electron energy of the hydrogen atom was inversely proportional to the square of the stationary state given by the equation

$$E = \frac{-2.1799 * 10^{-18}}{n^2}$$

where E = energy in joules per atom

n = energy state having values of 1, 2, 3, . . .

a. What is the energy of the $n = 3$ state of the hydrogen atom?

b. Determine the energy state of an excited hydrogen atom if the energy being emitted is $-5.4498 * 10^{-19}$ joules/atom.

Solution a.
$$E = \frac{-2.1799 * 10^{-18}}{n^2}$$

$$E = \frac{-2.1799 * 10^{-18}}{3^2} \quad \text{Substitute } n = 3.$$

$$E \approx {-2.4221} * 10^{-19}$$

The energy is $-2.4221 * 10^{-19}$ joules per atom.

b.
$$E = \frac{-2.1799 * 10^{-18}}{n^2}$$

$$-5.4498 * 10^{-19} = \frac{-2.1799 * 10^{-18}}{n^2} \quad \text{Substitute } E = {-5.4498} * 10^{-19}.$$

$$\boldsymbol{n^2} * -5.4498 * 10^{-19} = \frac{-2.1799 * 10^{-18}}{n^2} * \boldsymbol{n^2} \quad \text{Multiply both sides by } n^2 \text{ to eliminate the fraction.}$$

$$-5.4498 * 10^{-19} n^2 = -2.1799 * 10^{-18} \quad \text{Simplify each side of the equation.}$$

$$\frac{-5.4498 * 10^{-19} n^2}{-5.4498 * 10^{-19}} = \frac{-2.1799 * 10^{-18}}{-5.4498 * 10^{-19}} \quad \text{Divide both sides by the coefficient of } n^2 \text{ to isolate } n^2.$$

$$n^2 \approx 4.0000$$

$$n \approx \pm\sqrt{4.0000} \quad \text{The solution is the positive or negative square root of 4.0000.}$$

$$n \approx \pm 2.0000$$

The energy state is supposed to be a positive integer value, so the solution is $n = 2$. Therefore, the hydrogen atom has an energy state of $n = 2$.

Pythagorean Theorem

One property of right triangles, the Pythagorean theorem, is very important.

> *Definition*
> In a right triangle, the sides forming the right angle are called **legs,** and the side opposite the right angle is called the **hypotenuse.**

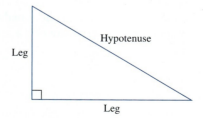

You may recall that we looked at the relationship between the lengths of the legs and the hypotenuse earlier when we explored the area of the following model.

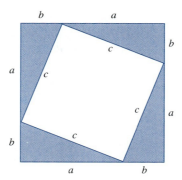

In the figure, the area of the large square can be expressed as the sum of the area of the smaller square and the areas of the four shaded triangles.

area of large square = area of small square + 4 ∗ area of triangle

$$(a + b)^2 = c^2 + 4 * (\tfrac{1}{2}ab)$$
$$(a + b)(a + b) = c^2 + 2ab$$
$$a^2 + 2ab + b^2 = c^2 + 2ab \quad \text{We can subtract } 2ab \text{ from both sides of the equation.}$$
$$a^2 + b^2 = c^2$$

In doing the proof of the Pythagorean theorem, we used the letters a and b to represent the legs of the right triangles and c to represent the hypotenuse. On another occasion, we might use a to represent the hypotenuse or use other letters altogether. Remember that the relationship between the legs and the hypotenuse is important, not which particular letters are used in any given application. It is also important to notice that this theorem applies only to right triangles.

These points are summarized in the following box.

The Pythagorean Theorem

The **Pythagorean theorem** states that in a right triangle, the sum of the squares of the lengths of the legs is equal to the square of the length of the hypotenuse, or

$$\text{leg}_1^2 + \text{leg}_2^2 = \text{hypotenuse}^2$$

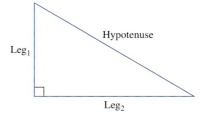

Conversely, it is also true that a triangle is a right triangle if the square of the length of the longest side is equal to the sum of the squares of the lengths of the other two sides. The right angle is opposite the longest side.

Example 4

A tree growing on level ground is to be anchored with 7.5 feet of guy wire. The wire will be attached to the tree at 5.5 feet off the ground as shown in the picture. How far away from the tree should the stake on the ground be placed? Assume no extra wire is used to tie off to the tree or the stake.

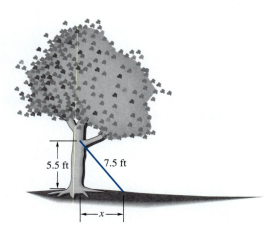

Solution Using the Pythagorean theorem, we can substitute 5.5 ft and x for the legs of the right triangle and 7.5 ft for the hypotenuse.

$$\text{leg}_1^2 + \text{leg}_2^2 = \text{hypotenuse}^2$$
$$(5.5 \text{ ft})^2 + x^2 = (7.5 \text{ ft})^2$$

Notice that when the lengths with units are substituted into the formula, we must put parentheses around the measurement. This is to indicate that both the number and the units are squared.

$$x^2 = (7.5 \text{ ft})^2 - (5.5 \text{ ft})^2$$ Subtract $(5.5 \text{ ft})^2$ from both sides to isolate x^2.

$$x = \pm\sqrt{(7.5 \text{ ft})^2 - (5.5 \text{ ft})^2}$$ The solution is the positive or negative square root of $(7.5 \text{ ft})^2 - (5.5 \text{ ft})^2$.

$$x = \pm\sqrt{26 \text{ ft}^2}$$ Subtracting ft^2 from ft^2 gives ft^2.

$$x \approx \pm 5.1 \text{ ft}$$ The square root of ft^2 is ft.

The negative answer does not make sense in this example, so the distance from the tree should be about 5.1 feet.

In this section, we looked at how to solve equations involving simple quadratics. Because the square of a positive *and* the square of a negative number both result in a positive result, we found that these equations have two solutions. However, in applications, both answers do not always make sense. Remember to find all possible solutions when solving equations, but use common sense to answer the question asked! Many applications result in simple quadratic equations. The Pythagorean theorem is an important application of simple quadratic equations.

Problem Set 6.3

1. Solve the following equations.
   a. $x^2 - 36 = 64$
   b. $4R^2 = 64$
   c. $\frac{3}{4}m^2 = 15$
   d. $T^2 = 0.81$
   e. $0.32T^2 + 0.72 = 1.36$
   f. $(y - 6)(y + 6) = 64$

2. Solve the following literal equations for the indicated variable.
   a. $J = kf^2$ for f
   b. $Ks^2 - K = 1$ for s
   c. $Ks^2 - K = 1$ for K
   d. $A = bc^2d$ for c
   e. $K = \dfrac{M^2 + 1}{P}$ for P
   f. $K = \dfrac{M^2 + 1}{P}$ for M

3. The volume of a cylinder is given by the formula
   $$V = \pi r^2 h$$
   where V = volume of the cylinder
   r = radius of the cylinder
   h = height of the cylinder

   Find the radius of a cylindrical tank that has a height of 35.0 ft and a volume of 74,330 ft³.

4. The capacity of a cylindrical tank can be determined by using the formula
   $$C = 0.0034\, D^2 L$$
   where C = capacity of the tank in gallons
   D = diameter of the tank in inches
   L = length of the tank in inches

   Suppose you want to mount a tank with a capacity of 120 gallons on a trailer that can hold a tank of no longer than $4\frac{1}{2}$ feet. Determine the diameter necessary for a tank to have the desired capacity.

5. **Kinetic Energy.** The kinetic energy (KE) of any object depends on the mass (m) and the velocity (v) of an object. The kinetic energy of an object is directly proportional to the mass of the object and the square of the velocity. This relationship is shown by the following equation.
   $$KE = \tfrac{1}{2}mv^2$$
   where KE = kinetic energy   (Note this is not K times E, KE represents a single variable.)
   m = mass of the object
   v = velocity of the object

   a. If a platform diver has a mass of 70.5 kg and is traveling at 9.8 meters per second in the middle of his dive from a platform, what is his kinetic energy?
   b. If the kinetic energy of the diver when he hit the water was 16,100 (kg m²/s²), what speed was he traveling when he hit the water?

6. Solve for the missing side of the following triangles.

a.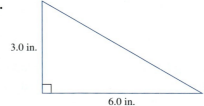

b.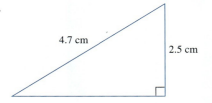

c.

d.

7. In the following right triangle, $a = 5.5$ cm and $b = 7.8$ cm. A student was asked to determine the length of side c. Below is the student's work. Is there anything wrong with the student's solution? If so, explain.

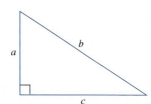

$$a^2 + b^2 = c^2$$
$$(5.5 \text{ cm})^2 + (7.8 \text{ cm})^2 = c^2$$
$$\sqrt{30.25 \text{ cm}^2 + 60.84 \text{ cm}^2} = c$$
$$9.5 \text{ cm} \approx c$$

The length of c is approximately 9.5 cm.

8. Two holes are drilled into a piece of metal. The measurements in the figure are from the centers of the holes. Determine the diagonal distance between the centers of the two holes.

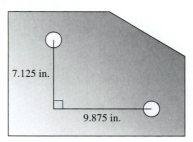

9. **The Ramp.** A handicap ramp is being built by the carpentry class. The ramp measures 7.50 meters. The horizontal distance for the ramp is 7.46 meters. What is the vertical height of the ramp? The specifications for a handicap ramp require the ratio of the vertical distance to the horizontal distance to be less than or equal to $\frac{1}{12}$ for commercial buildings. Does this ramp meet specifications? If not what adjustments would you make?

10. The hypotenuse of an isosceles right triangle is 10 inches long. Find the lengths of the two legs. Assume the measurement is accurate to two significant digits.

11. Find the length of the diagonal of the rectangular box whose dimensions are given in the following figure. The measurements are to the nearest tenth of an inch.

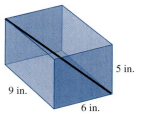

12. Determine the length of the diagonal of a cube with side length of 8.0 inches.

13. The area of a circle is 1.0 square foot. What is the radius of the circle in inches?

14. The area of a square plot of land is 5.00 hectares. Determine the length of the side of the plot in meters.

15. A bench for the corner of an outdoor deck is to be cut from a square piece of wood as shown in the following figure. Determine the length of the longest side of the bench.

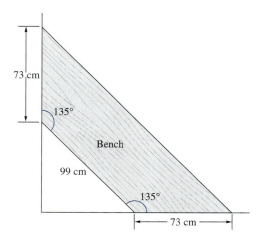

6.4 Rational Equations

We say that 12 is *evenly divisible* by 3 because 3 divides into 12 evenly, that is, $\frac{12}{3} = 4$. Similarly, $8x^2$ is evenly divisible by $4x$ because

$$\frac{8x^2}{4x} = 2x$$

1. Determine which of the following expressions are divisible by all of the factors, 3, h, and $2h$.

   a. $3h$ **b.** $3h^2$ **c.** $6h$ **d.** $2h$ **e.** 6

2. Determine which of the following expressions are divisible by all of the factors, 5, x, and $x - 5$.

   a. $5x$ **b.** $5x - 5$ **c.** $5(x - 5)$ **d.** $5x(x - 5)$ **e.** $25x^2$

3. Write an expression that is evenly divisible by $3T$, $4T$, and $T + 3$. Are there other possible expressions?

Discussion 6.4

Earlier you learned how to solve linear and simple quadratic equations. In this section you will be solving a specific type of equation called **rational equations.** Recall that rational numbers are the set of numbers that can be written as fractions in the form $\frac{a}{b}$, where a and b are both integers and $b \neq 0$. In solving rational equations, you will be solving equations that contain expressions written in the form of fractions. First let us review a strategy for solving equations in one variable.

Strategy Solving Linear Equations in One Variable

1. Simplify both sides of the equation by applying the distributive property or combining like terms. If the equation contains fractions, you may want to multiply both sides of the equation by the least common denominator to eliminate the fractions.

2. Add or subtract terms from both sides of the equation to collect all terms containing the variable on one side and all other terms on the opposite side. Again combine like terms if possible.

3. Divide both sides by the coefficient of the variable.

4. Check your solution in the original equation.

Step 1 of the strategy tells us that if an equation contains fractions, we need to multiply both sides of the equation by the **least common denominator.** A common denominator is an expression that is evenly divisible by all of the denominators; that is, the denominator of each term is a factor of the common denominator.

Example 1

Determine the least common denominator for each equation.

a. $\dfrac{m-2}{5m} - \dfrac{3}{2m} = \dfrac{5}{4}$ b. $\dfrac{2x-3}{x^2} = \dfrac{6}{5x}$ c. $\dfrac{1}{p+3} - \dfrac{1}{4p} = \dfrac{10}{p}$

Solution

a. There are three denominators in the equation $\dfrac{m-2}{5m} - \dfrac{3}{2m} = \dfrac{5}{4}$: $5m$, $2m$, and 4. The numerical factors in the denominators are 5, 2, and 4. Each of these numerical factors divides evenly into 20. Therefore, 20 is the numerical factor of the common denominator.

In addition, the denominators $5m$ and $2m$ both contain a factor of m, Therefore, m must be a factor of the common denominator. We now have two factors of our common denominator, 20 and m. Let's try $20m$ as our common denominator.

$5m$ divides evenly into $20m$.

$2m$ divides evenly into $20m$.

4 divides evenly into $20m$.

NOTE: Only one factor of m is necessary even though it occurred in two different denominators.

Because each denominator divides evenly into $20m$, $20m$ is our common denominator. How do we know it is our least common denominator?

b. The equation $\dfrac{2x-3}{x^2} = \dfrac{6}{5x}$ has two denominators: x^2 and $5x$. Because 5 is the only numerical factor in the denominators, 5 is the numerical factor of the common denominator. We need a factor of x^2 in our common denominator because it is a factor in the denominator on the left. Do we also need a factor of x? Because x goes evenly into x^2, we do not need another factor of x. Let's try $5x^2$ as our common denominator.

x^2 divides evenly into $5x^2$.

$5x$ divides evenly into $5x^2$.

Therefore, the least common denominator is $5x^2$.

c. The equation $\dfrac{1}{p+3} - \dfrac{1}{4p} = \dfrac{10}{p}$ has three denominators: $p+3$, $4p$, and p. Four is the only numerical factor in the denominators (the 3 is a term of its denominator, not a factor). Therefore, 4 is the numerical factor of the common denominator.

The denominators $4p$ and p contain the factor p, so p must be a factor in the common denominator. In addition, $p+3$ is a factor of the denominator $p+3$ and must also be included. We have three factors to include in our common denominator, 4, p, and $p+3$. Therefore, our common denominator is $4p(p+3)$.

$p+3$ divides evenly into $4p(p+3)$.

$4p$ divides evenly into $4p(p+3)$.

p divides evenly into $4p(p+3)$.

Therefore $4p(p+3)$ is the least common denominator.

Example 2

Solve the following equations.

a. $\dfrac{h+8}{h} - \dfrac{3}{2h} = \dfrac{5}{3}$ b. $\dfrac{x+8}{x-4} = 6$ c. $\dfrac{1}{M+4} - \dfrac{1}{M-4} = 5$

Solution

a. The least common denominator for h, $2h$, and 3 is $2*3*h = 6h$. The first step is to multiply both sides of the equation by the common denominator.

$$\dfrac{h+8}{h} - \dfrac{3}{2h} = \dfrac{5}{3}$$

$$\mathbf{6h} * \left(\dfrac{h+8}{h} - \dfrac{3}{2h}\right) = \left(\dfrac{5}{3}\right) * \mathbf{6h} \qquad \text{Multiply both sides of the equation by the least common denominator.}$$

The next step is to distribute the $6h$ to each of the terms on the right side of the equation. The *key to consistent success* in solving rational equations is to write out this distribution step!

$$\mathbf{6h} * \dfrac{h+8}{h} - \mathbf{6h} * \dfrac{3}{2h} = \mathbf{6h} * \dfrac{5}{3} \qquad \text{Distribute the } 6h \text{ over the difference.}$$

$$6h * \dfrac{h+8}{h} - \overset{3}{6h} * \dfrac{3}{2h} = \overset{2}{6h} * \dfrac{5}{3} \qquad \text{Cancel common factors.}$$

$$6(h+8) - 3(3) = 2h(5) \qquad \text{Simplify each side of the equation.}$$

$$6h + 48 - 9 = 10h$$

$$6h + 39 = 10h$$

$$6h + 39 - \mathbf{6h} = 10h - \mathbf{6h} \qquad \text{Collect all terms containing } h \text{ to one side of the equation.}$$

$$39 = 4h$$

$$\dfrac{39}{4} = \dfrac{4h}{4} \qquad \text{Divide both sides by 4, the coefficient of } h.$$

$$9.75 = h$$

CHECK Substitute 9.75 for h in the original equation.

$$\dfrac{h+8}{h} - \dfrac{3}{2h} = \dfrac{5}{3}$$

$$\dfrac{9.75+8}{9.75} - \dfrac{3}{2*9.75} \overset{?}{=} \dfrac{5}{3}$$

$$\dfrac{5}{3} = \dfrac{5}{3} \quad ✓$$

We conclude that $h = 9.75$.

b. The only denominator is $x - 4$; therefore, the common denominator is $x - 4$.

$$\frac{x+8}{x-4} = 6$$

$(x-4) * \dfrac{x+8}{x-4} = 6 * (x-4)$ Multiply both sides of the equation by $(x-4)$.

$\cancel{(x-4)} * \dfrac{x+8}{\cancel{x-4}} = 6 * (x-4)$ Cancel common factors.

$x + 8 = 6x - 24$ Simplify each side of the equation.

$x + 8 - x = 6x - 24 - x$ Collect all terms containing x on the right side of the equation.

$8 = 5x - 24$

$8 + 24 = 5x - 24 + 24$ Collect all other terms on the left side of the equation.

$32 = 5x$

$\dfrac{32}{5} = \dfrac{5x}{5}$ Divide by the coefficient of x.

$6.4 = x$

CHECK   Check the solution in the original equation.

$$\frac{x+8}{x-4} = 6$$

$$\frac{6.4 + 8}{6.4 - 4} \stackrel{?}{=} 6$$

$$6 = 6 \quad \checkmark$$

We conclude that $x = 6.4$.

c. The least common denominator for $M + 4$ and $M - 4$ is $(M + 4)(M - 4)$.

$$\frac{1}{M+4} - \frac{1}{M-4} = 5$$

Multiply both sides of the equation by the least common denominator.

$$(M+4)(M-4) * \left(\frac{1}{M+4} - \frac{1}{M-4}\right) = 5 * (M+4)(M-4)$$

Distribute $(M+4)(M-4)$ over the difference. Multiply $M+4$ and $M-4$ on the right-hand side.

$$(M+4)(M-4) * \frac{1}{M+4} - (M+4)(M-4) * \frac{1}{M-4} = 5 * (M^2 - 16)$$

Cancel common factors.

$$\cancel{(M+4)}(M-4) * \frac{1}{\cancel{M+4}} - (M+4)\cancel{(M-4)} * \frac{1}{\cancel{M-4}} = 5 * (M^2 - 16)$$

$$(M-4)-(M+4) = 5M^2 - 80$$

$$M + {}^-4 + {}^-M + {}^-4 = 5M^2 - 80 \qquad \text{Rewrite subtraction as adding the opposite on the left side of the equation.}$$

$$-8 = 5M^2 - 80$$

$$-8 + 80 = 5M^2 - 80 + 80$$

$$\frac{72}{5} = \frac{5M^2}{5}$$

$$14.4 = M^2$$

$$\pm\sqrt{14.4} = M \qquad \text{Recall that a simple quadratic has two solutions.}$$

$$\pm 3.795 \approx M$$

CHECK Check both of the solutions in the original equation.

$$\frac{1}{M+4} - \frac{1}{M-4} = 5$$

If $M \approx 3.795$:

$$\frac{1}{3.795 + 4} - \frac{1}{3.795 - 4} \stackrel{?}{=} 5$$

$$5.0063 \approx 5 \quad ✓$$

If $M \approx -3.795$:

$$\frac{1}{-3.795 + 4} - \frac{1}{-3.795 - 4} \stackrel{?}{=} 5$$

$$5.0063 \approx 5 \quad ✓$$

Our solution is $M \approx \pm 3.795$.

In the past, we solved simple rational equations that resulted from setting up a proportion. Remember, a **proportion** is an equation in which one ratio is equal to another ratio.

> Definition
>
> An equation is **proportional** if it equates two ratios,
>
> $\dfrac{a}{b} = \dfrac{c}{d}$, where $b \neq 0$ and $d \neq 0$.

Because proportional equations are a type of rational equation, all proportional equations can be solved using the techniques just discussed. However, there is a shortcut that works with proportional equations but not with other rational equations. You are very likely to encounter more proportional equations than other rational equations because there are many applications of proportions. Learning the shortcut will make solving these faster and easier.

Let's look at a proportion and see if we can discover the shortcut.

$$\frac{a}{b} = \frac{c}{d}$$

$$bd * \frac{a}{b} = \frac{c}{d} * bd \qquad \text{Multiply both sides by the lowest common denominator to clear the fractions.}$$

$$ad = bc \qquad \text{Cancel common factors in each term.}$$

Notice that after clearing the fractions and canceling the common factors we get a product on each side of the equation. To be more specific, the product of the numerator of the left side of the proportion and the denominator of the right side is equal to the product of the denominator of the right side and the numerator of the left side. The result is always the same when we are simplifying proportions, so we can skip the step in which we multiplied both sides by the lowest common denominator and just write the product. That is, if $\frac{a}{b} = \frac{c}{d}$, then $ad = bc$. This shortcut for solving proportional equations is called **cross multiplication** because we are multiplying the numerator of each side by the denominator of the other. If we connect these paths the lines "cross."

$$\frac{a}{b} \times \frac{c}{d}$$
$$ad = bc$$

Example 3

Solve the following equation.

$$\frac{10}{y} = \frac{6}{y-3}$$

Solution Because this equation is a proportion, we can start by cross multiplying.

$$\frac{10}{y} = \frac{6}{y-3}$$
$$10(y - 3) = 6y \quad \text{Cross multiply to solve the proportion.}$$
$$10y - 30 = 6y$$
$$10y - 30 - 10y = 6y - 10y$$
$$-30 = -4y$$
$$\frac{-30}{-4} = \frac{-4y}{-4}$$
$$7.5 = y$$

CHECK

$$\frac{10}{7.5} \stackrel{?}{=} \frac{6}{7.5 - 3}$$
$$\frac{4}{3} = \frac{4}{3} \quad ✔$$

We conclude that $y = 7.5$.

In the previous example we saw that if we are solving equations in which a single term occurs on each side, then we can use "cross multiplication" as a shortcut. However, if more than one term occurs on either side of the equation, this shortcut does not work.

Example 4

Equations similar to the following one occur in the study of chemistry. Solve the equation for T_1.

Solution

$$\frac{P_1}{P_2} = \frac{H}{R}\left(\frac{1}{T_2} - \frac{1}{T_1}\right)$$

$\dfrac{P_1}{P_2} = \dfrac{H}{R}\left(\dfrac{1}{T_2} - \dfrac{1}{T_1}\right)$	Because the right side of the equation is a product, not a simple fraction, we do not try to use "cross multiplication" to solve this equation.
$\dfrac{P_1}{P_2} = \dfrac{H}{R} * \dfrac{1}{T_2} - \dfrac{H}{R} * \dfrac{1}{T_1}$	Distribute $\dfrac{H}{R}$ over the difference on the right.
$\dfrac{P_1}{P_2} = \dfrac{H}{RT_2} - \dfrac{H}{RT_1}$	The least common denominator for P_2, RT_2, and RT_1 is $P_2RT_2T_1$.
$P_2RT_2T_1 * \left(\dfrac{P_1}{P_2}\right) = \left(\dfrac{H}{RT_2} - \dfrac{H}{RT_1}\right) * P_2RT_2T_1$	Multiply both sides of the equation by the least common denominator.
$P_2RT_2T_1 * \dfrac{P_1}{P_2} = P_2RT_2T_1 * \dfrac{H}{RT_2} - P_2RT_2T_1 * \dfrac{H}{RT_1}$	Distribute $P_2RT_2T_1$ over the difference.
$RT_2T_1P_1 = P_2T_1H - P_2T_2H$	Identify the terms containing T_1.
$RT_2T_1P_1 - P_2T_1H = P_2T_1H - P_2T_2H - P_2T_1H$	Collect terms containing T_1 to one side of the equation.
$RT_2T_1P_1 - P_2T_1H = -P_2T_2H$	
$T_1(RT_2P_1 - P_2H) = -P_2T_2H$	Factor out T_1.
$\dfrac{T_1(RT_2P_1 - P_2H)}{(RT_2P_1 - P_2H)} = \dfrac{-P_2T_2H}{(RT_2P_1 - P_2H)}$	Divide by the coefficient of T_1.
$T_1 = \dfrac{-P_2T_2H}{(RT_2P_1 - P_2H)}$	

CHECK To check our solution, we will choose $P_1 = 10$, $P_2 = 20$, $T_2 = 8$, $R = 5$, and $H = 4$ and find the value of T_1.

$$T_1 = \frac{-20 * 8 * 4}{(5 * 8 * 10 - 20 * 4)}$$

$$T_1 = -2$$

Now substitute these values in the original equation.

$$\frac{P_1}{P_2} = \frac{H}{R}\left(\frac{1}{T_2} - \frac{1}{T_1}\right)$$

$$\frac{10}{20} \stackrel{?}{=} \frac{4}{5}\left(\frac{1}{8} - \frac{1}{-2}\right)$$

$$0.5 = 0.5 \quad \checkmark$$

We conclude that

$$T_1 = \frac{-P_2T_2H}{(RT_2P_1 - P_2H)}$$

Example 5

Solve the following equation and check your solution.

$$4 - \frac{2}{x-3} = \frac{x-5}{x-3}$$

Solution

Because our only denominator is $x - 3$ it is also our common denominator.

$$4 - \frac{2}{x-3} = \frac{x-5}{x-3}$$

$$(x-3)\left(4 - \frac{2}{x-3}\right) = \left(\frac{x-5}{x-3}\right)(x-3) \quad \text{Multiply both sides by the common denominator.}$$

$$(x-3) * 4 - (x-3) * \frac{2}{(x-3)} = \frac{x-5}{(x-3)} * (x-3) \quad \text{Distribute the } (x-3) \text{ over the difference.}$$

$$(x-3) * 4 - \cancel{(x-3)} * \frac{2}{\cancel{(x-3)}} = \frac{x-5}{\cancel{(x-3)}} * \cancel{(x-3)} \quad \text{Cancel common factors.}$$

$$4x - 12 - 2 = x - 5$$
$$4x - 14 = x - 5$$
$$4x - 14 - x = x - 5 - x$$
$$3x - 14 = {}^-5$$
$$3x - 14 + 14 = {}^-5 + 14$$
$$3x = 9$$
$$x = 3$$

It appears our solution is $x = 3$. Let's substitute $x = 3$ into the original equation to check.

$$4 - \frac{2}{3-3} = \frac{3-5}{3-3}$$

$$4 - \frac{2}{0} = \frac{-2}{0}$$

Because division by zero is undefined, $x = 3$ is not a solution. In this example we followed a correct mathematical process; however, by checking our result we can see that our result is not a solution to the original equation. When this happens, we say there is *no solution* to the equation.

In this section, we reviewed solving equations, applying strategies we learned previously to solving rational equations. In the preceding example, we saw that even if we follow those strategies correctly and find a result, that result may not work when substituted back into the original equation. If your result does not make the original equation true, it is not a solution. Therefore the preceding example had no solution. Results that do not work in the original equation are called extraneous solutions. Because solving a rational equation can lead to answers that do not solve the original equation, it is *always necessary* to check our answers to rational equations.

Problem Set 6.4

1. For each rational equation, determine the least common denominator.

   a. $\dfrac{10}{x-4} = \dfrac{5}{6}$

   b. $\dfrac{12}{t} + \dfrac{3}{2} = 4$

   c. $\dfrac{1}{2m} = \dfrac{1}{5m} - 1$

   d. $\dfrac{4k-1}{k} - \dfrac{5}{10k} = 2$

   e. $\dfrac{5}{3R} - \dfrac{2}{R^2} = \dfrac{7-R}{R^2}$

   f. $\dfrac{1}{p} + \dfrac{2}{p+4} = \dfrac{5}{p}$

   g. $\dfrac{10}{2x-1} - \dfrac{1}{3} = 8$

2. Solve each equation. Check your solutions.

   a. $\dfrac{10}{x-4} = \dfrac{5}{6}$

   b. $\dfrac{12}{t} + \dfrac{3}{2} = 4$

   c. $\dfrac{1}{2m} = \dfrac{1}{5m} - 1$

   d. $\dfrac{4k-1}{k} - \dfrac{5}{10k} = 2$

   e. $\dfrac{5}{3R} - \dfrac{2}{R^2} = \dfrac{7-R}{R^2}$

   f. $\dfrac{1}{p} + \dfrac{2}{p+4} = \dfrac{5}{p}$

   g. $\dfrac{10}{2x-1} - \dfrac{1}{3} = 8$

3. Solve each equation. Check your solutions.

   a. $\dfrac{3}{t} = \dfrac{15}{8}$

   b. $\dfrac{5}{X-8} = 4$

   c. $\dfrac{z-7}{z+4} = 3$

   d. $\dfrac{m-2}{5m} - \dfrac{3}{2m} = \dfrac{5}{4}$

   e. $\dfrac{2x-3}{x^2} = \dfrac{6}{5x}$

   f. $\dfrac{1}{p+3} - \dfrac{1}{4p} = \dfrac{10}{p}$

   g. $\dfrac{5}{3} - \dfrac{3}{2x} = \dfrac{1}{6}$

   h. $\dfrac{2}{w} = \dfrac{1}{w+5}$

   i. $\dfrac{1}{M+2} + \dfrac{1}{M} = 1$

   j. $\dfrac{4}{b} - \dfrac{20}{b-2} = 8$

4. What is wrong with the following solution?

   $$\dfrac{4}{5h} - 7 = \dfrac{5}{h}$$

   $$5h\left(\dfrac{4}{5h} - 7\right) = \left(\dfrac{5}{h}\right)5h$$

   $$\cancel{5h}\left(\dfrac{4}{\cancel{5h}} - 7\right) = \left(\dfrac{5}{\cancel{h}}\right)5\cancel{h}$$

   $$4 - 7 = 25$$

   $$-3 \neq 25$$

   Therefore, there is no solution to this equation.

5. **Red Spruce.** In forestry management, the *top volume* of a tree is not useful, so it cannot be sold. Top volume is related to both diameter and height of a tree. The following formula can be used to determine the top volume of a red spruce.

$$V_T = 0.19 + 0.025 \frac{H}{D - 4}$$

where V_T = the top volume in cubic feet

H = height of tree in feet

D = breast height diameter in inches

a. Use the formula to complete the first three rows of the following table.

b. Solve the formula for D. Use this formula to complete the last three rows of the table.

V_T (ft³)	H (ft)	D (in.)
?	35	14
?	35	12
?	30	10
0.378	30	?
0.503	25	?
0.815	25	?

c. Use the partial table that follows along with the preceding completed table to answer the following questions. What is the percent of top volume waste compared with the total volume for a 35-foot-tall tree with 14-inch diameter at breast height? for a 25-foot-tall tree with a 5-inch diameter at breast height?

TOTAL VOLUME OF RED SPRUCE IN CUBIC FEET			
Diameter at Breast Height (in.)	Height (ft)		
	25	30	35
4	1.5	1.7	1.9
5	2.0	2.3	2.6
6	2.7	3.1	3.5
7	3.5	4.1	4.7
8		5.1	5.9
9		6.4	7.3
10		7.7	9.0
11			10.7
12			12.7
13			14.8
14			17.1

6. Solve the following equations for the given variable.

a. $A_v = \dfrac{R_L}{R_E}$ for R_E

b. $X = \dfrac{1}{2\pi f C}$ for f

c. $\dfrac{V_1}{V_2} = \dfrac{N_1}{N_2}$ for N_1

d. $R = \dfrac{AB}{A + B + C}$ for A

e. $\dfrac{Q}{f_1 - f_2} = R$ for f_2

f. $\dfrac{1}{R_{eq}} = \dfrac{1}{R_1} + \dfrac{1}{R_2}$ for R_1

g. $L = H\left(\dfrac{1}{P_a} - \dfrac{1}{P_b}\right)$ for P_b

h. $\dfrac{1}{2}v^2 - \dfrac{GM}{r} = \dfrac{-GM}{2a}$ for M

i. $\dfrac{1}{U} = \dfrac{x}{k} + \dfrac{1}{h}$ for x

j. $\dfrac{1}{M} = \dfrac{1}{M_2} + \dfrac{1}{M_1 + M_3}$ for M_1

7. The Ryberg Equation. Chemists have discovered that wavelengths, L in meters, in the visible spectrum of hydrogen atoms can be predicted from the equation

$$\dfrac{1}{L} = R\left(\dfrac{1}{4} - \dfrac{1}{n^2}\right) \quad \text{for } n > 2$$

This is called the Ryberg equation. R is called the Ryberg constant and has a value of about $1.097 * 10^7$. If $n = 3$, the wavelength of the red line of the hydrogen spectrum is obtained. If $n = 4$, the wavelength of the green line is obtained and if $n = 5$, we get the wavelength of the blue line. Solve the equation for L. Use your new equation to find the wavelengths for the red, green, and blue lines of the hydrogen spectrum.

Chapter Six Summary

In this chapter, we learned several new algebraic techniques for manipulating expressions and solving equations.

We began by extending the process of multiplying expressions and then looked at division of an expression by a single term. To **multiply multiterm expressions** we multiply each term in the first factor by each term in the second factor. In the special case of multiplying a two-term expression by another two-term expression, we use the mnemonic **FOIL**. FOIL stands for **first, outer, inner,** and **last**.

$$(a + b)(c + d) = ac + ad + bc + bd$$

with First, Last indicated on the top and Inner, Outer indicated on the bottom.

To divide an expression by a single term, we use a version of the distributive property, which states that each term of the numerator is divided by the denominator.

> ### Distributing Division over Addition
> Stated more generally, the quotient of the sum of a and b and c is equal to the sum of a divided by c and b divided by c. That is,
> $$\frac{a+b}{c} = \frac{a}{c} + \frac{b}{c}$$

Up to this chapter, we had only solved equations that could be rewritten in a linear form. In Section 6.3, we learned how to solve **simple quadratic equations** of the form $ax^2 + b = c$. To solve a simple quadratic equation, we first isolate x^2. Then we know the solution is plus or minus the square root of the other side. In many applications only one of the solutions is reasonable. However, in the future we will encounter situations in which both solutions are important, so we need to be sure to get in the habit of finding *both* solutions.

A specific application of simple quadratic equations is the **Pythagorean theorem,** which states that, in a right triangle, the sum of the squares of the lengths of the legs of the triangle is equal to the square of the length of the hypotenuse. We will encounter this extremely important theorem frequently throughout this course.

We ended this chapter by solving **rational equations.** The first step in solving a rational equation is to eliminate all fractions by multiplying both sides of the equation by a common denominator. The next step is to simplify both sides of the equation. This often involves distributing the common factor on one or both sides. The *key to consistent* success in solving rational equations is to write out this distribution step before you cancel common factors.

A special kind of rational equation is called a proportion. A proportion looks like a rational equation in which each side consists of a single term: $\frac{a}{b} = \frac{c}{d}$. To solve a proportion we can multiply each side by a common denominator. In Section 6.4, we learned that a shortcut for this process is called **cross multiplication.** To cross multiply, we multiply the denominator of one side by the numerator of the other.

$$\frac{a}{b} \times \frac{c}{d}$$
$$ad = bc$$

Chapter Review 4–6

1. For each of the numerical expressions do the following.
   i. Determine the number of terms in each expression.
   ii. Underline the terms.
   iii. Evaluate the expression showing all steps.
   iv. Evaluate the expression *in one step* on your calculator. Compare your results with what you got in part iii. If your results are different, find where you made an error.

   a. $\dfrac{-42}{6} * {-8} + 2$

   b. $\dfrac{12 + 10 - 8 + 1}{3 * 5}$

   c. $-52 + (12 - 8)^2 \div 8$

   d. $6 + 4^3$

   e. $12 + 2 * 5^2$

   f. $\sqrt{3^2 + 4^2}$

   g. $-10\sqrt{25} + 24$

   h. $-9^2 - 10(5 * 3)$

   i. $0.25(18 - 5 + 3) + 0.10(6 * 5)$

   j. $\dfrac{(-3)^4 + 9}{-2^2 + 1}$

2. Solve each equation. Check your solutions.

   a. $6 - 3(x - 5) = 4x$

   b. $m + 3(m - 3) = \dfrac{m}{2}$

   c. $(k - 1)(k + 2) = k + 14$

   d. $\dfrac{p}{p + 2} = \dfrac{p + 3}{p}$

   e. $3y + 5y(y - 3) = 5y^2 + 12$

   f. $3.2x + 0.2(x + 7) = 3.1$

   g. $(n + 2)^2 = n(n - 5) + 22$

3. Determine the slope of each of the following lines. In this problem, assume that dots are on integer coordinates.

   a.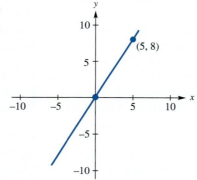

   b.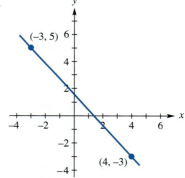

   c.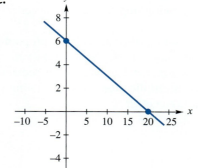

   d.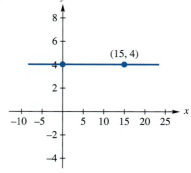

4. a. Identify the numerical coefficient of the term $-28pm^2t$.
   b. Identify the coefficient of t in the term $-28pm^2t$.
   c. Identify the coefficient of m^2 in the term $-28pm^2t$.

5. a. Identify the numerical coefficient of the term $\frac{3xy}{4}$.
   b. Identify the numerical coefficient of the term $\frac{-5m}{8}$.
   c. Identify the numerical coefficient of the term $\frac{k}{9}$.

6. The following equations are in slope–intercept form. Identify the slope and the vertical intercept.
   a. $y = 7 - 2x$ b. $y = 3x$ c. $y = \frac{x}{2}$ d. $y = -x$

7. Rewrite the equation $y = 3(x + 3) - 5(x - 1)$ in slope–intercept form. Determine the slope and vertical intercept of the line. Graph the line.

8. For each of the following expressions, determine the numerical value and units of the result.
   a. $\frac{3.5 \text{ cm}}{2}(24.2 \text{ cm})^2 - \pi(1.4 \text{ cm})^3$
   b. $42 \text{ in.}(280 \text{ in.}^2) * \frac{1 \text{ ft}}{12 \text{ in.}} * \frac{1 \text{ ft}}{12 \text{ in.}} * \frac{1 \text{ ft}}{12 \text{ in.}} * \frac{1 \text{ gal}}{7.481 \text{ ft}^3}$
   c. $\frac{52 \text{ mL/in.}^2}{2.0 \text{ in.}}$

9. a. A U.S. penny has a mass of 2.65 grams. What is its weight in ounces?
   b. How many cubic inches are in an engine that is rated at 4.0 liters?
   c. The density of air is $1.12 * 10^{-3}$ g/cm^3; convert this to ounces per cubic foot.
   d. Satellites that circle the earth in low orbits travel about 457,000 meters per minute. How fast is this in miles per hour?

10. A rectangular tank has dimensions 2.5 feet by 3 feet by 8 inches. How many liters can this tank hold?

11. Explain the difference between -5^2 and $(-5)^2$.

12. Evaluate each expression if $x = 3$. Show your substitution.
    a. x^2 b. x^3 c. $-x^4$ d. $(-x)^4$

13. Evaluate each expression if $x = -3$. Show your substitution.
    a. x^2 b. x^3 c. $-x^4$ d. $(-x)^4$

14. Solve the following equations. Round results to the nearest hundredth.
    a. $M^2 = 196$
    b. $3x^2 + 25 = 217$
    c. $y^2 - 10 = 0$
    d. $2\pi r^2 = 5.0$
    e. $10 = -4.9t^2 + 50$

15. **Sit-Ups.** Chris is signed up for a fitness PE class. On the first day of class all students are tested to determine their overall fitness. The instructor then creates a series of graphs for each student, which shows their target progress over the ten-week term. The following graph shows Chris's target for improvement in stomach strength measured by the number of sit-ups performed.

   a. How many sit-ups could Chris do at the beginning of the term?
   b. How many sit-ups should Chris be able to do at the end of ten weeks?
   c. What is the slope of the line in the graph? What are the units? Interpret the meaning of the slope in the context of this problem.

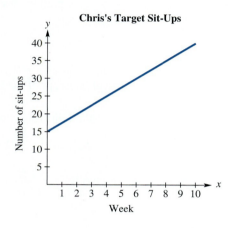

Chris's Target Sit-Ups

16. Solve the equation $\frac{v^2}{C} = R$ for v.

17. Write equations for the lines in the following graphs. Check that your equations are correct.

a.
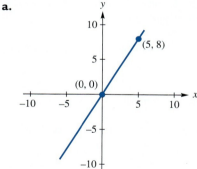

b.
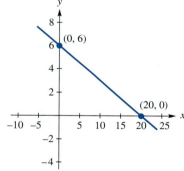

c.
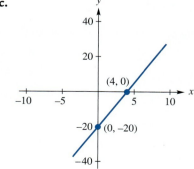

d.
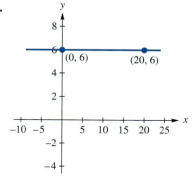

18. Without graphing, identify all of the following equations that graph as straight lines. Explain how you decided.

- **a.** $y = x$
- **b.** $y = \dfrac{1}{x}$
- **c.** $y = 4x - 5$
- **d.** $y = x^2$
- **e.** $y = 4(3 - 2x)$
- **f.** $y = \dfrac{x}{2}$
- **g.** $y = 2\sqrt{x} + 10$
- **h.** $y = 4.8 - 2.2x$
- **i.** $7x - 2y = 60$
- **j.** $y = 25$
- **k.** $y = -\dfrac{3}{8}x + 22$
- **l.** $xy = 4$

19. The New Patio. Gina and John are planning to prepare their backyard for a new patio. They received three bids for delivering sand to their home:

- Rivers Sand and Gravel is the most expensive per ton for sand, but does not charge a delivery fee.
- Better Building Supplies charges less per ton than Rivers Sand and Gravel, but more than Local Builders Emporium.
- Local Builders Emporium has the highest delivery cost of the three companies.

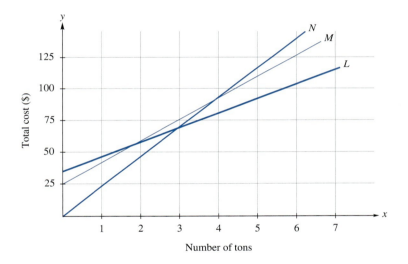

a. Match each of the lines L, M, and N with the company whose bid the line represents. Explain how you made your decision.

b. Approximately what is the delivery charge for Better Building Supplies? Explain how you made your decision.

c. Approximately what is the cost per ton for Local Builders Emporium? Explain how you made your decision.

d. Interpret the slope and vertical intercept for line L in the context of this problem. Include numerical values and units for each.

e. If Gina and John estimate that the patio will require 2.5 tons of sand, which company should they choose?

20. Suppose you are asked to graph each of the following equations. Without actually graphing them, rank them in order from easiest to hardest. (Label the easiest #1, second easiest #2, and so forth.)

   a. $y = -2x + 10$ _____

   b. $y = 3x^2 + 1$ _____

   c. $y = 5.25x - 82.50$ _____

   d. $y = 14$ _____

21. Graph each of the following equations.

   a. $y = -2x + 7$

   b. $y = \dfrac{x}{5} - 2$

   c. $y = 3x^2 - 10$

   d. $y = -x$

22. For what values of x does $\sqrt{-x}$ produce real values?

23. Evaluate the expression $a^4 - b^3$, if $a = -5$ and $b = 10$. Show your substitution step.

24. Evaluate the expression $3x^2 - y^4$ for $x = -5$ and $y = 2$. Show your substitution step.

25. Simplify the following expressions. Verify your results.

   a. $(20 - x)(30 + 5x)$

   b. $w(2w - 1) + 5w + 2w * 4w$

   c. $3(x)(2x) + (3x + 2)(x - 5)$

   d. $p^2 - (p + 2)(p - 2)$

   e. $(3p - 2)^2$

   f. $7x * 6x - \dfrac{1}{2} * 6x(x - 3)$

26. Solve each equation for the given variable. Verify your results.

   a. $L = \pi D + 2C$ for D

   b. $V = 0.7854\, LB^2$ for L

   c. $V = 0.7854\, LB^2$ for B

   d. $(m - a)(m + b) = a + b$ for a

   e. $V = 0.19 + 0.025 \dfrac{H}{D - 4}$ for H

   f. $f = \dfrac{Fc}{c - v}$ for F

   g. $f = \dfrac{Fc}{c - v}$ for c

27. Find the lengths of missing sides in each of the following triangles.

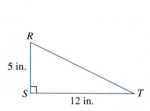
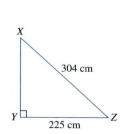
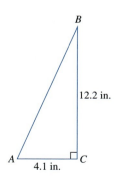

28. **Always True?** Determine if each statement is true for all values of the variables. If the statement is not true for all values, rewrite the right-hand side to make it true, if possible.
   a. $(AB)^2 = A^2B^2$
   b. $3AB^2 = (3AB)^2$
   c. $(A+B)^4 = A^4 + B^4$
   d. $\left(\dfrac{A}{B}\right)^3 = \dfrac{A^3}{B^3}$
   e. $A^7 + A^2 * A^5 = 2A^7$
   f. $(2A^2B)^3 = 6A^6B^3$
   g. $5 * 5^2 = 25^2$
   h. $B^6 + 9B^6 = 10B^{12}$

29. Simplify each of the following expressions. Assume that all variables are nonzero real numbers. Verify that your result is correct by numerical substitution. Record your results using only positive exponents.
   a. $b^5 * b^3$
   b. $\dfrac{x^8}{x^3}$
   c. $\dfrac{A^{13}}{A^{20}}$
   d. $(x^5)^4$
   e. $(5q^4)^3$
   f. $\dfrac{15m}{3m^3}$
   g. $m^3mm^2 - 5(m^2)^3$
   h. $3y^4(y^4 + 2y^2) + (y^4)^2$

30. a. Graph the equation $2x + 8y = {-}16$. What is the slope of this line?
    b. Graph the equation $9x - 4y = 6$. What is the slope of this line?
    c. Graph the equation ${-}5x + 2y = 10$. What is the slope of this line?

31. Multiply to write each product as a sum or difference. Verify your results numerically.
    a. $(4x - 2y)(x + 7y)$ b. $(3A - B)^2$

32. Solve the formula $P = \pi d + 2L + W + 2\pi d$ for d.

33. **Mile Run.** In 1942, the men's world record in the mile run was 4:06 (this means 4 minutes and 6 seconds). An exercise promoter claims that according to records since that year the times have been falling at the constant rate of 0.4 seconds per year. Assuming the promoter's claim is true, write an equation for the men's world record time in the mile run in terms of the number of years *after 1942*. Use your equation to determine the world record time for the mile run in 1997.

34. Divide to write the following quotients as a sum or difference.
    a. $\dfrac{3m^4 - 2m^2 + m}{m}$
    b. $\dfrac{3x^2y - 12x}{xy}$
    c. $\dfrac{5AB^3 + (3B^2)^3}{4B^3}$

35. Solve and check the following equations.
    a. $\dfrac{2}{x-3} = 4$
    b. $\dfrac{7}{m} + \dfrac{3}{2} = 5$
    c. $\dfrac{1}{2w} + \dfrac{1}{5w} = 10$
    d. $\dfrac{3}{A+2} = \dfrac{6}{A}$

36. Solve and check the following equations.

   a. $\dfrac{3x+2}{6x} - 3 = \dfrac{7}{3x}$

   b. $\dfrac{3}{x} + \dfrac{2}{5} = \dfrac{4}{2x}$

   c. $\dfrac{1}{y} + 4 = \dfrac{15}{4-y}$

   d. $\dfrac{2}{x+1} - \dfrac{5}{x} = \dfrac{7}{x+1}$

37. Solve the equation

   $$\dfrac{1}{R_1} + \dfrac{1}{2R_2} = \dfrac{1}{L} \quad \text{for } R_1$$

38. **Piano Music.** J. P. Dickson plays the piano at an outdoor cafe at Saturday Market. Near the piano is a display where he sells tape cassettes for $7.95 and CDs for $9.95 each.

   a. Assuming J. P. sold a total of 24 recordings last Saturday, complete a table similar to following.

Number of Cassettes Sold at $7.95	Number of CDs Sold at $9.95	Total Sales from the Recordings
0	24	
2	22	
4		
6		
10		
20		
•		
•		
•		
n		

   b. Graph the *total sales* from the recordings *in terms of the number of cassettes sold* by plotting points from your table.

   c. Let n represent the number of cassettes. Write an equation for the number of CDs sold in terms of n.

   d. Write an equation for the total sales from the recordings in terms of n.

   e. Last Saturday, J. P. had total sales of $216.80. Use your equation from part d to algebraically solve for the number of cassettes sold. How many CDs were sold? Explain why your answer is reasonable.

Chapter Seven

More Linear Equations

Previously, we studied linear equations in one variable and in two variables. In Chapter 3, we learned how to solve linear equations in one variable algebraically. In Chapter 4, we looked at the relationship between solutions to linear equations in two variables and their graphs. In this chapter, we will be extending both of these ideas. Then, we will move on to systems of linear equations and linear inequalities.

7.1 Solving Linear Equations Graphically

Activity Set 7.1

1. Use the following graph to answer parts a–c.

   a. Identify two different solutions to the equation

   $$y = 2.5x + 3.5$$

   b. Identify two different solutions to the equation

   $$y = \frac{-2x}{7} + \frac{5}{7}$$

   c. Identify a solution to the equation

   $$2.5x + 3.5 = \frac{-2x}{7} + \frac{5}{7}$$

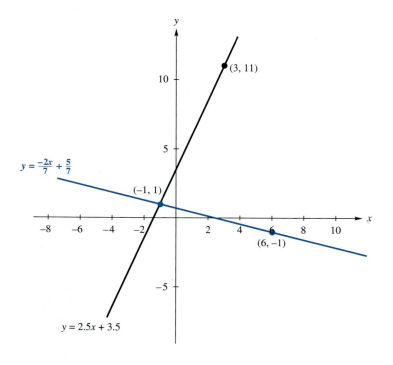

Discussion 7.1

Solving Equations Graphically

Previously we solved linear equations in one variable algebraically. In this section we will learn how to solve these same equations graphically.

Recall that the graph of any equation consists of all the points that are solutions to that equation. We used this fact in the activities to identify solutions to linear equations in two variables. For example, because the point (3, 11) is on the graph of $y = 2.5x + 3.5$, we know that $x = 3, y = 11$ is a solution to this equation.

Example 1

Solve the following equation graphically. Check your solution.

$$2x - 8 = \frac{1}{3}x + \frac{1}{3}$$

Solution To solve an equation graphically, we need to graph each side of the equation. To do this, we set each side of the equation equal to y and graph the two resulting equations in two variables. In this case, we graph

$$y = 2x - 8$$

and

$$y = \frac{1}{3}x + \frac{1}{3}$$

Both equations are in slope–intercept form. The graph of the first equation, $y = 2x - 8$, has a slope of 2 and crosses the vertical axis at -8. The graph of the second equation, $y = \frac{1}{3}x + \frac{1}{3}$, has a slope of $\frac{1}{3}$ and crosses the vertical axis at $\frac{1}{3}$. With this information, we know that both graphs fit in a standard viewing window. We let both the horizontal and vertical axes run from -10 to 10.

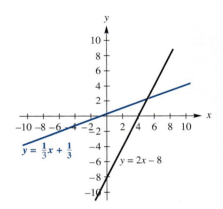

Because we are trying to solve the equation $2x - 8 = \frac{1}{3}x + \frac{1}{3}$, we want to find the value of x, where $2x - 8$ is equal to $\frac{1}{3}x + \frac{1}{3}$. From the graph, the intersection point appears to be (5, 2). Because we want the value of x, we only want the first coordinate of the ordered pair. Therefore, it appears that our solution is $x = 5$. Let's check this value in the original equation.

CHECK

$$2x - 8 = \frac{1}{3}x + \frac{1}{3}$$
$$2 * 5 - 8 \stackrel{?}{=} \frac{1}{3} * 5 + \frac{1}{3}$$
$$2 = 2 \quad ✔$$

Our solution checks, so we know that $x = 5$ is the solution to the equation $2x - 8 = \frac{1}{3}x + \frac{1}{3}$.

Did you notice that in checking our solution the resulting value on each side of the equation was 2, the same as the y-coordinate of the intersection point? Can you explain this?

Example 2

Solve the following equation graphically. Check your solution.

$$\frac{4w - 1}{7} + \frac{3w}{2} = 50$$

Solution To solve this equation graphically, we set each side of the equation equal to y and graph the resulting equations in two variables,

$$y = \frac{4w - 1}{7} + \frac{3w}{2}$$

and

$$y = 50$$

Because the second equation, $y = 50$, is a horizontal line through 50, these two equations do not fit in the standard viewing window. The vertical axis needs to reach to at least 50. In the first equation

$$y = \frac{4w - 1}{7} + \frac{3w}{2}$$

the coefficient of w is positive, so the graph has a positive slope. Let's try a window in which the horizontal axis runs from -10 to 30, and the vertical axis runs from -10 to 60.

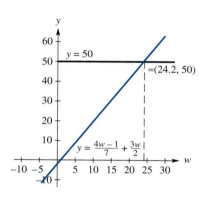

The intersection point appears to be the point $\approx (24.2, 50)$. Again, we solve the equation

$$\frac{4w - 1}{7} + \frac{3w}{2} = 50 \quad \text{for } w,$$

so we want only the first coordinate of the ordered pair, $w \approx 24.2$.

CHECK

$$\frac{4w - 1}{7} + \frac{3w}{2} = 50$$

$$\frac{4 * 24.2 - 1}{7} + \frac{3 * 24.2}{2} \stackrel{?}{=} 50$$

$$49.986 \approx 50 \quad \checkmark$$

Because our solution checks approximately, this tells us that $w \approx 24.2$ is an approximate solution to the equation.

You have now learned to solve linear equations in one variable algebraically, and graphically using a two-dimensional graph. The next step is for you to decide when to use which method. Will you solve $25 - 5x = 100$ graphically or algebraically? Will you solve $\frac{m}{3} + m + 6 = \frac{-3}{5}$ graphically or algebraically?

Problem Set 7.1

1. Identify the solution to each equation by inspection. Check your solution.
   a. $x + 2 = 15$ b. $p - 5 = 20$ c. $10 + w = 9$

2. List three solutions to each equation.
   a. $y = x + 8$ b. $y = 20 - 2x$ c. $y = x^2 + 1$ d. $y = \sqrt{x}$

3. Solve each equation graphically. Check your solution.
   a. $-0.2x + 6.4 = \dfrac{x}{2} - 2$

   b. $25 - 5k = 100$

   c. $\dfrac{m}{3} + m + 6 = \dfrac{-3}{5}$

   d. $\dfrac{4x + 60}{3} = 1 - \dfrac{31x}{5}$

4. Suppose you were asked to solve the equations from Problem 3 algebraically. Without actually solving them, rank them in order from easiest to hardest. (Label the easiest #1, second easiest #2, and so on.) Based on your experience solving these equations graphically and your ranking of solving them algebraically, compare the difficulty level of solving each equation graphically and solving the same equation algebraically.

   a. $-0.2x + 6.4 = \dfrac{x}{2} - 2$ _____

   b. $25 - 5k = 100$ _____

   c. $\dfrac{m}{3} + m + 6 = \dfrac{-3}{5}$ _____

   d. $\dfrac{4x + 60}{3} = 1 - \dfrac{31x}{5}$ _____

5. Solve each equation either algebraically or graphically. Use your experience from Problems 3 and 4 to make your choice. Check your solutions. Round approximate solutions to the nearest hundredth.
   a. $0.75p + 0.25(800 - p) = 1000$

   b. $\dfrac{10 - 7x}{8} + \dfrac{x}{2} = \dfrac{3x}{5}$

   c. $30 - 5(x - 2) = 15$

   d. $10.2 - 1.7(2.2x - 6.3) = 16.4$

6. *Use the following graph* to solve the equation $0.25(x - 2) + 2.5 = 4(3 - x) + 3x - 15$. Do not graph this on your own.

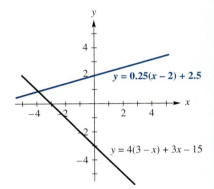

7. *Use the following graph* to solve the equation $14 - 2(4 - 0.65x) = 32$. Do not graph this on your own.

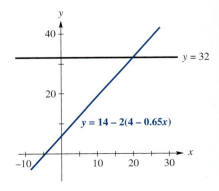

7.2 Writing Equations of Lines

Activity Set 7.2

1. Use the following graph to answer parts a–d.
   a. Determine the slope of the line.
   b. Estimate the vertical intercept.
   c. Use the results of parts a and b to write an equation of the line.
   d. Check both of the known points in the equation you wrote in part c. Did your equation check exactly?

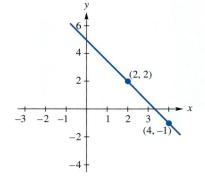

2. Use the following graph to answer parts a–d.
   a. Determine the slope of the line.
   b. Estimate the vertical intercept.
   c. Use the results of parts a and b to write an equation of the line.
   d. Check both of the known points in the equation you wrote in part c. Did your equation check exactly? Explain why or why not.

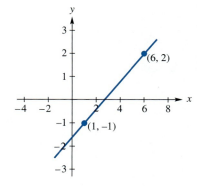

3. Two graphs are shown here. On each graph two lines are drawn.
   a. Determine the slopes of lines L and M. What relationship do you observe between the lines and the slopes of the lines?
   b. Determine the slopes of lines P and Q. What relationship do you observe between the lines and the slopes of the lines?

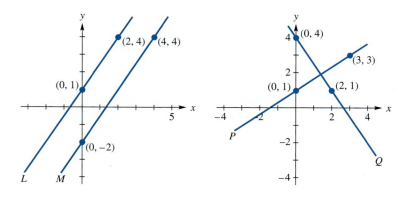

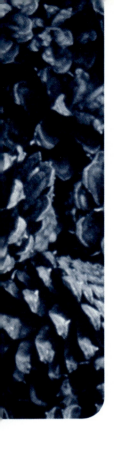

4. Recall that two lines are parallel if and only if the lines never intersect, no matter how far they extend. Two lines are perpendicular if and only if they form a right angle.

   For each of the following parts carefully *hand draw* the graphs of both lines on the same set of axes, using the same scale for both the vertical and horizontal axes.

   What can you say about the equations of parallel lines? What can you say about the equations of perpendicular lines?

   a. $\begin{cases} y = 2x - 1 \\ y = 2x + 3 \end{cases}$

   b. $\begin{cases} y = \frac{2}{3}x - 2 \\ y = \frac{3}{2}x - 4 \end{cases}$

   c. $\begin{cases} y = 2x + 1 \\ y = -\frac{1}{2}x - 3 \end{cases}$

   d. $\begin{cases} y = \frac{2}{3}x - 2 \\ y = -\frac{3}{2}x + 4 \end{cases}$

5. Given the equations of two lines, can you tell if the lines are parallel, perpendicular, or neither? If so, explain how you made your decision.

6. Redo parts c and d in Activity 4 using your graphing calculator. Set the calculator to its standard graphing window. What do you observe? Explain why these lines no longer appear perpendicular.

Discussion 7.2

Writing Linear Equations in Two Variables

In Section 4.2, we found that any equation of the form $y = mx + b$ graphs as a line. Furthermore, we learned that m, the coefficient of x, represents the slope of that line and that b, the constant term, is the vertical coordinate of the vertical intercept. For this reason we called the equation $y = mx + b$ the slope–intercept equation and wrote the equation as

$$y = slope * x + vertical\ intercept$$

With this information, we were able to write the equation of a line provided we were given the vertical intercept and enough information to determine the slope. In this section, we will see that it is not necessary to be given the vertical intercept to write the equation.

In Activity 2, you were given the graph of a line with two points on the line, $(1, {}^-1)$ and $(6, 2)$. Were you able to write the exact equation of the line? Most likely, you were not. Your equation was probably something like $y = \frac{3}{5}x - 1.5$. When we check this equation with our two points we find that it is close but not exact.

   CHECK   with $(1, {}^-1)$.   CHECK   with $(6, 2)$.

   $-1 \overset{?}{=} \frac{3}{5} * 1 - 1.5$ $2 \overset{?}{=} \frac{3}{5} * 6 - 1.5$

   $-1 \approx {}^-0.9$ $2 \approx 2.1$

Which part of our equation $y = \frac{3}{5}x - 1.5$ do we know and which part did we estimate? We found the slope from the two points given, so we know it is exact. However, the vertical intercept was an estimate from the graph. We cannot be sure that the graph crosses the vertical axis at exactly -1.5. Therefore, we know that our equation is $y = \frac{3}{5}x + b$, and if we can find the exact value of b, then we will have an exact equation.

Next, for our equation $y = \frac{3}{5}x + b$ to be correct, it must check exactly with each of the points on the graph. Therefore, if we substitute either point $(1, -1)$ or $(6, 2)$, our equation must check. Let's substitute the point $(1, -1)$ into our equation $y = \frac{3}{5}x + b$.

$$-1 = \frac{3}{5} * 1 + b \quad \text{Substitute } x = 1 \text{ and } y = -1 \text{ into the equation.}$$

Notice that we now have an equation with the variable b. We can solve this equation for b to determine the exact vertical intercept of the graph.

$$-1 = \frac{3}{5} * 1 + b$$
$$-1 = \frac{3}{5} + b$$
$$-1 - \frac{3}{5} = b$$
$$-1\frac{3}{5} = b \quad \text{or} \quad -1.6 = b$$

Therefore, we know that the graph crosses the vertical axis at exactly -1.6 rather than at -1.5. With this information we can write the exact equation, $y = \frac{3}{5}x - 1.6$. Because we used the point $(1, -1)$ to determine the value of b, let's use the other point $(6, 2)$ to check our equation.

CHECK with $(6, 2)$.

$$y = \frac{3}{5}x - 1.6$$
$$2 \stackrel{?}{=} \frac{3}{5} * 6 - 1.6$$
$$2 = 2 \quad ✓$$

Because the point $(6, 2)$ checks exactly in our equation, we conclude that $y = \frac{3}{5}x - 1.6$ is the exact equation of the line that passes through the points $(1, -1)$ and $(6, 2)$. The equation could also have been written as $y = 0.6x - 1.6$ or as $y = \frac{3}{5}x - 1\frac{3}{5}$.

Example 1

Write an equation for the following line. Check your equation.

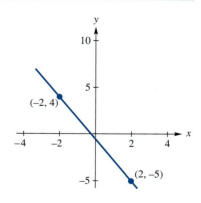

Solution First we need to determine the slope. Going from the point (−2, 4) to (2, −5) we go down 9 units and right 4 units. Therefore, the slope is $\frac{-9}{4}$. We can substitute this into the slope–intercept equation to obtain

$$y = -\frac{9}{4}x + b$$

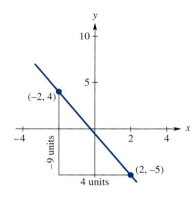

Next, we can substitute either of the points given into our equation, $y = -\frac{9}{4}x + b$ to obtain a value for b. Let's substitute (−2, 4).

$$4 = -\frac{9}{4} * {-2} + b$$
$$4 = 4.5 + b$$
$$-0.5 = b$$

Because b is −0.5, this means that the line crosses the vertical axis at −0.5. Is this reasonable based on the given graph?

Now that we have determined the slope and the vertical intercept we can write the equation of the line, $y = -\frac{9}{4}x - 0.5$. To check our equation, let's substitute the other known point. It is important to notice that we are substituting the point (2, −5) not (−2, 4). Because we used the point (−2, 4) to write the equation $y = -\frac{9}{4}x - 0.5$, we know (−2, 4) works in the equation. But, many lines pass through the point (−2, 4), so we need to verify that the equation we wrote goes through *both* points.

CHECK
$$y = -\frac{9}{4}x - 0.5$$
$$-5 \stackrel{?}{=} -\frac{9}{4} * 2 - 0.5$$
$$-5 = -5 \quad ✓$$

Therefore, the equation of the line is $y = -\frac{9}{4}x - 0.5$.

Example 2

The Blue Book Value. Diane bought a used car. It was two years old when she bought it and she paid $8550, which was the *Kelley Blue Book* trade-in value. Three years later, she looked up the value and found it was $5250.

a. Assuming that the trade-in value of this car decreases linearly, write an equation for the trade-in value in terms of the age of the car.

b. If Diane wants to sell her car before the trade-in value drops below $2000, when should she sell it?

c. What are the units associated with the slope? What does the slope tell you about the problem situation?

d. What do the vertical and horizontal intercepts tell you in this problem situation?

Solution a. First let's sketch a graph. To draw a graph we must first choose the independent and dependent variables. The trade-in value of a car depends on its age. Therefore, age is the independent variable and is graphed on the horizontal axis. Trade-in value is the dependent variable and is graphed on the vertical axis.

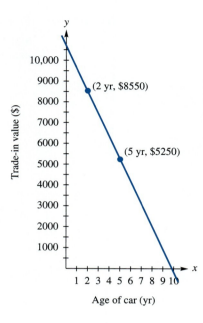

To write the equation of a line, we need to determine the slope. From the graph the slope is

$$\frac{\text{vertical change}}{\text{horizontal change}} = \frac{-\$3300}{3 \text{ years}} = -\$1100/\text{year}$$

If T represents the trade-in value in dollars and x represents the age of the car in years, we know that

$$T = -1100x + b$$

Next, we need to find a value for b. To do this we can substitute one of the known points into the equation and solve for b. Let's use (5, 5250).

$$5250 = -1100 * 5 + b$$
$$5250 = -5500 + b$$
$$10{,}750 = b$$

Now, we know that the vertical intercept is the point (0, 10,750). Is this reasonable based on our graph?

Now that we know the slope and the vertical intercept we can write the equation.

$$T = -1100x + 10{,}750$$

To check our equation, let's substitute the other known point.

CHECK Using the point (2, 8550), we have

$$8550 \stackrel{?}{=} -1100 * 2 + 10{,}750$$
$$8550 = 8550 \quad ✔$$

Our equation checks.

b. We can solve this problem in two different ways. We can approximate the solution using the graph, and we can use the linear equation that we wrote in part a.

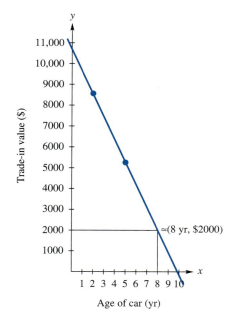

From the graph it appears that the car will be worth $2000 when it is about eight years old.

Using the equation from part a, $T = {-}1100x + 10{,}750$, we need to substitute 2000 for T and solve for x.

$$2000 = {-}1100x + 10{,}750$$
$${-}8750 = {-}1100x$$
$$7.95 \approx x$$

Both algebraically and graphically, we see that the car's trade-in value will be $2000 when it is about eight years old. Therefore, if Diane wants to sell it before its value drops below $2000, she should sell it before it is eight years old.

c. The slope, which we found in part a, is $\frac{-\$1100}{1 \text{ year}}$. This tells us that the yearly depreciation is $1100.

d. The vertical intercept is the point (0 years, $10,750). Therefore, based on our equation, the vertical intercept tells us that the initial value of the car was $10,750. In reality the trade-in value of a car varies depending on many criteria, including the condition of the car and its accessories.

From the graph the horizontal intercept appears to be the point $\approx$(10 years, $0). How can we find the horizontal intercept algebraically?

Algebraically, we can set T equal to zero and solve for x.

$$0 = {-}1100x + 10{,}750$$
$$1100x = 10{,}750$$
$$x = \frac{10750}{1100}$$
$$x \approx 9.77$$

Therefore, the horizontal intercept is $\approx$(9.77 years, $0).

The horizontal intercept represents the age when the trade-in value of the car is zero. Therefore, when the car is about ten years old the trade-in value of the car will be zero. However, in most cases a ten-year-old car will be worth something, though not very much.

As you found in Activities 3 and 4, parallel lines have the same slope and perpendicular lines have opposite reciprocal slopes. For example, if line L has a slope of $\frac{2}{5}$, then a line parallel to L will have a slope of $\frac{2}{5}$ and a line perpendicular to L will have a slope of $-\frac{5}{2}$.

> *Parallel and Perpendicular Lines*
>
> Two lines are **parallel** if and only if they have the same slope.
>
> Two nonhorizontal, nonvertical lines are **perpendicular** if and only if their slopes are opposite reciprocals. That is, if L is a line with slope m, then any line perpendicular to L will have slope $-\frac{1}{m}$.

Example 3

Soda Pop. At a local stadium, the person in charge of concessions keeps track of the amount of soda pop that is bought. The following data have been collected.

Write a linear equation that "fits" the data. Use your equation to determine the amount of soda pop that the concessionaires should order for the next event if the temperature is expected to reach 92°F.

Temperature (in °F)	Soda Pop Bought (in oz)
75	16,000
50	8,000
80	20,000
85	21,000
96	24,000
78	15,000
82	22,000
65	14,000

Solution First let's graph the data. The amount of soda pop that is bought depends on the temperature. Therefore, the temperature is the independent variable and is graphed on the horizontal axis. The amount of soda pop is the dependent variable and is graphed on the vertical axis.

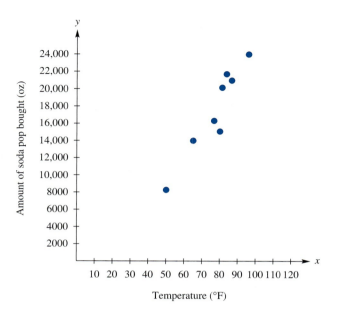

Notice that the data are clearly not linear. However, the points fall in a linear pattern. Therefore, we can approximate the data with a linear model. This is not a perfect fit but approximates the trend and can be used for making rough predictions. The first thing we need to do is draw a line that looks like it "fits" the data. We call this line an **eyeball fit line.**

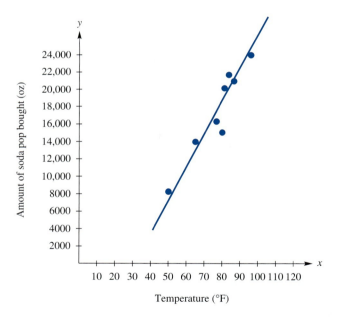

As you might imagine, several lines can be drawn that appear to fit the data. However, if all of the points lie on one side of the line then that line is a poor choice. You should use a straight edge and line it up until you have a "good fit."

Notice that the line drawn in the figure passes through a couple of the data points. As we shall see, this is helpful in writing the linear equation but is not necessary. The line we drew passes through the points (65°F, 14,000 oz) and (85°F, 21,000 oz). Using these points, we can determine the slope of our line.

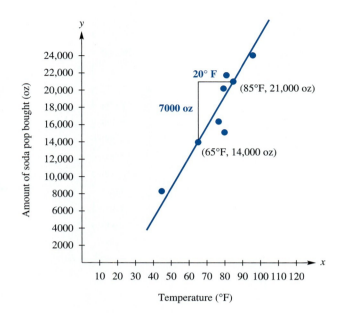

The slope of our eyeball fit line is

$$\frac{7000 \text{ oz}}{20°F} = \frac{350 \text{ oz}}{°F}$$

We can use this together with one of the data points that the line passes through to determine the vertical intercept, and thus the equation of the eyeball fit line. Let P be the amount of soda pop purchased and T be the temperature in degrees Fahrenheit. Then

$P = 350T + b$

$21{,}000 = 350 * 85 + b$ Substitute the point (85, 21,000) into the equation and solve for b.

$-8750 = b$

So the equation of our eyeball fit line is

$P = 350T - 8750$

To determine the amount of soda pop to order when the temperature is predicted to be 92°F, we substitute 92 for T in our equation.

$P = 350 * 92 - 8750$

$P = 23{,}450$

The concessionaires should order about 24,000 ounces of pop. The actual amount ordered depends on how the pop is distributed.

In Section 4.2, we were able to write the equation of a line only if we were given the vertical intercept. In this section, we extended that knowledge to being able to write the equation of a line, given any two points or the slope and any one point. This information might be given to us directly, through a graph, or in a numerical table. We also found that parallel lines have equal slopes and that perpendicular lines have opposite reciprocal slopes. Eyeball fit models for real data were introduced. If data appear to be linearly related when we graph it, we can approximate a linear equation to model the data.

Problem Set 7.2

1. Write the equation of each line pictured here. Check that your equation is correct.

   a.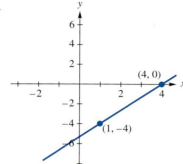

   b.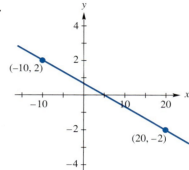

   c.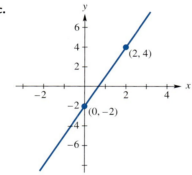

   d.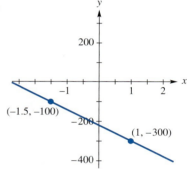

2. For each part, draw the graph of the line that satisfies the given conditions, label two points on the line, and write the equation of the line.
   a. Passing through the point $(-1, 4)$, with a slope of -2
   b. Passing through the point $(6, 0)$, with a slope of $\frac{3}{4}$
   c. Passing through the points $(3, -2)$ and $(-2, 3)$
   d. Passing through the points $(-5.7, 3.6)$ and $(4.8, 9.1)$ (Give coefficients to two significant digits.)

3. For each part, draw the graph and write the equation of the line satisfying the given conditions. Label two points on the line.
   a. Parallel to the line $y = 2x - 4$ and passing through the point $(0, 5)$
   b. Perpendicular to the line $y = 2x - 4$ and passing through the point $(-2, 1)$
   c. Parallel to the line $y = 5$ and passing through the point $(3.5, 6.5)$
   d. Parallel to the line $2x - 3y = 5$ and passing through the point $(0, 3)$
   e. Perpendicular to the line $2x - 3y = 5$ and passing through the point $(4, -1)$

4. Write the equation of each line pictured here. Check that your equation is correct.

a.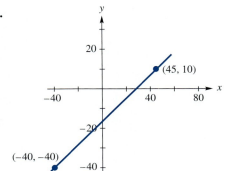

b.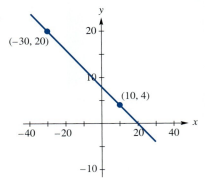

c.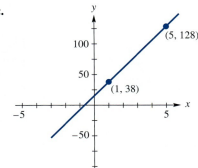

d.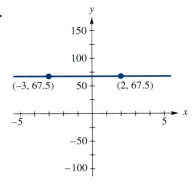

5. On each of the following problems, many equations are possible.
   a. Write the equations of three different lines that have a slope of 4.
   b. Write the equations of three different lines that pass through the point (1, 2).
   c. Write the equations of three different lines that are parallel to $2x + y = 6$.
   d. Write the equations of three different lines that have a vertical intercept of (0, −10).
   e. Write the equations of three different lines that have a horizontal intercept of (3, 0).
   f. Write the equations of three different lines that do not pass through quadrant I.

6. **Grain Production.** In 1980, the world produced 1447 million tons of grain, and in 1991, the world produced 1696 million tons of grain.
   a. Assuming this relationship is linear, write an equation for the amount of grain produced in terms of the number of years after 1980.
   b. Graph your equation.
   c. What is the slope, including units for the grain production equation? What does the slope mean?
   d. What does the vertical intercept tell you about grain production?
   e. Using your equation, determine the year in which the world grain production was about 1000 million tons.
   f. Use your equation to predict the world grain production for 2000. Find the actual data and compare.

7. **Penny's Call.** On Penny's long-distance telephone bill for last month were several calls from Vancouver, Washington, to Olympia, Washington, where one of her sons attends college. Among those calls were evening calls for 1 minute for $.29, 4 minutes for $.74, 6 minutes for $1.05, and 11 minutes for $1.80.

   a. Is this a linear relationship? Explain how you decide.

   b. Graph the data. What is the independent variable? Why?

   c. Assuming that this relationship is linear, choose two of the phone calls to write an equation. How do the other two phone calls check in your equation?

   d. What is the slope, including units, for the problem situation? What does the slope mean? Does this suggest that you made the correct choice in part b or that you should change your mind?

   e. What does the vertical intercept tell you?

   f. According to your equation, what is the cost of a 20-minute phone call?

   g. How long could an evening call be and still keep the cost under $5?

8. The following charts show the populations for two suburbs from 1980 to 1990.

Year	Suburb 1
1980	12,690
1982	13,071
1984	13,463
1986	13,867
1988	14,283
1990	14,711

Year	Suburb 2
1980	23,112
1982	23,693
1984	24,274
1986	24,855
1988	25,436
1990	26,017

   a. Analyze the data and decide which suburb's population growth can be modeled by a linear equation. Explain how you decided.

   b. Write an equation for the one that is linear.

   c. What does your model predict the population will be in the year 2010 if the current trends continue?

   d. When will the population be greater than 30,000?

9. **Classified Ads.** From the classified section of a newspaper, select one particular brand and model of used car that appears several times. Collect data on the asking price in terms of the age (number of years old) of the vehicle. You should have at least eight data points.

   a. Graph the asking price of the car in terms of the age of the car (Is the car two years old? five years old? and so forth).

   b. Draw an eyeball fit line through the data. (Your next task will be easier if the line goes through two of the data points.)

   c. Write an equation for the line you have drawn.

   d. What are the units associated with the slope of your line? What does the slope tell you about the problem situation?

   e. What does the vertical intercept tell you about the problem situation?

   f. If some data points do not seem to fit the overall linear pattern of the other data, try to explain why.

10. Go to your local supermarket and record the total cost and weight for at least six whole chickens of one particular brand (your choice).

    a. Plot your data.

    b. If the data appear to be linear, write an equation.

    c. What is the slope of your line? What are the units for the slope? What does the slope of the line tell us?

11. **A Diet Comparison Between Natural and Commercial Feed.** Students in a fisheries program compared the growth of rainbow trout who were fed natural feed with those fed commercial feed. The students weighed a group of fish at different intervals and computed the mean weight. Below are the data they collected.

    a. Plot the data for the fish fed natural feed in one color and the fish fed commercial feed in a different color.

    b. For each set of data, draw an eyeball fit line through the data. (Your next task will be easier if the line goes through two of the data points.)

    c. For each line you drew, write an equation of the line.

    d. Using your equations, predict the average weight of each set of fish after 60 days.

Days	Natural Feed Mean Weight of Fish (in grams)	Commercial Feed Mean Weight of Fish (in grams)
0	63.6	58.8
32	71.6	68.3
45	84.5	78.0
70	90.8	92.7
118	108.8	101.9

12. **Third World Debt.** The debt of all developing countries for years 1980–1992 is tabulated here.

    a. Plot the points given in the chart.

    b. Draw an eyeball fit line through the data. (Your next task will be easier if the line goes through two of the data points.)

    c. Write an equation for the line you have drawn.

    d. Interpret the meaning of the slope in the context of the problem. Include the numerical value and the units of the slope in your statement.

    e. If the debt continues to rise linearly, what do you predict the debt will be in the year 2000?

Year	1980	1982	1984	1986	1988	1990	1992
Debt (billions of dollars)	639	846	943	1147	1282	1355	1428

7.3 Systems of Linear Equations

Activity Set 7.3

1. What is the solution to $x + 3 = 5$?

2. What is a solution to $y = x + 3$? Is there another solution to $y = x + 3$?

3. *Use the following graph* to answer parts a–c. (Many people find it helpful to color the graphs and their corresponding equations in different colors as a way to help them see the relationships.)

    a. Identify two different solutions to the equation $y = -2.5x + 9.5$.
    b. Identify two different solutions to the equation $y = \frac{2}{9}x + \frac{4}{3}$.
    c. Identify the solution to the equation $-2.5x + 9.5 = \frac{2}{9}x + \frac{4}{3}$.

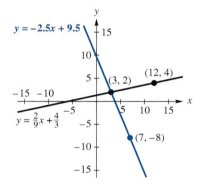

4. Following is a two-by-two linear system. To solve such a system means to find a solution that satisfies both equations. Use the graph to determine the solution to the system. How would you check your solution?

$$\begin{cases} y = -2.5x + 9.5 \\ y = \frac{2}{9}x + \frac{4}{3} \end{cases}$$

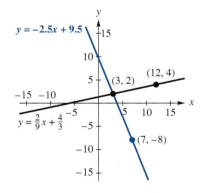

5. Observe the following two scales. In the diagram, each brick weighs the same and each prism weighs the same. When equal weights are on each side of a balance, the top bar is horizontal. For example, the first balance shows that five bricks and six prisms weigh the same as 23 pounds.

   Use the two scales to determine how much each brick and each prism weighs. Explain your process. (*Hint:* You might try *removing* objects from both sides of a scale. For example, if you removed one brick and two prisms from the left side of scale 1, what would you need to remove from the right-hand side?)

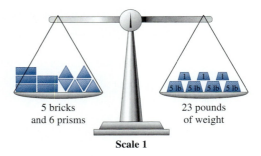

5 bricks and 6 prisms — 23 pounds of weight

Scale 1

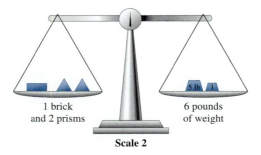

1 brick and 2 prisms — 6 pounds of weight

Scale 2

Discussion 7.3

Graphical Solutions to Systems of Equations

As we saw in the activities, a linear equation in two variables does not have a single solution. An infinite number of possible ordered pairs satisfy the equation, and these ordered pairs lie on the line that is the graph of the equation. Furthermore, we saw in the activities that two linear equations in two variables will intersect at one point (assuming the graphs of these equations are not parallel). *Two linear equations in two variables* are called a **2 × 2 linear system of equations.** To solve a system of equations, we must find a value for each of the two variables that will make *both* of the equations true. Graphically, the solution corresponds to the point of intersection of the graphs of the two equations. Because the solution to a 2 × 2 system must satisfy both equations, we can check our solution by substituting the values for the variables into the two original equations of the system.

Example 1

Solve the system of equations graphically. Check your solution.

$$\begin{cases} y = \frac{-1}{4}x + 12 \\ y = \frac{2}{3}x - 10 \end{cases}$$

Solution First, looking at the equations, we see that the equation $y = \frac{-1}{4}x + 12$ graphs as a line with a vertical intercept at $(0, 12)$ and a slope of $\frac{-1}{4}$. The equation $y = \frac{2}{3}x - 10$ graphs as line with a vertical intercept at $(0, -10)$ and a slope of $\frac{2}{3}$. Because the vertical intercepts are at $(0, 12)$ and $(0, -10)$, we want a slightly larger window than the standard window. We let the horizontal and vertical axes range from -20 to 20.

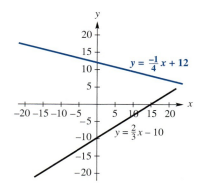

We can see that these two lines will intersect if we extend the graph just a little to the right. So we extend the positive x-axis to 35.

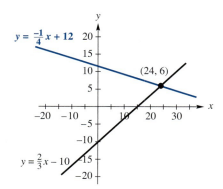

The two lines intersect at the point $(24, 6)$. So, our solution appears to be the point $(24, 6)$, or $x = 24$ and $y = 6$. To be a solution to a system means that the point must be a solution for both of the equations. Therefore, we can numerically check this solution by substituting these values into *both* of the original equations.

CHECK

$$y = \frac{-1}{4}x + 12 \qquad y = \frac{2}{3}x - 10$$
$$6 \stackrel{?}{=} \frac{-1}{4} * 24 + 12 \qquad 6 \stackrel{?}{=} \frac{2}{3} * 24 - 10$$
$$6 = 6 \quad ✔ \qquad 6 = 6 \quad ✔$$

The ordered pair $(24, 6)$ satisfies both equations, so our solution is $(24, 6)$, which can also be written as $x = 24$ and $y = 6$.

In the previous example, we solved a system of linear equations graphically. We will now look at two different methods for solving systems of equations algebraically.

Elimination Method to Solve Linear Systems of Equations

In the activities, you solved the problem of how much each prism and brick weighed by imagining removing equal amounts from both sides of scale 1. Using the information in scale 2, we know that two prisms and one brick together weigh 6 pounds. Therefore, we can remove two prisms and one brick from the left side of scale 1 as long as we remove 6 pounds from the right side of the scale.

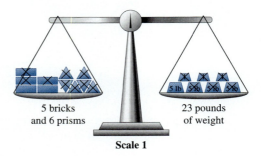

5 bricks and 6 prisms — 23 pounds of weight

Scale 1

By repeating this process three times, it was possible to get the balance down to just two bricks on the left and 5 pounds on the right. Then, two bricks are equal in weight to 5 pounds, so each brick must weigh 2.5 pounds. Knowing that one brick weighs 2.5 pounds, it was possible to determine the weight of each prism. Let's see how this process can be done algebraically.

We begin by writing an equation to model each of the balance scales. First define the variables.

Let p = weight of one prism in pounds

b = weight of one brick in pounds

Then, from the scales we know that

Ⓐ $\begin{cases} 5b + 6p = 23 \\ b + 2p = 6 \end{cases}$ Ⓑ

Notice that we have added labels Ⓐ and Ⓑ to the equations. This makes it easier to refer to each of the equations as we work through the solution process.

In the activities, you took away the amount of the second scale three times. We can model this by multiplying equation Ⓑ by $^-3$ and adding the result to equation Ⓐ. (*Note:* Adding negative three times a quantity is equivalent to subtracting three times a quantity and actually makes the arithmetic easier.)

The process is modeled as follows:

$$\begin{array}{r} \text{Ⓐ} \quad \{5b + 6p = 23 \\ -3 \text{ Ⓑ} \quad \{-3b - 6p = -18 \\ \hline 2b = 5 \end{array}$$

Multiply *both sides* of equation Ⓑ by $^-3$. Add the two equations.

We have *eliminated* the variable p and have one equation containing only the variable b. Now solve this equation.

$$2b = 5$$
$$\frac{2b}{2} = \frac{5}{2}$$
$$b = 2.5$$

Remember that to solve this problem we need to know how much each shape weighs. To solve for p, substitute $b = 2.5$ into any equation that contains both of the variables. We choose equation Ⓑ.

Ⓑ $b + 2p = 6$
   $2.5 + 2p = 6$ Substitute 2.5 for b.
   $2p = 3.5$
   $\frac{2p}{2} = \frac{3.5}{2}$
   $p = 1.75$

The solution appears to be $b = 2.5$ and $p = 1.75$. To check the solution, substitute the values for b and p into *both of the original* equations.

CHECK

Ⓐ $5b + 6p = 23$ Ⓑ $b + 2p = 6$
   $5 * 2.5 + 6 * 1.75 \stackrel{?}{=} 23$ $2.5 + 2 * 1.75 \stackrel{?}{=} 6$
   $23 = 23$ ✔ $6 = 6$ ✔

Therefore one brick weighs 2.5 pounds, and one prism weighs 1.75 pounds.

The algebraic method we just demonstrated is called the **elimination method.** Multiplying the first equation by -3 and adding it to the second equation eliminated the variable p. The result was an equation in one variable that we could then solve. The first step in the elimination method is determining a numerical factor, or sometimes two factors, that can be used to eliminate one of the variables.

Example 2

Solve the following system of equations using the elimination method. Check your solution.

Ⓐ $\begin{cases} 5x + 3y = 25 \\ 8x + 10y = 118 \end{cases}$
Ⓑ

Solution

First we need to find a factor that will eliminate one of the variables. You can't easily multiply 5 by anything and get 8, or 3 by anything and get 10. In this case, it is easier to multiply both of the equations by factors that result in the elimination of one of the variables. Suppose we choose to eliminate the terms containing y. By multiplying equation Ⓐ by 10 and equation Ⓑ by -3, the coefficients of y are 30 and -30. Adding these together eliminates the variable y.

$$\begin{array}{r} 10 \; Ⓐ \\ -3 \; Ⓑ \end{array} \begin{cases} 50x + 30y = 250 \\ \underline{-24x - 30y = -354} \end{cases}$$
$$26x = -104$$

Next, we solve this equation for x.

$$26x = {}^{-}104$$

$$\frac{26x}{26} = \frac{{}^{-}104}{26}$$

$$x = {}^{-}4$$

To find y, we substitute $x = {}^{-}4$ into any equation that contains both variables. Let's choose equation Ⓐ.

Ⓐ $\quad 5x + 3y = 25$

$$5 * {}^{-}4 + 3y = 25$$

$$-20 + 3y = 25$$

$$3y = 45$$

$$\frac{3y}{3} = \frac{45}{3}$$

$$y = 15$$

To check the solution, substitute these values for x and y into both of the original equations.

CHECK

Ⓐ $\quad 5x + 3y = 25 \quad\quad$ Ⓑ $\quad 8x + 10y = 118$

$\quad 5 * {}^{-}4 + 3 * 15 \stackrel{?}{=} 25 \quad\quad\quad 8 * {}^{-}4 + 10 * 15 \stackrel{?}{=} 118$

$\quad\quad\quad\quad 25 = 25 \quad ✔ \quad\quad\quad\quad\quad\quad 118 = 118 \quad ✔$

Our solution is $x = {}^{-}4$ and $y = 15$. This could also be written as the ordered pair $({}^{-}4, 15)$.

In this example, we chose to eliminate y as the first step. This resulted in our selection of 10 and ${}^{-}3$ as factors to multiply each of the equations. We could have decided to eliminate x as the first step. What factors could we use in this case?

Substitution Method to Solve Linear Systems of Equations

We have looked at solving systems of linear equations graphically and algebraically using the elimination method. The elimination method always works to solve linear systems of equations algebraically, although at times this method can be cumbersome.

A second algebraic method can be used to solve systems of equations: the **substitution method.** Both the elimination method and the substitution method produce the same solutions, even though the process is different. The method you choose will depend on the form of the system and your personal preference.

To solve a 2 × 2 system using the substitution method, we first solve one equation for a variable. Then, we substitute the resulting expression into the other equation.

Suppose we want to solve the following system of equations using the substitution method.

Ⓐ $\begin{cases} 5x + 3y = 37 \\ y = 4x - 75 \end{cases}$
Ⓑ

The first step is to solve one of the equations for one of the variables. In this system, equation Ⓑ is already solved for y. The next step is to *substitute* the expression $(4x - 75)$ for y in equation Ⓐ. This step is how the substitution method got its name. It is important when using the substitution method that you solve one of the equations for a variable and then substitute the resulting expression into *the other* equation.

Ⓐ $5x + 3(y) = 37$

Ⓐ $5x + 3(4x - 75) = 37$ Substitute $4x - 75$ for y in equation Ⓐ.

Notice that this equation now contains only one variable, x. We can now solve this equation.

$$5x + 3(4x - 75) = 37$$
$$5x + 12x - 225 = 37$$
$$17x - 225 = 37$$
$$17x = 262$$
$$\frac{17x}{17} = \frac{262}{17}$$
$$x \approx 15.4$$

We are not finished yet; we need to find y. To find y, we substitute 15.4 for x in any of the equations containing both x and y. Because the equation $y = 4x - 75$ is already solved for y, selecting this equation makes our next step easier.

$$y = 4x - 75$$
$$y \approx 4 * 15.4 - 75$$
$$y \approx -13.4$$

Finally, we check our solution using both of the original equations.

CHECK

Ⓐ $5x + 3y = 37$ Ⓑ $y = 4x - 75$

$5 * 15.4 + 3 * {-13.4} \stackrel{?}{=} 37$ ${-13.4} \stackrel{?}{=} 4 * 15.4 - 75$

$36.8 \approx 37$ ✔ ${-13.4} = {-13.4}$ ✔

Therefore, the solution to the system is $x \approx 15.4$ and $y \approx -13.4$, which can also be written as $\approx(15.4, -13.4)$.

Example 3

Creative Computing Company's Clutter. In the clutter of everyday work, Creative Computing Company's secretary has lost the price list for paper for their two fax machines. One of the machines uses plain paper and the other uses fax paper. The secretary did find the notes for the last two orders. Creative Computing ordered seven boxes of plain paper and ten boxes of fax paper at a total cost of $250.65 and six boxes of plain paper and one box of fax paper at a total cost of $120.20.

Write a 2 × 2 linear system of equations for this problem situation, and solve the system algebraically.

Solution First we must write two equations to model the last two paper orders. Let P represent the cost per box for plain paper and F represent the cost per box for fax paper.

Ⓐ $\begin{cases} 7P + 10F = 250.65 \\ 6P + F = 120.20 \end{cases}$
Ⓑ

Two methods can be used to solve this system algebraically: the elimination method and the substitution method. If we choose the elimination method, our first step is to decide which variable to eliminate. By multiplying equation Ⓑ by ⁻10, we eliminate the variable F. To use the substitution method, our first step is to solve one of the equations for one of the variables. Variable F has a numerical coefficient of 1 in equation Ⓑ. This is the easiest variable to solve for. We complete the solution using this method.

Ⓑ $\quad 6P + F = 120.20$

Ⓑ $\qquad F = 120.20 - 6P \quad$ Solve for F.

Next substitute the expression $120.20 - 6P$ for F into equation Ⓐ.

Ⓐ $\qquad 7P + 10(F) = 250.65$

$$7P + 10(120.20 - 6P) = 250.65$$
$$7P + 1202.0 - 60P = 250.65$$
$$-53P + 1202.0 = 250.65$$
$$-53P = -951.35$$
$$\frac{-53P}{-53} = \frac{-951.35}{-53}$$
$$P = 17.95$$

To find F, we need to substitute 17.95 for P into any equation containing both variables and solve for F. Our next step is made easier by selecting the equation $F = 120.20 - 6P$ because we already solved this equation for F.

$F = 120.20 - 6P$

$F = 120.20 - 6 * 17.95$

$F = 12.5$

The solution appears to be $(P, F) = (17.95, 12.50)$.

We now need to check if this answer works in the problem situation, not just the equations we have written, because we may have set up our equations incorrectly.

The first order was for seven boxes of plain paper and ten boxes of fax paper for a total of $250.65. The second order was for six boxes of plain paper and one box of fax paper for a total of $120.20.

First order: $\qquad$ 7 boxes ∗ $17.95 per box = $125.65

$\qquad\qquad\qquad\quad$ 10 boxes ∗ $12.50 per box = $125.00

Total order is $\qquad$ $125.65 + $125.00 = $250.65 ✔

Second order: $\qquad$ 6 boxes ∗ $17.95 per box = $107.70

$\qquad\qquad\qquad\quad$ 1 box ∗ $12.50 per box = $12.50

Total order is $\qquad$ $107.70 + $12.50 = $120.20 ✔

We can now say that the cost of the plain paper is $17.95 per box and the cost of the fax paper is $12.50 per box.

Although it may seem confusing to introduce two methods for solving a system algebraically, you may find one method easier to use on a given problem. Try solving the linear system in the preceding example using the method of elimination. Did you find one method easier?

The Number of Solutions to a System of Linear Equations

Example 4 Solve each of the following linear systems.

a. $\begin{cases} y = -5x + 2 \\ 1.25x + 0.25y = 0 \end{cases}$ b. $\begin{cases} 3x - 12y = 4 \\ -6x + 24y + 8 = 0 \end{cases}$

Solution a. Because the first equation is already solved for y, let's solve the system in part a by substitution. We can substitute the expression $(-5x + 2)$ for y in the second equation.

$$1.25x + 0.25(y) = 0$$
$$1.25x + 0.25(-5x + 2) = 0$$
$$1.25x - 1.25x + 0.5 = 0$$
$$0.5 = 0$$

This statement is clearly not true! But, what does this tell us about the solution to the system of equations in part a? Let's try looking at the graphs. Rewrite each equation in slope–intercept form. The equation $1.25x + 0.25y = 0$ becomes $y = -5x$. What do you notice about the two equations?

$\begin{cases} y = -5x + 2 \\ y = -5x \end{cases}$

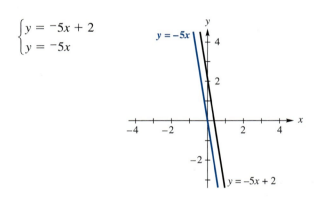

The graph shows what we expected, that these lines are parallel. The solution to a system of linear equations is the point of intersection of the two lines. Because parallel lines never intersect, this system has no solution. This is why our algebraic solutions resulted in a **false statement,** a statement that is always false regardless of the input. Our final answer must be that no solutions exist to the system of equations in part a.

b. Solving for x or y in either of the equations in the system in part b results in fractional coefficients. Therefore, the method of elimination looks like the best algebraic method for solving this system. To use the elimination method, the variables and constants must "line up." So, the first step is to rewrite the second equation with variables to one side of the equal sign and constants on the other. We rewrite $-6x + 24y + 8 = 0$ as $-6x + 24y = -8$.

Ⓐ $\begin{cases} 3x - 12y = 4 \\ -6x + 24y = -8 \end{cases}$
Ⓑ

We can see that by multiplying equation Ⓐ by 2 and adding it to equation Ⓑ, x is eliminated.

2 Ⓐ $\begin{cases} 6x - 24y = 8 \\ -6x + 24y = -8 \end{cases}$
Ⓑ
$$0 = 0$$

This equation is definitely true, but does not seem to help solve the system. Again, we explore the possible solution by looking at the graphs of the two lines. To graph the equations we must first solve each equation for y. We can rewrite $3x - 12y = 4$ as $y = \frac{1}{4}x - \frac{1}{3}$ and $^-6x + 24y + 8 = 0$ as $y = \frac{1}{4}x - \frac{1}{3}$. What do you observe about these two equations?

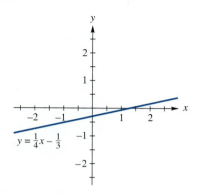

The two equations are identical and, therefore, graph as the same line. Two lines that have the same graph and the same set of solutions are called **coincidental.** The solutions for x and y are the coordinates of the points of intersection of the lines. Therefore all points lying on the line $y = \frac{1}{4}x - \frac{1}{3}$ are solutions to this system. Note that the solution is not "all real numbers." Every point in the Cartesian coordinate system does not satisfy these two equations, only those that lie on the line. Nevertheless, the solution is an infinite number of points. This means we cannot list all possible solutions and therefore, we describe the solution by writing the equation of the line. Any equivalent form of the equation can be used in the statement of the solution. For example, the solution could have been written as, "all points lying on the line $3x - 12y = 4$."

The previous two examples are special cases. In most applications the lines are nonparallel and noncoincidental. In other words, they intersect at just one point. When lines are nonparallel and noncoincidental, exactly one point of intersection exists, and, therefore, one ordered pair is a solution to the system of equations.

We learned three techniques for solving systems of equations: one graphical method and two algebraic methods. We looked specifically at solving **2 × 2 linear systems. Solving a system of equations** means finding values of the variables that simultaneously satisfy all equations in the system. The **graphical method** involves graphing the two linear equations and looking for the point of intersection. The other two methods we explored were algebraic. Both the **elimination method** and the **substitution method** give us techniques to reduce the two equations in two variables to a single equation in one variable that we can easily solve. It is important to remember that *checking* a system of equations requires that we substitute the values for the variables into *both* equations. An ordered pair that works in one equation and not the other is not a solution to the system. How would such an ordered pair show up in a graphical representation of the system?

Problem Set 7.3

1. For each part, without solving the system, determine whether $x = -2$ and $y = 5$ is a solution. (*Hint:* If you substitute values into an equation, have you solved the equation?)

   a. $\begin{cases} 2x - 3y = -19 \\ 5x + 2y = 0 \end{cases}$

   b. $\begin{cases} y = -3x - 1 \\ 5x - 2y = 11 \end{cases}$

2. For each part, determine whether $x = 3$ and $y = -2.1$ is a solution. Explain how you can decide *without* solving the systems either algebraically or graphically.

   a. $\begin{cases} x + y = 5.1 \\ 4x - 3y = 18.3 \end{cases}$

   b. $\begin{cases} y = 3x - 11.1 \\ x = 2.1y + 7.41 \end{cases}$

3. Solve each 2×2 linear system algebraically using either the elimination or the substitution method. Numerically check all solutions.

   a. $\begin{cases} y = 2x - 5 \\ x + y = 7 \end{cases}$

   b. $\begin{cases} 4x - 5y = 9 \\ 2x + 3y = -12 \end{cases}$

   c. $\begin{cases} 9x - y = 21 \\ 3x + 4y = 150 \end{cases}$

   d. $\begin{cases} 1.5x - 3y = 45 \\ y = 1.5x - 25 \end{cases}$

   e. $\begin{cases} y = 25 + 7.5x \\ y = 6.5x + 30 \end{cases}$

   f. $\begin{cases} 1.5x - y = 45 \\ y = 1.5x + 25 \end{cases}$

4. Solve each 2×2 linear system graphically. Sketch the graph used to solve the system. Check all solutions.

   a. $\begin{cases} y = 2x - 7 \\ x + y = 8 \end{cases}$

   b. $\begin{cases} 0.8x + y = 4 \\ y = 2x - 3 \end{cases}$

   c. $\begin{cases} 1.5x - 3y = 45 \\ y = 1.5x - 25 \end{cases}$

   d. $\begin{cases} 9x - y = 21 \\ 3x + 4y = 150 \end{cases}$

5. Which of the following graphs could be a graph of the system shown? Explain how you arrived at your decision.

   $\begin{cases} y = \frac{3}{5}x - 27 \\ y = \frac{3}{5}x - 52 \end{cases}$

   a.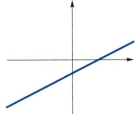

   b.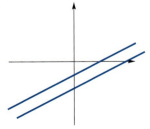

   c.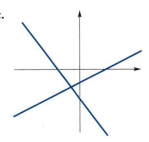

6. In Problem 3, you solved each system algebraically. In Problem 4, you solved systems graphically. In this problem you are to solve each 2 × 2 linear system either graphically or algebraically. Your experience in Problems 3 and 4 may help you decide when to choose a particular method. Regardless of what method you choose, check all solutions. Round any approximate solutions to the nearest hundredth.

   a. $\begin{cases} 2x + y = 6 \\ x - 3y = 10 \end{cases}$

   b. $\begin{cases} 7x - 3y = 4 \\ 5x + 4y = 52 \end{cases}$

   c. $\begin{cases} 2x - 105 = y \\ x = 6 - 0.5y \end{cases}$

   d. $\begin{cases} -x + 5y = 19 \\ x = 2.1y + 11.45 \end{cases}$

   e. $\begin{cases} x + y = 20 \\ 0.25x + 0.75y = 7 \end{cases}$

   f. $\begin{cases} x = 3x - 2y + 25 \\ y = x - 3y + 14 \end{cases}$

   g. $\begin{cases} 2x - 15y = 180 \\ y - 6x = 125 \end{cases}$

   h. $\begin{cases} 0.25x + 1.65y = 42 \\ x + y = 200 \end{cases}$

   i. $\begin{cases} 3.5x - 4.4y = 8.6 \\ 7.2x - 5.9y = 17.0 \end{cases}$

7. Solve the following systems of equations using either the graphical method or one of the two algebraic methods. Round approximate solutions to the hundredths place. Numerically check all solutions.

   a. $\begin{cases} 0.75x + 0.25y = 293.75 \\ 0.80x + 0.20y = 280 \end{cases}$

   b. $\begin{cases} y = 3x + 50 \\ 2x + 3y = 105 \end{cases}$

   c. $\begin{cases} A = 1.25B + 32 \\ A = 0.8B - 25.6 \end{cases}$

8. Estimate the solution to the system of equations that is graphed here.

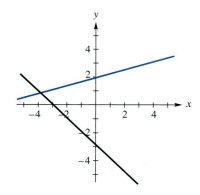

9. Use the following graph (do not graph your own) to solve the following system of equations.

$\begin{cases} y = 3(x + 3) - (x + 14) \\ y = \dfrac{x}{2} - \dfrac{10x - 25}{10} \end{cases}$

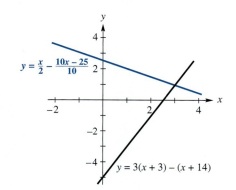

10. **The Paint Sprayer.** The Radical Rental Emporium charges $10.00 plus 75¢ per hour to rent a paint sprayer. A competitor, Truly Temporary Equipment Service, rents the same sprayer for $2.95 plus $1.22 per hour.

   a. Write an equation for the cost in terms of the number of rental hours for each business.

   b. Determine algebraically and graphically the number of hours for which the cost of renting a paint sprayer is the same from Radical Rental or Truly Temporary.

   c. If you need to rent a paint sprayer, how will you decide which company to rent from?

11. **The Investment.** Cathy just won $10,000 in the Oregon Lottery. She wants to put some of the money in an account where it is accessible and to commit the rest to a long-term investment. She has been considering a savings account that has been paying 2.64% APR and a mutual fund that has been paying 8.9% APR. At the end of one year, Cathy hopes to earn $750. How much money must she invest in each type of account?

   a. Write a 2 × 2 system to model this situation.

   b. Determine the solution.

   c. Is there any way that Cathy can invest the money to earn $1000 in interest in one year? Explain how you arrived at your conclusion.

12. **TGIF.** Chris and Terry have been trying to live within their budget, but miss going out to dinner on Friday nights. They decide not to spend any nickels or quarters they receive as change for a month and save these in a jar. At the end of the second week they have a total of 90 coins in the jar. The value of the coins totals $15.10. How many nickels and how many quarters were in the jar?

13. **Jazz Concert.** Last spring, several of the Mt. Hood Community College (MHCC) jazz groups held a joint concert. Tickets sold for $5.50 to the general public and for $3.00 for students and staff. Ticket sales brought in $1607.50. If 390 people attended the concert, how many were MHCC students and staff and how many were from the general public?

14. **Chemistry Lab.** The chemistry lab aide has been told that tomorrow's lab experiments require 2 liters of 17% sulfuric acid. When she checks the storeroom she finds only 10% and 30% solutions of sulfuric acid in stock. She realizes that she can mix these two to get what is needed. How much of the 10% acid and how much of the 30% acid should she mix together to create 2 liters of 17% sulfuric acid solution?

7.4 Solving Linear Inequalities in One Variable

Activity Set 7.4

1.
   a. Is $-3 < 8$ true or false?
   b. Is $4 > 10$ true or false?
   c. Is $-3 > -5$ true or false?

2.
   a. Is $x = 4$ a solution to $x + 1 < 5$?
   b. Is $x = 3$ a solution to $x + 1 < 5$?
   c. Is $x = 2$ a solution to $x + 1 < 5$?
   d. Is $x = 5$ a solution to $x + 1 < 5$?
   e. Describe all the values of x that are solutions to $x + 1 < 5$.

3. Follow the directions in parts a–e. Does the inequality remain true?
   a. Add 5 to both sides of the inequality. $3 < 7$
   b. Add 5 to both sides of the inequality. $3 > -7$
   c. Add -5 to both sides of the inequality. $3 < 7$
   d. Add -5 to both sides of the inequality. $3 > -7$
   e. If you add or subtract the same amount from both sides of an inequality, does the inequality remain true?

4. Follow the directions in parts a–e. Does the inequality remain true?
   a. Multiply both sides of the inequality by 5. $25 < 40$
   b. Multiply both sides of the inequality by $\frac{1}{5}$. $25 > -40$
   c. Multiply both sides of the inequality by -5. $25 < 40$
   d. Multiply both sides of the inequality by $\frac{-1}{5}$. $25 > -40$
   e. If you multiply or divide both sides of an inequality by a given number, does the inequality always remain true? If not, describe when it remains true and when it does not.

5. Using what you learned in Activities 3 and 4, solve the inequality $17 - 25x > 177$.

Discussion 7.4

Graphing Inequalities on a Number Line.

In this section we will be looking at the solutions to inequalities. Unlike an equation in one variable, many values make an inequality in one variable true. Any value that makes the inequality statement true is part of the solution of the inequality. A number line can be used as a visual model of a solution for an inequality. There are four basic types of number line graphs of inequalities. These are seen in the following table.

Verbal Phrase	Inequality	Graph
All real numbers less than 4	$x < 4$	
All real numbers greater than $^-3$	$x > {^-3}$	
All real numbers less than or equal to $^-2$	$x \leq {^-2}$	
All real numbers greater than or equal to 1	$x \geq 1$	

In the table, we see that if x is *greater than* or *greater than or equal to* a number, the graph is shaded to the right. Similarly, if x is *less than* or *less than or equal to* a number, the graph is shaded to the left. The graphs in the first two rows are drawn with open dots; that is, the dots are not shaded in. This is because 4 is not in the solution of $x < 4$ and $^-3$ is not in the solution of $x > {^-3}$. In the last two rows, the graphs are drawn with solid dots; the dots are shaded. This means that $^-2$ is a solution of $x \leq {^-2}$ and 1 is a solution of $x \geq 1$.

Example 1

Graph the following inequalities on a number line.

   a. $R > {^-2}$ b. $y \leq 0$ c. $5 > M$

Solution a. The graph of $R > {^-2}$ is all real numbers greater than $^-2$.

b. The graph of $y \leq 0$ is all real number less than or equal to 0.

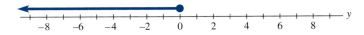

c. In reading an inequality like $5 > M$, we often read the variable first. Because the smaller part of the inequality sign is pointing to the M, we know that M is less than 5. The graph is then all the numbers less than 5. Writing $5 > M$ as $M < 5$ avoids confusion.

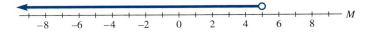

You may have noticed in the examples of graphing inequalities, that if the *variable is written on the left-hand side,* the inequality symbol "points" in the direction that the graph is shaded.

Solving Linear Inequalities in One Variable

In the activities, we found that adding or subtracting the same amount from both sides of an inequality allows the inequality to remain true. However, multiplying or dividing it is not as simple. In this case we need to know the sign of the number that we are multiplying or dividing by. If we multiply or divide both sides of an inequality by a positive number, then the inequality remains true. However, if we multiply or divide by a negative number, then the inequality is not true until we reverse its direction.

These ideas are the properties we will use to solve linear inequalities algebraically. The properties are summarized in the following box.

> *Properties of Inequalities*
>
> 1. *Addition and Subtraction* If you add (subtract) the same real number to (from) both sides of an inequality, the direction of the inequality does not change.
> 2. *Multiplication (Division) by a Positive Number* If you multiply or divide both sides of an inequality by a positive number, the direction of the inequality does not change.
> 3. *Multiplication (Division) by a Negative Number* If you multiply or divide both sides of an inequality by a negative number, the direction of the inequality must be reversed.

Notice that we can perform the same operations on a linear inequality that we can on a linear equation, except that when we multiply or divide by a negative number we must remember to reverse the direction of the inequality.

Example 2 For each of the following inequalities, solve the inequality, graph the solution on a number line, and check your solution.

  a. $8x - 24 > 80$ b. $13 - 7w \geq {}^{-}491$ c. $6T + 14 \leq 10T + 24$

Solution a. $8x - 24 > 80$

$8x - 24 + 24 > 80 + 24$ Add 24 to both sides of the inequality.

$8x > 104$

$\dfrac{8x}{8} > \dfrac{104}{8}$

$x > 13$ Divide both sides of the inequality by 8.

Because 8 is a positive number, the direction of the inequality remains the same.

Next we draw a number line graph of the solution. Because *x* is *greater than* 13, we use an open dot, which means that 13 is not part of the solution. The arrow is directed to the right, which means that all values to the right of 13 are solutions.

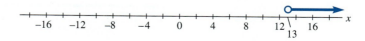

However, to be sure we must check our solution. Checking a solution to an inequality is somewhat different from checking the solution to an equation. Typically, the solution to an inequality consists of a range of values. It is not possible to check every value, so we check values in different regions.

In this example, we call 13 the **border point** because it borders the shaded region and the unshaded region. To check a solution to an inequality, we check the border point, a point in the solution region (shaded) and a point that is not in the solution region (unshaded). When we check the border point, both sides of the inequality should be equal. When we check a point in the solution region, the inequality should be true, and when we check a point that is not in the solution region the inequality should be false.

CHECK   Even though 13 is not part of the solution, we must check to see if it is the correct border point.

Border point: $x = 13$ (Both sides of the inequality should be *equal*.)
$$8x - 24 = 80$$
$$8 * 13 - 24 \stackrel{?}{=} 80$$
$$80 = 80 \quad ✔ \quad \text{The border point makes the two sides equal, as it should.}$$

Point in Region: $x = 15$ (The inequality should be *true*.)
$$8 * 15 - 24 \stackrel{?}{>} 80$$
$$96 > 80 \quad ✔ \quad \text{This is a true statement, as it should be.}$$

Point not in region: $x = 0$ (The inequality should be *false*.)
$$8 * 0 - 4 \stackrel{?}{>} 80$$
$$-4 \not> 80 \quad ✔ \quad \text{This is a false statement, as it should be.}$$

We conclude that our solution is $x > 13$.

b.
$$13 - 7w \geq -491$$
$$13 - 7w - \mathbf{13} \geq -491 - \mathbf{13}$$
$$-7w \geq -504$$
$$\frac{-7w}{\mathbf{-7}} \leq \frac{-504}{\mathbf{-7}} \quad \text{Divide both sides by } -7. \text{ Because we are dividing by a negative number, reverse the direction of the inequality.}$$
$$w \leq 72$$

Next, graph the solution on a number line.

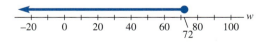

CHECK   To verify our solution, we again check three points: the border point, a point in the solution region, and a point not in the solution region.

Border point: $w = 72$ (Both sides of the inequality should be *equal*.)
$$13 - 7w \stackrel{?}{=} -491$$
$$13 - 7 * 72 \stackrel{?}{=} -491$$
$$-491 = -491 \quad ✔ \quad \text{Both sides are equal, as they should be.}$$

Point in Region: $w = 0$ (The inequality should be *true*.)
$$13 - 7 * 0 \stackrel{?}{\geq} -491$$
$$13 \geq -491 \quad ✔$$

Point not in Region: $w = 75$ (The inequality should be *false.*)

$$13 - 7 * 75 \stackrel{?}{\geq} {}^-491$$
$$^-512 \not\geq {}^-491 \quad ✔ \quad \text{This is a false statement, which is what we want.}$$

We conclude that our solution is $w \leq 72$.

c.
$$6T + 14 \leq 10T + 24$$
$$6T + 14 - \mathbf{6T} \leq 10T + 24 - \mathbf{6T}$$
$$14 \leq 4T + 24$$
$$14 - \mathbf{24} \leq 4T + 24 - \mathbf{24}$$
$$^-10 \leq 4T$$
$$\frac{^-10}{4} \leq \frac{4T}{4}$$
$$^-2.5 \leq T$$
$$T \geq {}^-2.5 \quad \text{For clarity, we can rewrite the inequality with the variable on the left.}$$

Next, graph the solution on a number line.

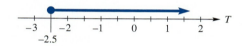

CHECK To verify our solution, we again check three points: the border point, a point in the solution region, and a point not in the solution region.

Border point: $T = {}^-2.5$ (Both sides of the inequality should be *equal.*)
$$6 * {}^-2.5 + 14 \stackrel{?}{=} 10 * {}^-2.5 + 24$$
$$^-1 = {}^-1 \quad ✔ \quad \text{The border point makes the two sides equal, as it should.}$$

Point in region: $T = 0$ (The inequality should be *true.*)
$$6 * 0 + 14 \stackrel{?}{\leq} 10 * 0 + 24$$
$$14 \leq 24 \quad ✔$$

Point not in region: $T = {}^-3$ (The inequality should be *false.*)
$$6 * {}^-3 + 14 \stackrel{?}{\leq} 10 * {}^-3 + 24$$
$$^-4 \not\leq {}^-6 \quad ✔ \quad \text{This is a false statement, which is what we want.}$$

We conclude that the solution is $T \geq {}^-2.5$.

Example 3

Abe is planning for a microbrewery festival. He wants to encourage lots of people to try his beer, but he does not want to lose money. He has arranged for ten volunteers to staff his booth on Saturday. Each volunteer will receive a free T-shirt. The T-shirts will cost Abe $12.50 each. He has figured that the cost of a 12-ounce beer and cup is $0.75. He plans to charge $1.25 per 12-ounce serving. How many servings of beer must Abe's booth sell to make money on Saturday?

Solution Revenue is the money Abe earns by selling beer. Abe's costs are the cost of the beer and cups plus the cost of the T-shirts for the volunteers. His profit is the difference of the revenue and the costs.

Let n represent the number of servings of beer sold. Then

$$\text{Revenue} = 1.25 * n$$
$$\text{Cost of beer} = 0.75 * n$$
$$\text{Cost of T-shirts} = 10 * 12.50$$
$$\text{Profit} = \text{Revenue} - \text{Cost}$$
$$\text{Profit} = 1.25 * n - (0.75 * n + 10 * 12.50)$$

For Abe to make money, his profit must be greater than zero. This can be written as an inequality.

$$1.25 * n - (0.75 * n + 10 * 12.50) > 0$$

$1.25n + {}^-(0.75n + 125.0) > 0$	Simplify, and write subtraction as addition of the opposite.
$1.25n + {}^-0.75n + {}^-125.0 > 0$	Distribute $^-1$ over the sum.
$0.5n + {}^-125.0 > 0$	
$0.5n > 125.0$	Add 125.0 to both sides.
$n > 250$	Divide both sides by 0.5.

Let's see if our answer makes sense. If Abe sells 250 beers he makes $1.25 * 250 = 312.50$ in revenue. His costs are $0.75 * 250 = 187.50$ for the beer and cups and $10 * 12.50 = 125$ for the volunteer's T-shirts. Therefore, Abe's profit for selling 250 servings is

$$312.50 - (187.50 + 125) = 0$$

We conclude that Abe will make money if he sells more than 250 servings on Saturday.

In this section, we saw how to graph inequalities on a number line. We discovered the properties of inequalities and used these to solve linear inequalities. We found that the properties of inequalities are the same as properties for equations *except* when we multiply or divide both sides of an inequality by a negative number. In this case the direction of the inequality must be reversed. In checking our solutions, we tested three values: the border point, a value in the solution region, and a value outside the solution region. In the problem set we will see why checking only one point is not sufficient.

Problem Set 7.4

1. Write an inequality to describe each of the following graphs.

   a. [number line showing open circle at 0, shaded to the right, variable x]

   b. [number line showing closed circle at 2, shaded to the right, variable k]

   c. [number line showing closed circle at 8, shaded to the left, variable M]

2. Graph the following inequalities on a number line.

   a. $x \geq 5$ b. $w < {}^-3$ c. $10 \leq m$ d. $0 \geq K$ e. $T > {}^-4.5$ f. $^-7.5 \geq p$

3. Write an inequality to describe the following situations.
   a. temperatures below 40°F
   b. lengths under 48 inches
   c. speeds that exceed 55 mph
   d. heart rates of at least 40 beats per minute

4. Solve the following inequalities. Graph the solutions on a number line. Check your solutions.
   a. $5x + 43 > 98$
   b. $700 \leq 55 + 8w$
   c. $-50 - 3y \geq 10$
   d. $7m + 24 < 3m - 15$
   e. $-2(4x - 32) \leq -40$
   f. $-81 > 27 - 9H$
   g. $10 - x \leq -6$
   h. $5(2.3B - 5.5) - 2(8.7B - 2.4) > 98$

5. In the following problem, an error occurs in both the solution and the check. Find both errors.

   $$5P - (10 - 3P) > 15$$
   $$5P - 10 - 3P > 15$$
   $$2P - 10 > 15$$
   $$2P > 25$$
   $$\frac{2P}{2} > \frac{25}{2}$$
   $$P > 12.5$$

   CHECK   Let $P = 15$.

   $$5P - (10 - 3P) > 15$$
   $$5 * 15 - (10 - 3 * 15) \stackrel{?}{>} 15$$
   $$110 > 15 \quad ✔$$

   The solution is $P > 12.5$.

6. **Friday Night at the Movies.** Three friends want to go to the movies this Friday. They estimate that they will need at least $5 for gas. Pooling their funds they find that they have $21.50. Use this information to answer the following questions.
   a. How much can the friends afford to spend on the movie tickets?
   b. The movie they want to see is $6.50 for regular tickets or $3.50 for economy-hour tickets. Which tickets can they afford?
   c. Write an inequality to determine the price of movie tickets that the friends can afford. What is the solution to this inequality?

7. **Municipal Water Supply.** Currently 620,000 gallons of water are in one of the city's emergency reservoirs. Water consumption for the area served by this reservoir is approximately 22,500 gallons per day. Could this reservoir be used for ten days? two weeks? How many days can the city predict that this reservoir could be used as a source of emergency water?

8. The following pattern is given to a group of students. They are also given 200 blocks. Can they build the 50th structure? Which structures can they build? Write an inequality statement to model this problem. (Assume that they are building *just* the last structure.)

Structure 1

Structure 2

Structure 3

9. **The Jog-a-Thon.** Students at a local elementary school are holding a jog-a-thon to purchase instruments for their music program. Nicole has been collecting pledges. Her pledge sheet is shown here. (*Note:* Some people pledge a flat amount, others pledge an amount per lap. The total will be computed after Nicole runs the jog-a-thon.) The jog-a-thon will be run on a track where each lap is one quarter mile. Nicole wants to earn at least $30 from the jog-a-thon. How far should Nicole run?

	Pledges per Lap	Number of Laps	Total Pledge
1.	25¢		
2.			$5.00
3.			$2.50
4.	10¢		
5.	$1.00		
6.	50¢		

10. **Garbage Transfer.** In a certain metro region, daily garbage is collected at two transfer stations and then transported by truck to a landfill in the eastern part of the state. Each of the two transfer stations can hold 3200 tons of garbage. In 1996, a severe ice storm hit this region, closing the highway to the east. The average person in this metro region of 1.3 million people makes 4.2 pounds of garbage per day. How many days can the roads remain closed before the transfer stations have to stop accepting garbage?

Chapter Seven Summary

In this chapter, we extended our understanding and skills in working with linear equations and their graphs. In Chapter 3 we learned how to solve linear equations in one variable algebraically. We began Chapter 7 by solving these same equations graphically.

To **solve a linear equation in one variable graphically,** we graph each side of the equation. Then, we find the intersection point of these two graphs. Usually this is done using a graphing calculator. Because the original equation contains only one variable, we know that our solution is only the *first coordinate* of the intersection point. We check our solution by substituting this value into the original equation.

Even though it may seem redundant to solve these equations graphically because we already have an algebraic technique, there are benefits. First, a graphical solution gives us a visual representation. Second, the algebra required to solve some linear equations may be quite cumbersome. Third, as you proceed in mathematics you will find that some equations in one variable cannot be solved algebraically. However, the graphical technique we learned in Section 7.1 can be used to solve these equations.

In Section 4.2, we learned that the graph of the equation $y = mx + b$ is a line with slope m and vertical intercept $(0, b)$. With this information we were able to graph linear equations and write equations, provided we were given the vertical intercept and enough information to determine the slope. However, there are many situations that can be modeled with lines in which we do not know the vertical intercept. In Section 7.2 we learned how to write the equation of any line, given two points or given the slope and one point.

> *Writing the Equation of a Line, Given Two Points*
> 1. Find the slope of the line using the two points given.
> 2. Substitute the value for the slope in the slope–intercept equation.
> 3. Find the value of b by substituting the coordinates of one of the given points into the slope–intercept equation. Solve the resulting equation for b.
> 4. Write the equation of the line by substituting the value for m and b into the slope–intercept equation.
> 5. Check your equation by substituting the coordinates of the other point into your equation, or graph your equation to see that it passes through the two given points.

To **write the equation for a line, given the slope and one point,** follow the same process as shown in the box, except skip Step 1, because you already know the slope.

Many situations are best modeled using two variables. If two variables are required, then we need two equations to solve the situation. In Section 7.3 we looked at 2×2 linear systems of equations. Three methods are available to solve a 2×2 linear system. We can solve a 2×2 linear system graphically or algebraically using either the elimination method or the substitution method.

To **solve a 2×2 linear system graphically,** we graph both of the equations on a calculator. It is helpful to consider the slope and vertical intercept of each equation to determine a reasonable window. The intersection of the two graphs is the solution to the system. We can check our solution by substituting the values for both variables into the *two original equations.*

To solve a 2 × 2 linear system using the **elimination method,** we choose factors to multiply one or both equations so that when the equations are added, one of the variables is eliminated. This produces an equation with only one variable. We solve this equation. Substitute the value for this variable into any of the equations containing two variables to determine the value of the other variable. Once we have the value of both variables, we check our solution by substituting these values into the *two original equations.*

To solve a 2 × 2 linear system using the **substitution method,** we solve one of the equations for one of the variables. Then we substitute the resulting expression into the other equation. This produces an equation with only one variable. We solve this equation, and proceed as we did in the elimination method.

We concluded this chapter by looking at **linear inequalities.** We found that the algebraic process for solving linear inequalities is the same as the process for solving equations except that if we multiply or divide both sides of an inequality by a negative number we must reverse the direction of the inequality.

The process used to **check the solution to an inequality** is more involved. We must check the border point, a point in the solution region, and a point that is not in the solution region. The **border point** should always make both sides of the inequality equal. A **point in the solution region** should make the inequality true. A **point that is not in the solution region** should make the inequality false.

Chapter Eight

Zero and Negative Integer Exponents

We have seen the importance and use of positive integer exponents in many applications. We know that they are shorthand notation for repeated multiplication. But what do negative integer exponents mean? In this section, we will learn the meaning of negative integer exponents and see how they can be used in problem situations. We will see that the properties of positive integer exponents will also apply to negative integer exponents.

8.1 Language and Evaluation

Activity Set 8.1

1. In biology we often study bacteria that reproduce by splitting. In ideal situations a population can double over a specific time. Suppose we are studying a population of bacteria that can double every hour. It has been determined that Monday at 8:00 A.M. 400 bacteria were in the population.

   a. How many bacteria were in the population Monday at 9:00 A.M.? At 10:00 A.M.? How many bacteria were in the population Monday at 7:00 A.M.? At 6:00 A.M.?

   b. Complete the following table for this situation.

Number of Hours after 8:00 A.M.	−3	−2	−1	0	1	2	3	24
Number of Bacteria in Population				400				

   c. If n represents *number of hours after 8:00 A.M.* and A represents *number of bacteria in the population*, write an equation for A in terms of n.

   d. Substitute $n = 2$ hours and $n = 3$ hours into your equation in part c. Do the values you get match the values in the table? Substitute $n = -2$ hours and $n = -3$ hours into your equation in part c. Do the values you get match the values in the table? If the values do not match, modify your equation and check the values again.

   e. Substitute $n = 0$ into your equation in part c. What is the value of 2^0?

   f. What fraction is equivalent to 2^{-1}? 2^{-2}? 2^{-3}?

2. Complete the following diagram. Express numbers as integers or fractions (not decimals).

   $2^4 = 16$
   $\quad\searrow$ Divide by 2.
   $2^3 = 8$
   $\quad\searrow$ Divide by 2.
   $2^2 = 4$
   $\quad\searrow$ Divide by 2.
   $2^1 =$
   $2^0 =$
   $2^{-1} =$
   $2^{-2} =$
   $2^{-2} =$

   How would you define x^{-n}, where n is a positive integer?

3. Evaluate each of the following expressions using a calculator.

   $\dfrac{1}{2^{-2}}$ $\qquad$ $\dfrac{1}{2^{-1}}$

   $\dfrac{1}{3^{-2}}$ $\qquad$ $\dfrac{1}{2^{-2}}$

   $\dfrac{1}{4^{-2}}$ $\qquad$ $\dfrac{1}{2^{-3}}$

   $\dfrac{1}{5^{-2}}$ $\qquad$ $\dfrac{1}{2^{-4}}$

   Determine a rule for $\dfrac{1}{x^{-n}}$, where x is a nonzero real number and n is a positive integer.

Discussion 8.1

Negative Integer Exponents: Language and Evaluation

In Section 5.1 we worked with positive integer exponents. In this section we will review expressions with positive integer exponents and extend the ideas to negative integer exponents. Reading the expressions to reflect their exact meanings will help us to interpret them precisely. The following examples will give us some practice reading expressions that include positive integer exponents.

Example 1

Evaluate each expression for the given value of the variable.

a. x^2 for $x = {}^-4$ b. $({}^-y)^2$ for $y = {}^-3$ c. ${}^-k^2$ for $k = 5$

Solution

a. The expression x^2 is *the square of* x. Because we are substituting a negative value for the variable, we must enclose the $^-4$ in parentheses when doing the substitution and the exponent is placed just outside of the parentheses. The expression becomes the square of $^-4$ or $(^-4)^2$.

$$(^-4)^2 = 16$$

b. The expression $(^-y)^2$ is *the square of the opposite of* y. Because y is $^-3$, the expression becomes the square of the opposite of $^-3$ or $(^{--}3)^2$. We have substituted $^-3$ into the expression for y.

$$(^{--}3)^2 = 3^2 = 9$$

c. The expression $^-k^2$ is *the opposite of the square of* k. Because k is 5, the expression becomes the opposite of the square of 5, or $^-5^2$. In this case, we do not need parentheses, but the expression could also be written as $^-(5)^2$ to show very clearly that it is only the 5 that is to be squared.

$$^-5^2 = ^-25 \quad \text{or} \quad ^-(5)^2 = ^-25$$

In the activities, you may have discovered how to interpret zero as an exponent. The equation $A = 400 * 2^0$ represented the number of bacteria in the population zero hours after 8:00 A.M.; that is, at 8:00 A.M. We know that at 8:00 A.M. 400 bacteria were in the population; that means 2^0 must equal 1. This agrees with the following definition.

Definition

If x is a nonzero real number, then the **zero power** of x is 1, or $x^0 = 1$.

You may have noticed that, in the definition, x is a nonzero number. What if we want to know what the expression 0^0 equals? Using the definition, we can create a pattern of numbers raised to the zero power,

$$3^0 = 1$$
$$2^0 = 1$$
$$1^0 = 1$$

and it seems logical that the next step should be $0^0 = 1$. But, if we look at a pattern of powers of zero,

$$0^3 = 0$$
$$0^2 = 0$$
$$0^1 = 0$$

it seems reasonable to think that $0^0 = 0$. Because of the difficulty in deciding which pattern to use, 0^0 is undefined.

> The expression 0^0 is undefined.

Next we look at negative integer exponents. Consider the expression $2^1 * 2^{-1}$. The properties of exponents should hold for negative exponents, so

$2^1 * 2^{-1}$
$= 2^{1+-1}$ Add exponents.
$= 2^0$
$= 1$

We see that $2^1 * 2^{-1} = 1$. This means that 2^{-1} must be equivalent to $\frac{1}{2}$, because $2 * \frac{1}{2} = 1$. We say that 2 and $\frac{1}{2}$ are multiplicative inverses because their product is 1. We conclude that $2^{-1} = \frac{1}{2}$.

Using this idea and the property $(x^m)^n = x^{m*n}$, we determine that $2^{-3} = (2^{-1})^3 = \left(\frac{1}{2}\right)^3$. That is, the negative third power of two is the cube of the reciprocal of two. This motivates the following definition.

> **Definition**
>
> If n is a positive integer and x is a nonzero real number then $x^{-n} = \left(\frac{1}{x}\right)^n$.

In words we say that x^{-n} is the nth power of the reciprocal of x.

If we then apply the property

$$\left(\frac{x}{y}\right)^m = \frac{x^m}{y^m}$$

to this result we obtain

$$x^{-n} = \left(\frac{1}{x}\right)^n = \frac{1^n}{x^n} = \frac{1}{x^n}$$

That is, x^{-n} is also the reciprocal of the nth power of x.

> If x is a nonzero real number,
>
> $$x^{-n} = \left(\frac{1}{x}\right)^n = \frac{1}{x^n}$$

Now that we know what zero and negative integer exponents mean, we need to return to the careful reading and evaluating of expressions. We are combining what we know of positive integer exponent expressions with our new definitions of zero and negative integer exponents. Remember that exponents apply only to what is to their immediate left in an expression.

Example 2

Decide how each expression is read, and evaluate the expression.

a. $(^-5)^0$ b. $(^-5)^2$ c. 5^{-2} d. $^-5^{-2}$ e. $(^-5)^{-2}$

Solution

a. The expression $(^-5)^0$ can be read as *the zero power of $^-5$*. The zero power of any nonzero number is 1. Therefore,

$$(^-5)^0 = 1$$

b. The expression $(^-5)^2$ can be read as *the square of $^-5$*. The square of negative 5 is $^-5 * {}^-5 = 25$. Therefore,

$$(^-5)^2 = 25$$

c. The expression 5^{-2} can be read as *the negative second power of 5*. The negative second power of 5 can be rewritten as $5^{-2} = \frac{1}{5^2}$ and $\frac{1}{5^2} = \frac{1}{25}$. Therefore,

$$5^{-2}$$
$$= \frac{1}{5^2}$$
$$= \frac{1}{25}$$

d. The expression $^-5^{-2}$ can be read as *the opposite of the negative second power of 5*. The negative second power of 5 can be rewritten as the reciprocal of the second power of 5, $\frac{1}{5^2} = \frac{1}{25}$. Therefore the opposite of the negative second power of 5 is

$$^-5^{-2}$$
$$= -\frac{1}{5^2}$$ Notice that the power applies only to the number 5 and not to the negative sign, which indicates that we are taking the opposite of the whole fraction.
$$= -\frac{1}{25}$$

e. The expression $(^-5)^{-2}$ can be read as *the negative second power of $^-5$*. The negative second power of $^-5$ can be rewritten as

$$(^-5)^{-2}$$
$$= \frac{1}{(^-5)^2}$$ Rewrite using a positive exponent. Notice that in this expression the power applies to both the 5 and the negative sign because the $^-5$ is enclosed in parentheses. Therefore, when we rewrite the expression, we must continue to group the $^-5$ in the denominator.
$$= \frac{1}{25}$$

Example 3

Evaluate each expression for the given value of the variable.

a. x^{-2} for $x = -4$ b. $(-y)^{-2}$ for $y = 3$ c. $-k^{-2}$ for $k = -5$

Solution

a. Evaluate the expression x^{-2} when $x = -4$.

The expression x^{-2} is *the negative second power of* x. Remember that we must enclose any negative number in parentheses when we substitute into an expression with a power.

$$x^{-2}$$
$$= (-4)^{-2} \quad \text{Substitute } -4 \text{ for the } x. \text{ Remember parentheses are necessary.}$$
$$= \frac{1}{(-4)^2} \quad \text{Rewrite using a positive exponent.}$$
$$= \frac{1}{16}$$

b. Evaluate the expression $(-y)^{-2}$ when $y = 3$.

The expression $(-y)^{-2}$ is *the negative second power of the opposite of* y.

$$(-y)^{-2}$$
$$= (-3)^{-2} \quad \text{Substitute 3 for } y.$$
$$= \frac{1}{(-3)^2} \quad \text{Rewrite using a positive exponent.}$$
$$= \frac{1}{9}$$

c. Evaluate the expression $-k^{-2}$ when $k = -5$.

The expression $-k^{-2}$ is *the opposite of the negative second power of* k. Because we are substituting a negative number for k, we must enclose it in parentheses. This gives us $-(-5)^{-2}$. Notice that the negative sign at the beginning of the expression is not inside the parentheses because the power does not apply to it. Therefore,

$$-k^{-2}$$
$$= -(-5)^{-2} \quad \text{Substitute } -5 \text{ for } k. \text{ Put the parentheses only around the } -5.$$
$$= -\frac{1}{(-5)^2} \quad \text{Rewrite using a positive exponent.}$$
$$= -\frac{1}{25}$$

In this section, we defined the result of raising a nonzero base to the **zero power** or to a **negative integer exponent.** Emphasis was placed on how to read expressions in a way that makes their meaning clear. We used the meaning and careful reading to help in evaluating expressions correctly.

Problem Set 8.1

1. Explain the difference between $(-3)^4$ and -3^4. How is each expression read?

2. Translate each of the following English phrases into correct symbolic mathematics. Do not simplify.
   a. the square of negative four
   b. the opposite of the square of four
   c. the square of the opposite of x
   d. the opposite of the square of x
   e. twice the square of three
   f. the square of the product of two and three
   g. the negative second power of negative three
   h. the negative fourth power of twice n

3. Translate each algebraic expression into an English phrase.
   a. $(-x)^2$ b. $-x^6$ c. $(xy)^2$ d. $(x+y)^3$

4. Without using a calculator, evaluate the following expressions.
   a. -2^2 b. -2^3 c. -2^4 d. -2^5

5. Without using a calculator, evaluate the following expressions.
   a. $(-2)^2$ b. $(-2)^3$ c. $(-2)^4$ d. $(-2)^5$

6. Complete the following table. Show your substitutions.

x	x^2	$-x^2$	$(-x)^2$
-4	$(-4)^2 = 16$		
0			
4			

Without using a calculator, evaluate the following numerical expressions.

7. a. 5^3 b. 5^{-3} c. 5^0

8. a. -2^3 b. -2^{-3} c. $(-8)^2$ d. -8^{-2} e. $(-8)^0$

9. a. $(-3)^3$ b. -3^{-3} c. -3^4 d. $(-3)^{-4}$

10. a. Show your substitution of $x = -3$ into the expression x^{-2}.
    b. Without evaluating the expression, decide whether the expression is positive or negative for the given substitution.
    c. Evaluate the expression.

11. a. Show your substitution of $x = -3$ into the expression x^{-3}.
    b. Without evaluating the expression, decide whether the expression is positive or negative for the given substitution.
    c. Evaluate the expression.

12. a. Show your substitution of $x = 0$ into the expression x^{-2}.
    b. Without evaluating the expression, decide whether the expression is positive, negative, zero, or none of these for the given substitution.
    c. Evaluate the expression.

13. Evaluate the following expressions for the given values of the variables. Show your substitution step. Express any approximate results to the nearest hundredth.
    a. $a^2 + b^3$ for $a = {^-}3$ and $b = {^-}5$
    b. $R^3 - T^2$ for $R = 8$ and $T = {^-}7$
    c. $(2kp)^4 + {^-}5kp^3$ for $k = {^-}6$ and $p = 9$
    d. $\dfrac{11(m^2n)^3}{4n}$ for $m = 2.32$ and $n = {^-}0.41$
    e. $3x^3y - (7xy^2)^3$ for $x = {^-}14.13$ and $y = {^-}0.27$
    f. x^{-2} for $x = {^-}2$
    g. $a^{-2} + b^{-4}$ for $a = {^-}4$ and $b = 2$
    h. $\dfrac{1}{a^2 + b^4}$ for $a = {^-}4$ and $b = 2$

14. Without using a calculator, decide which of the following numbers are equivalent.
    a. $3.4 * 10^{-2}$ b. 0.034 c. $3.4 * 10^2$ d. 3400 e. $0.34 * 10^{-1}$

15. Which of the following numbers are written in scientific notation?
    a. $0.56 * 10^{-2}$ b. $5.6 * 10^{-2}$ c. $7 * 10^2$ d. $56 * 10^2$ e. $3.700 * 10^{-1}$

16. Write the following numbers in scientific notation.
    a. $1900 * 10^2$ b. $0.19 * 10^{-2}$ c. $19 * 10^4$ d. $0.0019 * 10^6$ e. $19 * 10^{-6}$

8.2 Properties and Simplification

Negative exponents are defined so that the properties of positive integer exponents remain true. In each of the following activities, *apply the properties of integer exponents* to simplify each expression.

Activity Set 8.2

1. Simplify each expression using the properties of integer exponents. Verify by numerical substitution.
   a. $k^3 * k^5$ b. $m^5 * m^{-2}$

2. Simplify each expression using the properties of integer exponents. Verify by numerical substitution.
   a. $\dfrac{x^6}{x^2}$ b. $\dfrac{p^6}{p^{-4}}$

3. Simplify each expression using the properties of integer exponents. Verify by numerical substitution.
   a. $(x^3)^5$ b. $(k^{-2})^3$

4. Simplify each expression using the properties of integer exponents. Verify by numerical substitution.
   a. $(2y^4)^3$ b. $(2m^4)^{-2}$

Discussion 8.2

Recall that negative exponents are defined in the following way.

> If x is a nonzero real number,
> $$x^{-n} = \left(\dfrac{1}{x}\right)^n = \dfrac{1}{x^n}$$

Because this definition was created so that the properties of exponents remain true, we can restate the properties with the condition that m and n are integers. As you apply the properties, it is important to understand what each one says and how it is different from the others. Reading the verbal description of each property, as well as the symbolic description, should make it easier to decide which property is appropriate in a given situation.

> **Properties of Integer Exponents**
>
> Let x and y be nonzero real numbers. Let n and m be integers. Then the following are true:
>
> 1. $x^m x^n = x^{m+n}$ When *multiplying powers of* x, add the exponents.
> 2. $\dfrac{x^m}{x^n} = x^{m-n}$ When *dividing powers of* x, subtract the exponents.
> 3. $(xy)^m = x^m y^m$ The *power of a product* can be simplified by applying the power to each factor.
> 4. $\left(\dfrac{x}{y}\right)^m = \dfrac{x^m}{y^m}$ The *power of a quotient* can be simplified by applying the power to each factor in the numerator and to each factor in the denominator.
> 5. $(x^m)^n = x^{m*n}$ When simplifying a *power of a power of* x, multiply the exponents.

Some people have difficulty remembering when they need to multiply exponents and when they need to add them. Some people create their own *incorrect* rule that says if an expression includes parentheses then exponents should be multiplied. Consider the following four expressions.

$x^3 * x^4$

$x^3 x^4$

$x^3 \cdot x^4$

$x^3(x^4)$

All four of these expressions represent the product of x^3 and x^4. When we are *multiplying powers* of x we need to add the exponents. Therefore, all four of the expressions simplify to $x^{3+4} = x^7$. Notice that the last expression contains parentheses, yet we still add the exponents.

We only multiply the exponents when we are simplifying a *power of a power*. A *power of a power*, like $(x^3)^4$, always contains parentheses and simplifies to $x^{3*4} = x^{12}$. However, other expressions that are not powers of powers may also contain parentheses.

Next, we look at simplifying some expressions similar to those you did in the activities. When working with positive integer exponents, you have seen that you can often simplify either by thinking about what the powers mean or by applying the properties of exponents. In working with negative exponents, you will find that the properties will be extremely useful.

Example 1

Simplify each of the following expressions using the properties of integer exponents. Assume all variables are nonzero real numbers. Write results using only positive exponents.

a. $y^4 * y^{-6}$ b. $\dfrac{x^4}{x^{-3}}$ c. $(3M^{-4})^{-2}$

Solution

a. The expression $y^4 * y^{-6}$ is the product of powers of y. The properties tell us that to simplify the product of powers, we add the exponents.

$$y^4 * y^{-6}$$
$$= y^{4+{-6}} \quad \text{Add the exponents when finding the product of powers.}$$
$$= y^{-2}$$
$$= \frac{1}{y^2} \quad \text{Use the definition of negative exponents to rewrite with a positive exponent.}$$

VERIFY Let $x = 5$.

$$y^4 * y^{-6} \stackrel{?}{=} \frac{1}{y^2}$$
$$5^4 * 5^{-6} \stackrel{?}{=} \frac{1}{5^2}$$
$$0.04 = 0.04 \quad \checkmark$$

We conclude that $y^4 * y^{-6}$ simplifies to $\frac{1}{y^2}$.

b. The expression $\frac{x^4}{x^{-3}}$ is the quotient of powers of x. The properties of exponents tell us to subtract the exponents when dividing powers of x.

$$\frac{x^4}{x^{-3}}$$
$$= x^{4-(-3)} \quad \text{Subtract exponents when dividing powers of } x.$$
$$= x^7$$

VERIFY Let $x = 5$.

$$\frac{x^4}{x^{-3}} \stackrel{?}{=} x^7$$
$$\frac{5^4}{5^{-3}} \stackrel{?}{=} 5^7$$
$$78,125 = 78,125 \quad \checkmark$$

We conclude that $\frac{x^4}{x^{-3}}$ simplifies to x^7.

c. The expression $(3M^{-4})^{-2}$ is the power of a product. The properties of exponents tell us that we need to apply the power to each of the factors.

$$(3M^{-4})^{-2}$$
$$= 3^{-2} * (M^{-4})^{-2} \quad \text{Apply the power of } {-2} \text{ to both factors of the product.}$$
$$= \frac{1}{9}M^8 \quad \text{Use the definition of negative exponents. Multiply exponents to find the power of a power.}$$

VERIFY Let $M = 6$.

$$(3M^{-4})^{-2} \stackrel{?}{=} \frac{1}{9}M^8$$
$$(3 * 6^{-4})^{-2} \stackrel{?}{=} \frac{1}{9} * 6^8$$
$$186,624 = 186,624 \quad \checkmark$$

We conclude that $(3M^{-4})^{-2}$ simplifies to $\frac{1}{9}M^8$.

Before simplifying more algebraic expressions with negative exponents, we look at some true/false statements to help us clarify the properties of exponents.

Example 2

First decide whether each of the following equations is true for all values of the variable. Then substitute numerical values into the equation to verify your conjecture. If the statement is not true for all values, rewrite the right-hand side of the equation to make it true. Assume that all variables are nonzero.

a. $(3m^2)^{-3} = -9m^{-6}$ b. $\dfrac{k^5}{k^{-3}} = k^8$ c. $2x^{-3}(x^2 + 5) = 2x^{-6} + 10x^{-3}$

Solution

a. Many simplified results may "look correct." In this first example it *appears* that the negative third power has been applied to each of the factors in the parentheses, so we might assume that this statement is true. However, it is always a good idea to substitute numerical values into an expression to determine if your guess was correct.

VERIFY Let $m = 4$.

$(3m^2)^{-3} \stackrel{?}{=} -9m^{-6}$

$(3 * 4^2)^{-3} \stackrel{?}{=} -9 * 4^{-6}$

$0.000009 \neq -0.002$ ✗

When we substitute $m = 4$ into the statement, we see that the statement is *false*. So our guess was incorrect! Let's see if we can determine why. From the properties of exponents we know that in the expression $(3m^2)^{-3}$ we can apply the power to each factor in the parentheses.

$(3m^2)^{-3}$

$= 3^{-3}(m^2)^{-3}$ Apply the power to each factor.

$= \dfrac{1}{27}m^{2*-3}$ Simplify 3^{-3}, and multiply exponents.

$= \dfrac{1}{27}m^{-6}$

or $\approx 0.037m^{-6}$ Write $\dfrac{1}{27}$ as an approximate decimal.

In the second step, notice that we take the negative third power of the coefficient 3. We do not multiply the base of 3 and the exponent of -3. We conclude that the statement $(3m^2)^{-3} = -9m^{-6}$ is *false*, and one correct statement is $(3m^2)^{-3} = \dfrac{1}{27}m^{-6}$.

b. We might initially think that the statement $\dfrac{k^5}{k^{-3}} = k^8$ is false. We do not expect division to give us a larger exponent. Let's substitute a number for k to see if it is true or false.

VERIFY Let $k = 4$.

$\dfrac{k^5}{k^{-3}} \stackrel{?}{=} k^8$

$\dfrac{4^5}{4^{-3}} \stackrel{?}{=} 4^8$

$65{,}536 = 65{,}536$ ✓

The statement is true when $k = 4$. Remember that when we are dividing powers, we subtract the exponents. Therefore,

$$\frac{k^5}{k^{-3}}$$
$$= k^{5--3}$$
$$= k^8$$

We conclude that the statement $\frac{k^5}{k^{-3}} = k^8$ is true.

c. The statement $2x^{-3}(x^2 + 5) = 2x^{-6} + 10x^{-3}$ may look true. It appears that we have distributed the $2x^{-3}$ over the sum. We substitute numbers into the statement to check its correctness.

VERIFY Let $x = 4$.

$$2x^{-3}(x^2 + 5) \stackrel{?}{=} 2x^{-6} + 10x^{-3}$$
$$2 * 4^{-3}(4^2 + 5) \stackrel{?}{=} 2 * 4^{-6} + 10 * 4^{-3}$$
$$0.65625 \neq 0.15674 \quad \text{✗}$$

The statement is false when we substitute a value for x. Let's look more carefully at simplifying the expression.

$$2x^{-3}(x^2 + 5)$$
$$= 2x^{-3} * x^2 + 2x^{-3} * 5 \quad \text{Distribute the } 2x^{-3} \text{ over the sum.}$$
$$= 2x^{-3+2} + 10x^{-3} \quad \text{Add the exponents in the first term. Multiply the numerical factors in the second term.}$$
$$= 2x^{-1} + 10x^{-3}$$

The mistake was to multiply the exponents when *multiplying powers of* x. The exponents should have been added. We conclude that the statement $2x^{-3}(x^2 + 5) = 2x^{-6} + 10x^{-3}$ is *false*, and the true statement is $2x^{-3}(x^2 + 5) = 2x^{-1} + 10x^{-3}$.

Example 3

Simplify the following expressions. Express the results using only positive exponents. Assume that all variables are nonzero real numbers. Verify your results numerically.

a. $2k^{-4} * 5k^7$ b. $\frac{k^{-4}}{k^2} + (3k^{-3})^2$ c. $2x^{-3}(x^5 - 4)$

Solution a. $2k^{-4} * 5k^7$

$= 2 * 5 * k^{-4} * k^7$ Re-order the multiplication.
$= 10k^{-4+7}$ Add exponents.
$= 10k^3$

VERIFY Let $k = 4$.

$$2k^{-4} * 5k^7 \stackrel{?}{=} 10k^3$$
$$2 * 4^{-4} * 5 * 4^7 \stackrel{?}{=} 10 * 4^3$$
$$640 = 640 \quad \checkmark$$

We conclude that $2k^{-4} * 5k^7$ simplifies to $10k^3$.

b. $$\frac{k^{-4}}{k^2} + (3k^{-3})^2$$

$= k^{-4-2} + 3^2 * (k^{-3})^2$ Subtract exponents in the first term. Apply the exponent to each of the factors in the second term.

$= k^{-6} + 9k^{-6}$ Simplify each term.

$= 10k^{-6}$ Combine like terms.

$= \dfrac{10}{k^6}$ Rewrite using a positive exponent.

Notice that in the expression $10k^{-6}$ the exponent -6 applied only to k and not to 10, so 10 remains in the numerator.

VERIFY   Let $k = 5$.

$$\frac{k^{-4}}{k^2} + (3k^{-3})^2 \stackrel{?}{=} \frac{10}{k^6}$$

$$\frac{5^{-4}}{5^2} + (3 * 5^{-3})^2 \stackrel{?}{=} \frac{10}{5^6}$$

$$0.00064 = 0.00064 \quad \checkmark$$

We conclude that $\dfrac{k^{-4}}{k^2} + (3k^{-3})^2$ simplifies to $\dfrac{10}{k^6}$.

c. We can use the distributive property and properties of exponents to simplify this expression.

$$2x^{-3}(x^5 - 4)$$

$= 2x^{-3} * x^5 - 2x^{-3} * 4$ Distribute the $2x^{-3}$ over the difference.

$= 2x^{-3+5} - 8x^{-3}$ Add the exponents when multiplying powers.

$= 2x^2 - \dfrac{8}{x^3}$ Rewrite the second term using a positive exponent.

Notice that in the expression $8x^{-3}$ the exponent -3 applied only to x and not to 8, so 8 remains in the numerator.

VERIFY   Let $x = 5$.

$$2x^{-3}(x^5 - 4) \stackrel{?}{=} 2x^2 - \frac{8}{x^3}$$

$$2 * 5^{-3}(5^5 - 4) \stackrel{?}{=} 2 * 5^2 - \frac{8}{5^3}$$

$$49.936 = 49.936 \quad \checkmark$$

We conclude that $2x^{-3}(x^5 - 4)$ simplifies to $2x^2 - \dfrac{8}{x^3}$.

Example 4

In chemistry, we are sometimes asked to evaluate expressions similar to the following one. To verify that an expression is evaluated correctly when using technology, it is helpful to estimate the result using paper, pencil, and mental arithmetic techniques prior to using the technology to evaluate the expression.

$$\frac{200.59}{1} * \frac{1}{6.0221 * 10^{23}} * \frac{0.035242}{1}$$

For the preceding numerical expression, do the following.

a. Estimate the result by rewriting each of the factors in the expression in scientific notation, rounding the numbers, and applying the properties of exponents to the powers of 10.

b. Evaluate the expression, in one step, using a calculator, and round your result to five significant digits.

c. Compare your results from parts a and b.

Solution a.

$$\frac{200.59}{1} * \frac{1}{6.0221 * 10^{23}} * \frac{0.035242}{1}$$

$$= \frac{2.0059 * 10^2}{1} * \frac{1}{6.0221 * 10^{23}} * \frac{3.5242 * 10^{-2}}{1}$$

Rewrite each number in scientific notation.

$$\approx \frac{2 * 10^2}{1} * \frac{1}{6 * 10^{23}} * \frac{4 * 10^{-2}}{1}$$

Round each number to a single digit.

$$= \frac{2 * 4}{6} * \frac{10^2 * 10^{-2}}{10^{23}}$$

Collect the coefficients and powers of 10.

$$= \frac{8}{6} * 10^{2+-2-23}$$

Apply the properties of exponents to the powers of 10.

$$\approx \frac{6}{6} * 10^{-23}$$

Because it is difficult to divide 6 into 8, we round 8 to 6 to make the division easy.

$$= 1 * 10^{-23}$$

Our estimate is

$$\frac{200.59}{1} * \frac{1}{6.0221 * 10^{23}} * \frac{0.035242}{1} \approx 1 * 10^{-23}$$

b. Next we enter the original expression in a calculator and evaluate it.

$$\frac{200.59}{1} * \frac{1}{6.0221 * 10^{23}} * \frac{0.035242}{1} \approx 1.1739 * 10^{-23}$$

c. In comparing our results, we see that $1 \approx 1.1739$ and the magnitudes (powers of 10) agree. The results are close enough to feel confident with the computed value.

In this section, we extended simplifying of expressions to include negative integer exponents. We saw that the properties of exponents still apply. When using negative exponents, we need to rely more on the properties than we did when simplifying expressions involving only positive integer exponents. Estimation of complex numerical expressions involving scientific notation was introduced as a way of verifying our calculator results.

Problem Set 8.2

1. Simplify each expression. Assume that all variables are nonzero real numbers. Express the results using only positive exponents. Verify that your result is correct by numerical substitution.

   a. $a^4 \cdot a^7$

   b. $3x^2 \cdot 5x^3$

   c. $b^5 * b^{-3}$

   d. $(5v^{-2})^3$

   e. $(z^{-5})^2$

   f. $\left(\dfrac{5}{x^2}\right)^3$

   g. $2w^2 \cdot 5w^{-5}$

2. **Always True?** First decide whether each of the following equations is true for all values of the variable. Then substitute numerical values into the equation to verify your conjecture. If the statement is not true for all values, rewrite the right-hand side of the equation to make it true. Assume that all variables are nonzero.

   a. $A^3(A)^4 = A^{12}$

   b. $(A + B)^3 = A^3 + B^3$

   c. $y^3 + y^3 = 2y^3$

   d. $4A^{-1} = \dfrac{1}{4A}$

   e. $3^2 * 4^3 = 12^5$

3. Simplify each expression. Assume that all variables are nonzero real numbers. Express the results using only positive exponents. Round any approximate numerical coefficients to the nearest hundredth. Verify that your result is correct by numerical substitution.

   a. $\left(\dfrac{6m^3}{n}\right)^2$

   b. $2k^4 + (2k)^4$

   c. $p^2(p^6) - 3(p^4)^2$

   d. $\dfrac{12a^6}{(-5a^2)^2}$

   e. $3k(4 - k) + (3k)^2$

   f. $\dfrac{x^3}{x^{-3}}$

   g. $4y^5(3y^2 + 5x)$

   h. $\dfrac{x^3 + x^4}{x^{-3}}$

   i. $\dfrac{15m}{3m^{-3}}$

   j. $\dfrac{x^3 * x^4}{x^{-3}}$

   k. $(4x)^2 + x^5 * x^{-3} - \dfrac{x^2}{x^4}$

   l. $3x^{-2}(5x^5 + 2) - x^3$

4. **Always True?** First decide whether each of the following equations is true for all values of the variable. Then substitute numerical values into the equation to verify your conjecture. If the statement is not true for all values, rewrite the right-hand side of the equation to make it true. Assume that all variables are nonzero.

   a. $k^3 + k^3 = k^6$

   b. $(A + B)^{-2} = \dfrac{1}{A^2} + \dfrac{1}{B^2}$

   c. $\dfrac{m^4}{m^{-3}} = m^7$

   d. $B^{5x} * B^{-4x} = B^x$

   e. $w^5 + w^{-4} = w$

   f. $\dfrac{B^{-5}}{B^3} = B^2$

   g. $x^{m+1} * x^{m-1} = x^{2m}$

5. Estimate the value of the following expressions. (*Hint:* First round each number to one significant digit, rewrite in scientific notation, and apply the properties of exponents to the powers of 10.) Next, evaluate the original expression, in one step, using your calculator. Round results to three significant digits. Compare this result to your estimate.

   a. $\dfrac{5.97 * 10^{-4}}{6.022 * 10^{-23}}$

   b. $(6.91 * 10^{-11})(1.87 * 10^9)^2$

   c. $(0.00285)(5.98 * 10^{24})(6{,}370{,}000)$

   d. $\dfrac{0.08206}{5.00 * 10^{-2}} * \dfrac{2.21 * 10^{18}}{1} * 298$

6. You want to purchase a new car. In a business math book you find the formula for the amount of money you can borrow in terms of the monthly payment, the interest rate, and the total number of payments. The formula is

$$A = R * \frac{1 - (1 + i)^{-n}}{i}$$

where A = amount of the loan

R = monthly payment

i = interest rate *per month,* as a decimal

n = total number of months of the loan

a. You are looking at a four-year loan with an annual interest rate of 9%. If you can afford monthly payments of $200, how much money can you borrow?

b. If you paid $200 a month for four years, what is the total amount you paid? How much of the total amount paid was for the interest on the loan?

c. About how much do you need to pay monthly for the same type of loan if you want to borrow $9000 for a car?

7. a. A graduated cylinder contains about 14.2 milliliters of mercury. If the density of mercury, under laboratory conditions, is about 13.534 g/cm^3, how many moles of mercury are in the cylinder? About 200.59 g of mercury is in one mole.

b. If one mole contains about $6.0221 * 10^{23}$ atoms, how many atoms of mercury are in the cylinder?

8. a. Determine the perimeter of each of the following figures.

Figure 1   Figure 2   Figure 3

b. Using your results and continuing the pattern, complete a table similar to the following.

Figure Number	Perimeter of Figure
1	
2	
3	
4	
5	
n	

c. Graph the perimeter of the figures in terms of the figure number.

Chapter Eight Summary

We know that a *positive exponent* is used as a shorthand notation for *repeated multiplication*. For example, x^4 means $x*x*x*x$. Because a negative number is the opposite of a positive number and division is the inverse of multiplication, a *negative exponent* is used as a shorthand notation for *repeated division*. That is, x^{-4} represents $\frac{1}{x^4}$ or, equivalently

$$\frac{1}{x*x*x*x}$$

Through applications and the properties of exponents we discovered the following definition.

$$\boxed{\begin{array}{c} \text{If } x \text{ is a nonzero real number,} \\[4pt] x^{-n} = \left(\frac{1}{x}\right)^n = \frac{1}{x^n} \end{array}}$$

The following are key ideas to use when working with negative exponents.

- Remember that exponents apply to the item that is the exponent's immediate left. For example, $-5^{-2} = \frac{-1}{5^2}$ and $(-5)^{-2} = \frac{1}{(-5)^2}$.

- When evaluating expressions that include negative exponents, rewrite the expression so that the exponent is positive using the preceding definition.

- When substituting a negative number for a variable that is raised to a power, include parentheses around the negative number and place the exponent just to the right of the parentheses.

- When simplifying algebraic expressions involving negative exponents, it is *not* necessary to rewrite the exponent so that it is positive. Instead, apply the properties of exponents using the exponents given. Then, verify your result numerically.

- Remember that there are *no* properties where an exponent is applied to the terms of a sum or difference. The only way to rewrite a power of a sum or difference is to write out the meaning of the exponent. Then perform the resulting multiplication. For example,

$$= (x+5)^2$$
$$= (x+5)(x+5)$$
$$= x^2 + 10x + 25$$

Chapter Nine

Geometry

The world we live in contains objects of many different shapes and sizes. Geometry is the study of points, lines, surfaces, solids, and angles. In other words, when we study geometry we are learning about relationships between many of the objects we use every day. For this reason, you will find that the applied geometry you learn in this chapter is useful in many situations in daily life.

9.1 Angles and Triangles

Activity Set 9.1

Throughout the remainder of the text, assume that all figures are *not* drawn to scale unless otherwise indicated or unless you are directed to measure the figure.

1. Measure each angle to the nearest degree.

   a.

   b.

   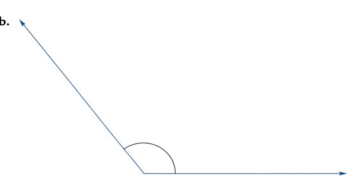

2. Use a ruler and protractor to carefully draw a triangle ABC with m $\angle A = 30°$, m $\angle B = 100°$, $AC = 7.7$ cm, and $AB = 6$ cm. Measure side BC and $\angle C$.

3. Use a ruler and protractor to carefully draw a square, each side of which has length 5.2 cm.

4. Use a ruler and protractor to carefully draw a trapezoid with bases $3\frac{3}{8}$ inches and $4\frac{3}{4}$ inches. The nonparallel sides of a trapezoid may have different lengths.

5. Compare your figures in Activities 2–4 with those of another team. Are they the same? Should they be?

6. Draw a quadrilateral $PQRS$ with $PQ = 5.5$ cm, m $\angle Q = 138°$, m $\angle P = 71°$, $QR = 4.0$ cm, and m $\angle R = 67°$. Measure side PS and $\angle S$. Compare your results with those of other teams.

7. a. Draw two parallel lines. Draw a third line that crosses the two parallel lines, but is *not* perpendicular. This third line is called a **transversal.**
   b. Measure the eight resulting angles.
   c. Which angles have the same measurement? Which angles add up to 180°?
   d. Compare your results with those of another team member. Are your angle measurements the same? Are the angle relationships you found the same as those your teammates found?
   e. Organize the angles with equal measurements into as many *pairs* as possible. Did you get the same pairs as your teammates?

8. **a.** Draw two nonparallel lines similar to the following figure.
   b. Draw a third line that crosses the two lines.
   c. Measure the eight resulting angles. What conclusions can be made about the resulting angles if no two lines are parallel?

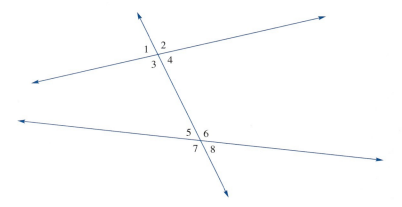

9. **a.** Carefully draw any triangle. Measure each angle in your triangle. What is the sum of the measures of the angles in your triangle?
   b. Cut out your triangle. Shade the tip of each vertex so that you can identify the vertices later. Then cut off the corners of your triangle as indicated in the following picture. Place the three vertices of the triangle together. What do you notice?

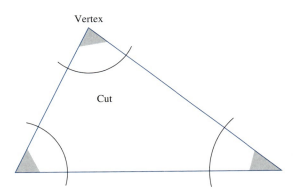

10. **a.** An **obtuse triangle** has one angle whose measure is greater than 90°. Draw an obtuse triangle. Measure each angle in your triangle. What is the sum of the measures of the angles in your triangle?
    b. Cut out your triangle. Then cut off the corners of your triangle as you did in Activity 9. Place the three vertices of the triangle together. What do you notice?

11. What is the sum of the measures of the angles in any triangle?

12. Measure each angle and each side in the following triangle. Match the smallest angle with the shortest side, the second smallest angle with the second shortest side, and the largest angle with the longest side.

Smallest Angle	Shortest Side	Second Smallest Angle	Second Shortest Side	Largest Angle	Longest Side

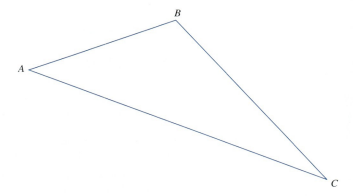

13. Repeat Activity 12 for the following triangle.

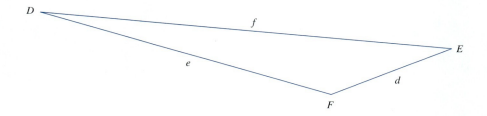

14. In Activities 12 and 13, what relationship did you find between the measures of the angles and the lengths of the sides?

15. In an isosceles triangle the angles opposite the sides of equal length are called base angles. Based on your observations from Activities 12 and 13, what can you say about the measure of the base angles in an isosceles triangle? Test your conjecture.

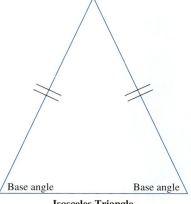

Isosceles Triangle

Discussion 9.1

Angles

In this section we will be discussing angles and relationships between angles. To begin this discussion we need to introduce some vocabulary.

> **Definition**
>
> An **angle** is formed by two rays with a common initial point or two line segments with a common endpoint. This common point is known as the **vertex** of the angle.

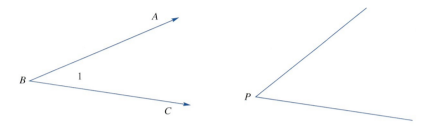

We can refer to the angle in the left-hand figure as ∠1, ∠B, or ∠ABC. If we use three letters to denote an angle, the vertex must be the middle letter.

An angle is measured in degrees. A protractor is a device for measuring angles. To measure an angle, place the center of the protractor on the vertex of the angle. Line up one side of the angle with the baseline. A protractor has two scales, and you use the baseline to determine which scale to use. Because we measure from the baseline, the baseline must correspond to 0°. In the following figure the top scale gives us 0° for the baseline, so we measure our angle using the top scale. The measure of ∠FGH in the figure is 40°. In symbols, we could write m ∠FGH = 40°. It is always a good idea to make sure your answer is reasonable. Clearly the measure of the angle is less than 90°; therefore, it must measure 40° rather than 140°.

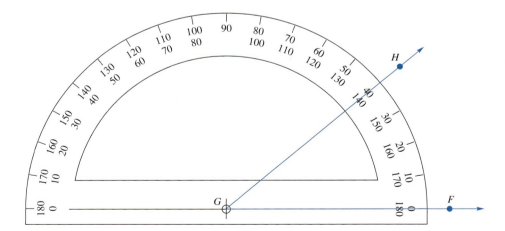

> **Definition**
>
> An angle that measures 90° is called a **right angle.** An angle that measures 180° is called a **straight angle.** An angle that measures between 0° and 90° is called an **acute angle,** and an angle that measures more than 90° and less than 180° is called an **obtuse angle.**

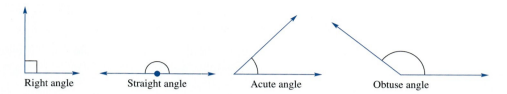

| Right angle | Straight angle | Acute angle | Obtuse angle |

NOTE: A small square in the corner of an angle indicates that it measures 90°.

> **Definition**
> Two geometrical figures are **congruent** if they have the same shape and the same measurements.
> - Two angles are congruent if they have the same measure.
> - Two line segments are congruent if they are the same length.
> - Two polygons are congruent if all of their corresponding sides and all of their corresponding angles are congruent.

Relationships Between Pairs of Angles

> **Definition**
> When two intersecting lines are drawn, four angles are formed. Pairs of angles that share a common side between them are called **adjacent angles.** Angles that are on opposite sides of the point of intersection are called **vertical angles** or **opposite angles.**

In the following figure, ∠1 and ∠2 are adjacent angles, while ∠2 and ∠4 are vertical angles. We saw in the activities that vertical angles are congruent, that is, their measures are equal.

> **Vertical Angles Theorem**
> If two angles are vertical angles, then they are congruent.

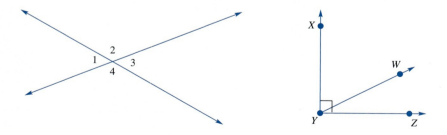

What is the sum of the measures of ∠1 and ∠2? What is the sum of the measures of ∠XYW and ∠WYZ in the preceding figure? We give special names to these pairs of angles.

> **Definition**
> If the sum of measures of two angles is 180°, then the angles are **supplementary.** If the sum of measures of two angles is 90°, then the angles are **complementary.**

In the preceding figure, ∠1 and ∠2 are supplementary angles, and ∠XYW and ∠WYZ are complementary angles. These represent special cases of supplementary and complementary angles because the angles are also adjacent. Supplementary and complementary angles do not need to be adjacent.

Example 1

In the following figure, assume that the angles are drawn to scale.

a. Identify two obtuse angles in the figure.
b. Identify four acute angles in the figure.
c. Identify one right angle in the figure.
d. Identify a pair of adjacent complementary angles.
e. Identify two pairs of adjacent supplementary angles.
f. Identify a pair of nonadjacent complementary angles.

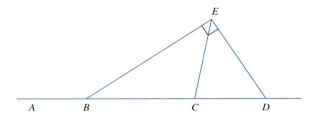

Solution

a. ∠ABE and ∠BCE are obtuse.
b. ∠EBC, ∠ECD, ∠BEC, ∠CED, and ∠EDC are acute angles.
c. ∠BED is a right angle.
d. ∠BEC and ∠CED are adjacent complementary angles.
e. ∠ABE and ∠EBC are a pair of adjacent supplementary angles, as are ∠BCE and ∠DCE.
f. In the activities, we found that the sum of the measures of the angles in a triangle is 180°. Because the measure of ∠BED is 90°, the sum of the measures of ∠EBD and ∠EDB is 90°, for a total of 180° in the triangle. Therefore, ∠EBD and ∠EDB are nonadjacent complementary angles.

In the activities, we discovered some relationships between pairs of angles formed when two parallel lines intersect a third line called a **transversal**. To begin discussing these relationships, we introduce some vocabulary related to angles created in this situation.

In the following picture, line L is parallel to line M, and line T is a transversal cutting through L and M. Angles 1, 2, 7, and 8 are called **exterior angles**. Angles 3, 4, 5, and 6 are called **interior angles**. Because ∠3 and ∠6 are interior angles lying on opposite sides of the transversal, they are called **alternate interior angles**. Similarly, ∠4 and ∠5 are alternate interior angles.

Next, consider ∠1, ∠2, ∠3, ∠4, and ∠5, ∠6, ∠7, ∠8. Because ∠4 and ∠8 are both in the lower right-hand corner of their respective group, we call these angles **corresponding angles**. Similarly, ∠1 and ∠5 are corresponding angles because they are in the upper left-hand corner of their respective group. Can you name any more pairs of corresponding angles?

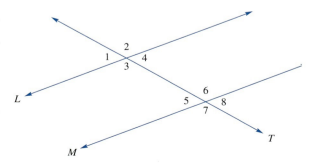

In the activities, we found that when two parallel lines are cut by a transversal, alternate interior angles are equal and corresponding angles are equal.

Using these relationships, we know that the following angles in the preceding figure are equal.

m ∠3 = m ∠6   Alternate interior angles
m ∠4 = m ∠8   Corresponding angles
m ∠5 = m ∠8   Vertical angles

Can you identify a different pair of alternate interior angles? Corresponding angles? Vertical angles?

To summarize, if two parallel lines are cut by a transversal, then alternate interior angles are equal and corresponding angles are equal. The converse of this statement is also true: If two lines are cut by a transversal and the alternate interior angles have the same measure, then the lines are parallel.

> **Alternate Interior Angles Theorem**
> Two lines are parallel if and only if alternate interior angles are congruent.

> **Corresponding Angles Theorem**
> Two lines are parallel if and only if corresponding angles are congruent.

Relationships Between Angles in Triangles

As mentioned previously, the sum of the measures of the angles in a triangle is 180°. We can see why this is always true by using relationships we found previously. Consider the following triangle.

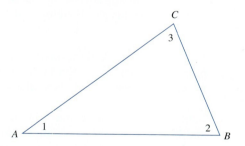

Draw a line parallel to one of the sides of the triangle. Line *L* is parallel to side *AB*.

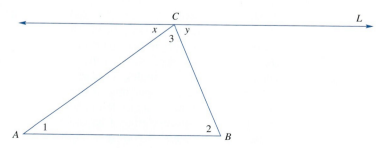

Side AC and side BC are transversals of lines L and AB. Because line L is parallel to side AB, alternate interior angles are equal. In this case,

m ∠1 = m ∠x
m ∠2 = m ∠y

Then, m ∠x + m ∠3 + m ∠y = 180° because they form a straight angle. Therefore, by substitution

m ∠1 + m ∠3 + m ∠2 = 180°

So, we can see that the sum of the measures of the angles in any triangle is 180°. In the problem set you will solve problems using the relationships between angles and triangles that we discovered in this section.

> **Sum of the Interior Angles of a Triangle**
> The sum of the interior angles of any triangle is 180°.

In this section we looked at angles, vocabulary of angles, and relationships between angles. We found that **vertical angles** are always equal, the measures of **complementary angles** sum to 90°, and the measures of **supplementary angles** sum to 180°. When two lines are parallel and cut by a transversal we discovered relationships between many pairs of angles, including **alternate interior angles** and **corresponding angles**. And finally, we found that the sum of the measures of the angles in a triangle is 180°.

Assume that all figures are *not* drawn to scale unless otherwise indicated or unless you are directed to measure the figure.

1. Determine the measure of ∠BDC in the following figure.

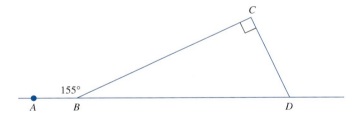

2. Determine the measures of angles 1 and 2 in the following rectangle.

3. If lines *L* and *M* are parallel, find *x*, *y*, and *z*.

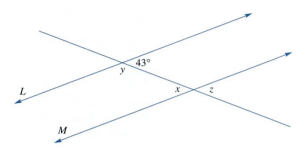

4. Determine the measures of the missing angles in the following figure.

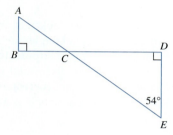

5. In the following figure, line *L* is parallel to line *M*, and angle 3 is a right angle.
   a. Name a pair of alternate interior angles.
   b. Name a pair of corresponding angles.
   c. Name a pair of nonadjacent complementary angles.

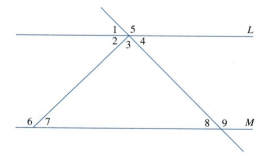

6. Assume lines *L* and *M* are parallel. Explain why each pair of angles given have the same measure. Use the vocabulary vertical angles, corresponding angles, and so forth in your explanation.

   a. ∠1 and ∠4   **b.** ∠3 and ∠6   **c.** ∠3 and ∠7   **d.** ∠2 and ∠7

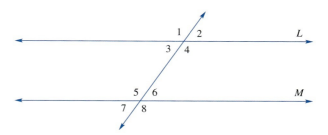

7. Determine the value of *x*.

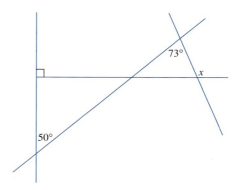

8. Determine the measure of ∠DBC.

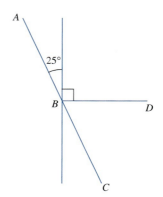

9. The line through points *A, B, C,* and *D* forms the deck of a bridge. Lines *BE* and *CF* represent pole supports, and they form right angles with the bridge deck. Line *EF* is parallel to line *AD*, line *EC* is parallel to line *FD*, and line *FB* is parallel to line *EA*. All other lines are support cables. Find the measures of the following angles.

   a. m ∠FCH =

   b. m ∠EFH =

   c. m ∠CHF =

   d. m ∠CDF =

   e. m ∠DCF =

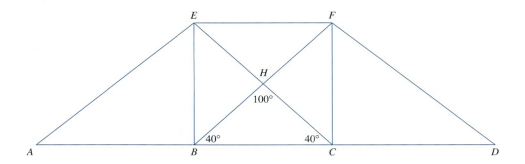

10. The following figure represents a roof truss. Line *BF* is parallel to line *DG*, and line *BG* is parallel to *DH*. The measure of angle *GDC* = 51° and m ∠HGF = 135°. Triangles *DGB* and *AGE* are isosceles triangles, with *BG* = *DG* and *AG* = *EG*. Find the measures of the following angles.

    a. m ∠FBA =

    b. m ∠DBG =

    c. m ∠HDG =

    d. m ∠DGB =

    e. m ∠BAF =

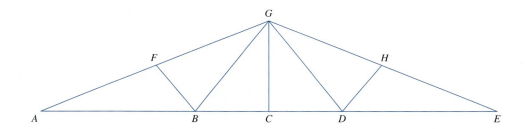

11. It can be difficult to determine whether two lines are parallel by inspection. Use your knowledge and properties related to parallel lines to determine whether lines *P* and *Q* are parallel. Explain how you made your decision.

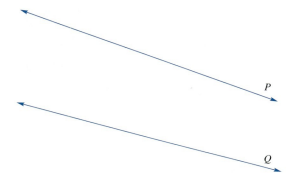

12. Determine the measures of the missing angles.

   a.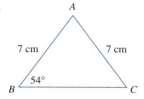

   b.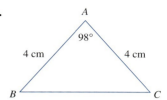

13. **Andres' Ramp.** Andres needs to make a wheelchair ramp for his niece. The ground is level where the ramp is to be built, and he has decided that the ramp should form a 4.7° angle with the ground. Both ends of the board used to make the ramp need to be cut at an angle. Determine the measure of these angles.

   The figure shows two representations of the ramp. The first picture shows the finished ramp, and the second picture shows the ramp before the board is cut. The second picture indicates the angles to be cut.

   Finished ramp

   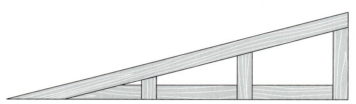

   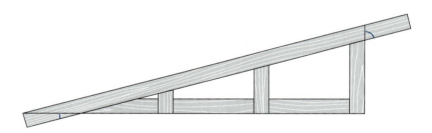

Activity Set 9.2

9.2 Polygons and Circles

1. The following pie chart shows a breakdown of how the average student uses time. What percent of the average student's life is spent on each activity? Explain how you decided.

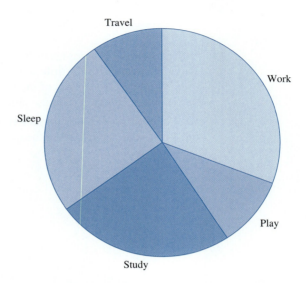

Activities for an Average Person

2. In a circle, an angle with its vertex at the center is called a **central angle**. An angle formed by two chords with the vertex on the circle is called an **inscribed angle**. A **chord** is any line segment connecting two points on a circle.

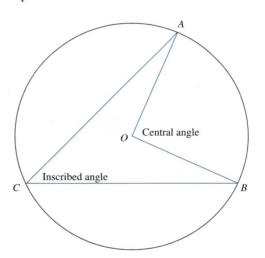

   a. Measure the central angle $\angle AOB$ and inscribed angle $\angle ACB$ in the figure. What conjecture might you make about the relationship between the central and inscribed angles in a circle?

   b. Mark a point D on the circle between points A and C. Draw $\angle ADB$. Compare the measures of angles $\angle ADB$ and $\angle AOB$. Does your conjecture from part a hold true?

9.2 Polygons and Circles 359

3. a. A quadrilateral is a four-sided polygon. For each of the quadrilaterals in the figure, measure each angle and determine the sum of the measures of the angles.

  b. What is the sum of the measures of the angles in any convex quadrilateral?

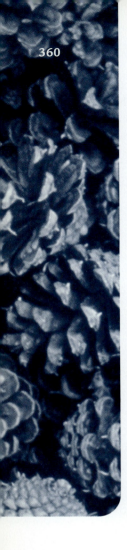

4. a. A pentagon is a five-sided polygon. For each pentagon in the figure, measure each angle and determine the sum of the measures of the angles.

   b. What is the sum of the measures of the angles in any convex pentagon?

5. A hexagon is a six-sided polygon. What do you think the sum of the measures of the angles in any convex hexagon is? Test your conjecture by drawing any convex hexagon and measuring the angles.

Discussion 9.2

In this section we will be discussing angle relationships in polygons and circles. When working with circles, the convention is to label the center of a circle with the letter O. Therefore, if a point is labeled O and is near the center of the circle, assume that this point is the center unless otherwise stated.

Definition

An **arc** of a circle is two points on the circle and the continuous part of the circle between the two points. The two points are the endpoints of the arc. A **minor arc** is an arc that is smaller than the semicircle, and a **major arc** is an arc that is larger than the semicircle.

Minor arcs are named with the letters of the two endpoints of the arc. Major arcs and semicircles are named with three letters; the first and last letters are the endpoints of the arc, and the middle letter is any other point on the arc.

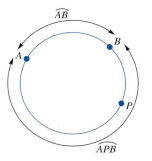

$\overset{\frown}{AB}$ is a minor arc of the circle.

$\overset{\frown}{APB}$ is a major arc.

Definition

In a circle, an angle formed with its vertex at the center is called a **central angle**. An angle formed by two chords with the vertex on the circle is called an **inscribed angle**. A **chord** is any line segment connecting two points on a circle.

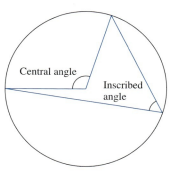

A central angle has its vertex at the center of the circle and intersects the circle in two points. The arc between these two points is called the **intercepted arc of the central angle**. In the following figure $\overset{\frown}{AB}$ is the intercepted arc of central angle AOB. We can also say that $\overset{\frown}{AB}$ **subtends** $\angle AOB$.

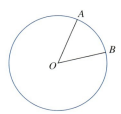

An inscribed angle has its vertex on the circle and intersects the circle in two other points. The arc formed by the inscribed angle is **the intercepted arc of the inscribed angle**. In the following figure, $\overset{\frown}{RS}$ is the intercepted arc of inscribed angle RPS and $\overset{\frown}{DEF}$ is the intercepted arc of inscribed angle DGF. We can also say that $\overset{\frown}{RS}$ subtends $\angle RPS$, and $\overset{\frown}{DEF}$ subtends $\angle DGF$.

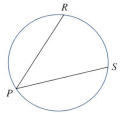 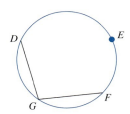

> **Definition**
> The **measure of a minor arc** is equal to the measure of its central angle. The **measure of a major arc** is equal to 360° minus the measure of the central angle.

In the following circle, $\overarc{AB}$ is the minor arc and $\overarc{ACB}$ is the major arc. The measure of $\overarc{AB}$ is equal to the measure of $\angle AOB$. The measure of $\overarc{ACB}$ is equal to 360° minus the measure of $\angle AOB$.

Measure of arc $AB = m \angle AOB$

Measure of arc $ACB = 360° - m \angle AOB$

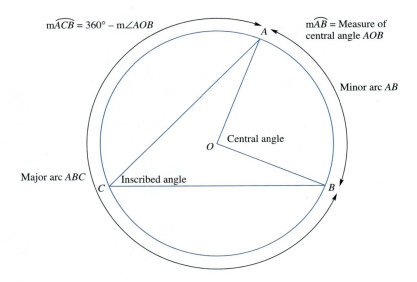

In the activities, you found the measure of an inscribed angle is equal to one half the measure of its intercepted arc.

> **Central and Inscribed Angles**
> - In a circle the measure of a central angle is equal to the measure of its intercepted arc.
> - In a circle the measure of an inscribed angle is equal to one half the measure of its intercepted arc.

Example 1

Find the measure of each angle in the following figure. Segment *BC* is the diameter of the circle, and the measure of $\widehat{AC}$ is 120°.

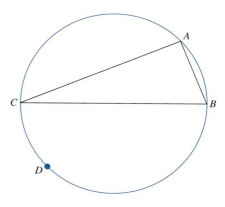

Solution Angle *CBA* is an inscribed angle that is subtended by arc *AC*. The measure of arc *AC* is 120°. Therefore,

$$\text{Measure of } \angle CBA = \tfrac{1}{2} * \text{Measure of } \widehat{AC}$$
$$= \tfrac{1}{2} * 120°$$
$$= 60°$$

Arc *BDC* is a semicircle. Therefore the measure of $\widehat{BDC}$ is 180°. Next, ∠*BAC* is the inscribed angle that is subtended by $\widehat{BDC}$. Therefore,

$$\text{Measure of } \angle BAC = \tfrac{1}{2} * \text{Measure of } \widehat{BCD}$$
$$= \tfrac{1}{2} * 180°$$
$$= 90°$$

The sum of the measures of the angles in a triangle is 180°. So,

$$\text{Measure of } \angle ACB = 180° - m\angle CBA - m\angle BAC$$
$$= 180° - 60° - 90°$$
$$= 30°$$

Let's look at a specific circle to see why the measure of an inscribed angle is equal to one half the measure of its intercepted arc. In the following figure we want to show that the measure of $\angle 3 = \tfrac{1}{2}(\text{measure of arc } AB)$. Because measure of ∠1 = measure of arc *AB*, we want to show that the

$$m\angle 3 = \tfrac{1}{2}(m\angle 1).$$

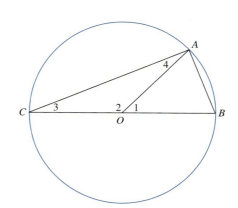

In the figure, ∠1 and ∠2 are supplementary angles. So,

m ∠2 = 180° − m ∠1

In △AOC, sides AO and OC are equal because both are radii of the circle. The angles opposite equal sides must also be equal. Therefore, ∠3 and ∠4 are equal.

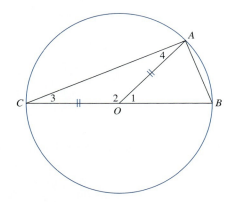

Then, because ∠2, ∠3, and ∠4 are angles in △AOC,

m ∠2 + m ∠3 + m ∠4 = 180°

(180° − m ∠1) + m ∠3 + m ∠4 = 180°   Substitute (180° − m ∠1) for m ∠2.

(180° − m ∠1) + m ∠3 + m ∠4 − **(180° − m ∠1)** = 180° − **(180° − m ∠1)**

Subtract 180° − m ∠1 from both sides.

m ∠3 + m ∠4 = m ∠1   Simplify both sides.

Now, because m ∠3 = m ∠4, m ∠3 = $\frac{1}{2}$(m ∠1).

So, we can see in this case that the measure of the inscribed angle is one half the measure of its intercepted arc.

Let's look at another situation involving circles and angles. We begin with circle O and a point P on the circle. Next, we draw line PT that is perpendicular to the radius OP as seen in the following figure.

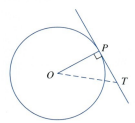

It appears that line PT is **tangent** to circle O; that is, the line intersects the circle in only one point. Because T is any point on the line distinct from P, if we can show that T cannot be on the circle, then we know that PT is tangent to the circle. The line segment OP is a radius of the circle, and any other point on the circle is equidistant from O. But △OPT is a right triangle, and OT is the hypotenuse of the triangle. Therefore, OT is greater than OP so the point T cannot be on the circle. We showed that a line perpendicular to the radius of a circle at its outer endpoint is tangent to the circle.

It can also be shown that any line tangent to a circle is perpendicular to a radius of the circle at the point of intersection. Because this proof is more difficult, we did not include it.

> *Tangent Line to a Circle Theorem*
>
> A line perpendicular to a radius of a circle at its outer endpoint is tangent to the circle.
>
> Conversely, if a line is tangent to a circle, it is perpendicular to a radius of the circle at the point of intersection.

In this section we continued our study of angles by looking at angles in polygons and circles. We looked at major and minor arcs of a circle and how they are named. We found that the measure of a central angle and its intercepted arc are equal and that the measure of an inscribed angle in a circle is equal to one half the measure of its intercepted arc. Additionally we saw that a line tangent to a circle is perpendicular to the radius at the point of tangency.

In the activities for this section you may have discovered that the sum of the measures of the interior angles of a quadrilateral is 360° and the sum of the interior angles of a pentagon is 540°. In the problem set, you will have the opportunity to create a formula for the sum of the measures of the interior angles of any convex polygon.

Problem Set 9.2

Assume that all figures are not drawn to scale unless otherwise indicated or unless you are directed to measure the figure.

1. Draw a pie chart (circle graph), similar to the one in Activity 1, to represent the following data.

Rent	$700
Utilities	$150
Phone	$ 80
Insurance	$100
Other	$350
Savings	$200

2. a. Complete the following table for yourself.
   b. Draw a pie chart to represent the data in your table. Label your pie chart with the appropriate percents.
   c. Determine the ratio of your study time to your class time.

Activity	Number of Hours Spent Each Week on Activity
Working	
Sleeping	
In class	
Studying	
Exercising	
Commuting	
Other	

3. In the following figure, the measure of angle T is 28°. Determine the measures of the other angles.

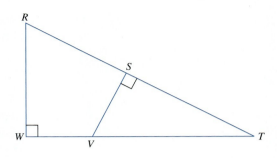

4. Determine the measures of the missing angles.

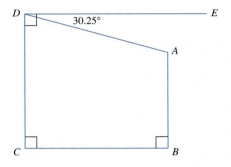

5. a. Identify four different central angles in the following figure.
   b. Identify four different inscribed angles in the figure.
   c. Identify four pairs of supplementary angles.
   d. Identify a pair of complementary angles.
   e. Assuming AE is parallel to BD, identify two pairs of equal angles that are not vertical angles.

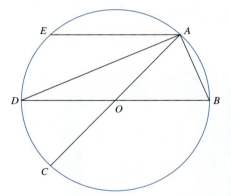

6. In the following circle, m ∠COB = 40°. Find the measures of the following angles.

   a. ∠COE
   b. ∠CEO
   c. ∠CEA
   d. ∠OAE

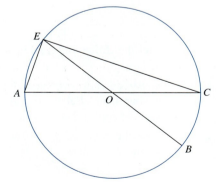

7. Draw a circle.
   a. In your circle, draw a central angle whose measure is less than 90°.
   b. Draw a central angle whose measure is between 90° and 180°.
   c. Draw an inscribed angle.
   d. Draw an inscribed angle whose measure is equal to 90°.

8. In this problem, when we refer to polygons, we mean convex polygons. The goal of this problem is to determine the sum of the measures of the interior angles in a convex polygon.
   a. A **diagonal** in a polygon is a line segment connecting two nonadjacent vertices. Start at one vertex in a quadrilateral. How many diagonals can be drawn from that one vertex? How many triangles does this create?
   b. Start at one vertex in a pentagon. How many diagonals can be drawn from that one vertex? How many triangles does this create? How does the sum of the measures of the angles in the triangles relate to the sum of the measures of the angles in the pentagon?
   c. A polygon with n sides is called an n-gon. How many diagonals can be drawn from one vertex in an n-gon? How many triangles does this create?
   d. Write a formula for the sum of the measures of the angles in an n-gon. Verify that your formula is correct for triangles, quadrilaterals, and pentagons.

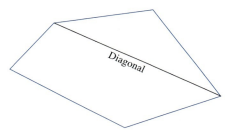

9. A concave polygon is a polygon in which at least one segment connecting two vertices lies outside the polygon. The following figure shows three concave polygons. Draw a concave pentagon and a concave hexagon. Determine the sum of the interior angles of your polygons. Are the results the same as for convex polygons?

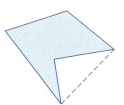 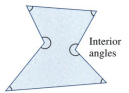 

10. A regular polygon is a polygon in which all sides are congruent and all angles are congruent. What is the measure of each angle in a regular octagon (eight sides)?

11. In a circle, a secant line is a line intersecting the circle in two points. Two secant lines are drawn as shown in the following figure. The measure of $\overset{\frown}{ED}$ is 100°, and the measure of $\overset{\frown}{BC}$ is 40°. Determine the measure of the angle formed at the intersection of the two secant lines, ∠BAC.

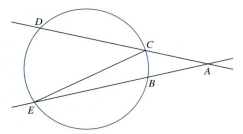

12. Line *PS* is tangent to the following circle at *P*, and line *RS* is tangent to the circle at *R*.
    a. Determine the measure of ∠OPS.
    b. Determine the length of *OP* if *PS* = 8 and *OS* = 12.
    c. Determine the length of *SR*.

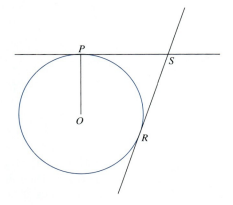

9.3 Similar Triangles

Activity Set 9.3

Definition
A pair of triangles are called **similar triangles** if the measures of each of the angles of one triangle are equal to the measures of each of the angles in the other triangle. The angles whose measures are equal are called **corresponding angles.**

Definition
In a pair of similar triangles, the sides opposite corresponding angles are called **corresponding sides.**

In the following figure, triangles I and II are similar and triangles III and IV are similar.

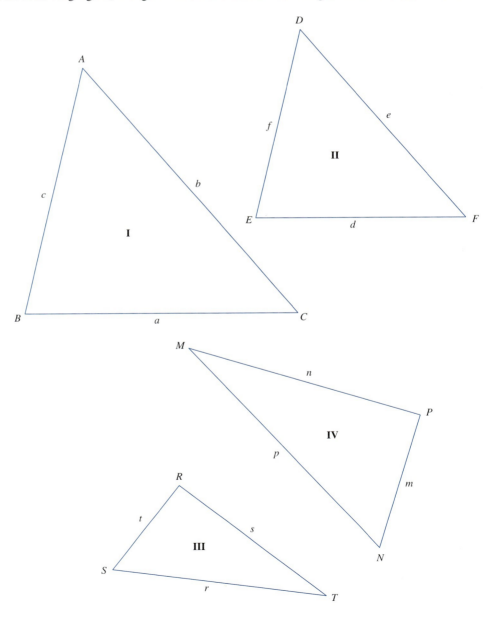

1. Using a ruler and protractor, carefully measure each of the angles and sides in the triangles in the preceding figure. Do the triangles fit the definition of similar triangles?

2. a. Calculate the ratios of the lengths of corresponding sides for each pair of similar triangles to fill in the following tables; that is, find the value of $\frac{a}{d}$, and so on. Write your results as decimals.

TRIANGLES I AND II	
Corresponding Sides	$\dfrac{\text{Side of Triangle I}}{\text{Side of Triangle II}}$
a and d	

TRIANGLES III AND IV	
Corresponding Sides	$\dfrac{\text{Side of Triangle III}}{\text{Side of Triangle IV}}$

b. Compare your results with those of your team. What do you observe about the ratios of corresponding sides in a pair of similar triangles?

Discussion 9.3

In the activities, we were introduced to the definitions of similar triangles, corresponding angles, and corresponding sides.

> **Definition**
>
> A pair of triangles are called **similar triangles** if the measures of each of the angles of one triangle are equal to the measures of each of the angles in the other triangle. The angles whose measures are equal are called **corresponding angles.**

> **Definition**
>
> In a pair of similar triangles, the sides opposite corresponding angles are called **corresponding sides.**

From these definitions, if two triangles are similar, their corresponding angles are equal. It is also true that if corresponding angles are equal in two triangles, then the triangles are similar. This can also be said as "two triangles are similar if and only if their corresponding angles are equal."

Some new notation will make it easier to talk about similar triangles. In the activities, you may have noticed that the vertices of the triangles were all labeled with capital letters and the sides labeled with lower-case letters. This is a common convention.

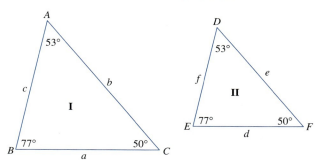

In talking about triangles I and II in the preceding figure, we can say that △ABC ~ △DEF. The symbol ~ is read as "is similar to." The order in which the vertices of the triangles are given indicates corresponding angles. Because m ∠A is equal to m ∠D, m ∠B = m ∠E, and m ∠C = m ∠F, we would say △ABC ~ △DEF. The position of each vertex in the label tells us which vertex it corresponds to in the other triangle. Once we know corresponding angles, we automatically know corresponding sides. For the triangles shown in the figure, can we say △BAC ~ △EDF? Can we say △ACB ~ △DEF?

In Activity 2, you may have found that the ratios of corresponding sides in triangles ABC and DEF were all approximately equal to 1.3. This ratio is sometimes called a **scale factor.** In this case, it indicates that each side in △ABC is approximately 1.3 times as long as the corresponding side in △DEF. For example, the length of side a is approximately 1.3 times the length of side d. What was the scale factor you identified for triangles MPN and TRS?

The following box summarizes the properties of similar triangles.

> **Properties of Similar Triangles**
>
> 1. Similar triangles have equal angles. Conversely, triangles with equal angles are similar.
> 2. The ratios of corresponding sides in similar triangles are a constant. Conversely, if the ratios of the corresponding sides are constant, the triangles are similar.

Example 1

In the following figure, △GHJ ~ △KPN. Find the lengths of the missing sides.

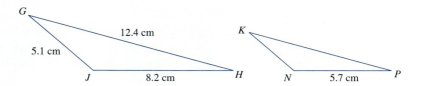

Solution Because the two triangles are similar, we know that the ratios of corresponding sides are equal. The following figures show the corresponding sides visually differentiated by the way the lines are drawn. You can do the same by using different colors to draw the lines.

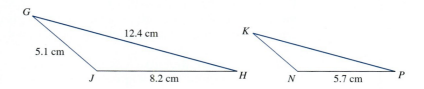

Side NP corresponds to side JH, and side KN corresponds to GJ. Therefore, we can write ratios using these sides. When we write our ratios we need to be sure and be consistent. For example, in the following proportion, each numerator is the length of a side in the larger triangle, and each denominator is the length of the corresponding side in the smaller triangle.

$$\frac{JH}{NP} = \frac{GJ}{KN}$$

$$\frac{8.2 \text{ cm}}{5.7 \text{ cm}} = \frac{5.1 \text{ cm}}{KN}$$ Substitute the values that we know.

$$\frac{8.2}{5.7} = \frac{5.1 \text{ cm}}{KN}$$ Cancel the units on the left-hand side.

$$5.7 * 5.1 \text{ cm} = 8.2 * KN$$ Cross multiply.

$$\frac{5.7 * 5.1 \text{ cm}}{8.2} = KN$$ Divide both sides by 8.2.

$$3.5 \text{ cm} \approx KN$$

Similarly,

$$\frac{JH}{NP} = \frac{GH}{KP}$$

$$\frac{8.2 \text{ cm}}{5.7 \text{ cm}} = \frac{12.4 \text{ cm}}{KN}$$ Substitute the values that we know.

$$\frac{8.2}{5.7} = \frac{12.4 \text{ cm}}{KN}$$ Cancel the units on the left-hand side.

$$5.7 * 12.4 \text{ cm} = 8.2 * KP$$ Cross multiply.

$$\frac{5.7 * 12.4 \text{ cm}}{8.2} = KP$$ Divide both sides by 8.2.

$$8.6 \text{ cm} \approx KP$$

We conclude that KN is approximately 3.5 cm and KP is approximately 8.6 cm.

Example 2

a. In the following figure, segment *DE* is parallel to segment *CB*. Show that triangle *ABC* is similar to triangle *AED*, that is, △*ABC* ~ △*AED*.

b. Find the length of segment *CD* if *DE* = 8 ft, *CB* = 12 ft, and *AD* = 9 ft.

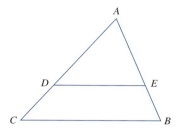

Solution

a. If we can show that the measures of corresponding angles in the two triangles are equal, we know that the triangles are similar.

We can see that ∠*A* in △*ABC* corresponds to ∠*A* in △*AED*. Clearly, m ∠*A* = m ∠*A*. Because *DE* is parallel to *CB* and cut by the transversal *AC*, ∠*ACB* and ∠*ADE* are corresponding angles. Therefore, m ∠*ACB* = m ∠*ADE*. Similarly, because *DE* is parallel to *CB* and cut by the transversal *AB*, ∠*ABC* and ∠*AED* are corresponding angles. Therefore, m ∠*ABC* = m ∠*AED*.

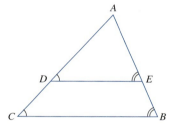

Because the measures of each of the angles of one triangle are equal to the measures of each of the corresponding angles of the other triangle, we conclude that △*ABC* ~ △*AED*; the two triangles are similar.

Notice that we talk about corresponding angles not only when a transversal cuts parallel lines but also in similar triangles. The context should make it clear which case we mean.

b. The following figure shows the corresponding angles marked and the given lengths labeled. Angle *A* corresponds to itself in both triangles.

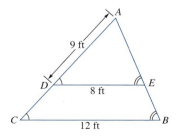

Separating the two similar triangles can make it easier to see corresponding sides. Again corresponding sides are indicated with the same style line.

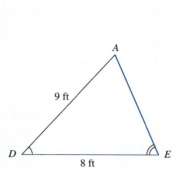
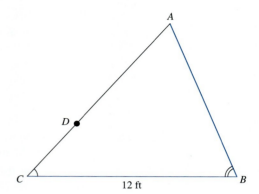

Because the two triangles are similar, we know that ratios of corresponding sides are equal. Therefore, we can write a proportion using corresponding sides.

$$\frac{CB}{DE} = \frac{AC}{AD}$$

$$\frac{12 \text{ ft}}{8 \text{ ft}} = \frac{AC}{9 \text{ ft}}$$

$$9 \text{ ft} * \frac{12 \text{ ft}}{8 \text{ ft}} = \frac{AC}{9 \text{ ft}} * 9 \text{ ft} \qquad \text{Multiply both sides by 9 ft to solve for } AC.$$

$$\frac{9 \text{ ft} * 12}{8} = AC \qquad \text{Rewrite the left-hand side.}$$

$$13.5 \text{ ft} = AC$$

The length we want to determine is CD. Because $CD = AC - AD$,

$$CD = 13.5 \text{ ft} - 9 \text{ ft} = 4.5 \text{ ft}$$

We conclude that the length of CD is 4.5 ft.

In the previous example, we could have solved the problem by cross multiplying. Because the variable was in the numerator, we were able to solve the problem by multiplying both sides by the denominator of the right-hand side instead.

We saw that some of the angle relationships of the previous section can be useful in showing that pairs of angles in triangles are equal. This in turn can help us to show that the triangles are similar. If two triangles are similar, we can form proportions using pairs of corresponding sides. As long as we know three of the lengths in a proportion, we can solve for the unknown length.

In this section, we were introduced to **similar triangles.** We saw how to label similar figures to make it easier to identify the **corresponding angles and sides.** Similar triangles were used to solve problems that we previously could not solve.

Problem Set 9.3

Assume that all figures are *not* drawn to scale unless otherwise indicated or unless you are directed to measure the figure.

1. In the following figure, $\triangle ABC \sim \triangle TUV$. Find the lengths of the missing sides.

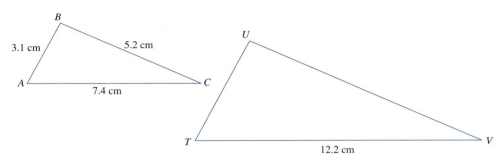

2. In the following figure, $\triangle DEF \sim \triangle NPL$. Find the lengths of the missing sides.

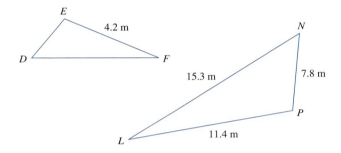

3. Simplify each expression, and verify your results.

    a. $y^5 y^2 y$ **b.** $(4m)^2$ **c.** $\dfrac{8n^4}{2n}$ **d.** $p^3(p^5)$ **e.** $3x^2 * {}^-8x$ **f.** $(p^3)^5$

4. a. Which of the following triangles are similar? Explain your reasoning.
   b. Determine the length of side *DF*.

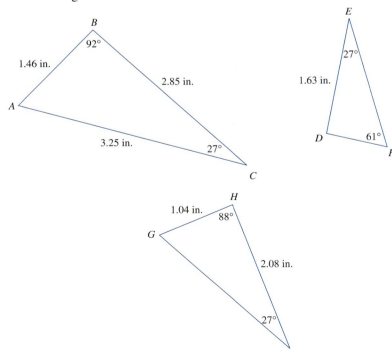

5. In the following figure if CD is parallel to AB, find the length of CD.

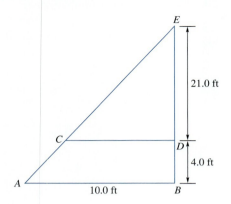

6. In the following figure, line L is parallel to line M.
   a. Name one pair of vertical angles.
   b. Name one pair of alternate interior angles.
   c. Name one pair of corresponding angles.
   d. Name one pair of adjacent angles.
   e. Name one pair of nonadjacent supplementary angles.

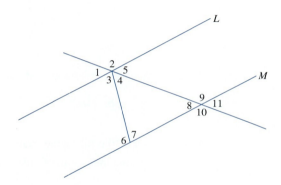

7. a. Which of the following triangles are similar? Explain your reasoning.
   b. Determine the length of sides WX and RT.

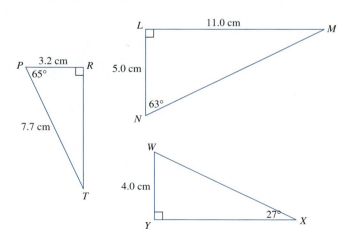

8. Given that segment *AE* is parallel to segment *DC*, identify the similar triangles in the following figure. Indicate corresponding sides by the way you name the triangles. Explain how you know the triangles are similar.

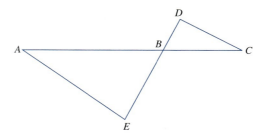

9. In the following figure △*GHJ* ~ △*MNK*. Find the lengths of the missing sides.

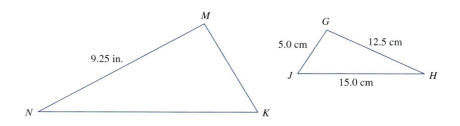

10. In the following figure, *QR* is parallel to *MP*. If you know that *QR* = 6.0 meters, *NR* = 5.4 meters, and *RP* = 4.6 meters, find the length of *MP*.

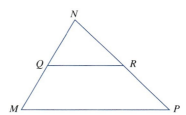

11. Handicapped Ramp. Jeff is building a ramp that has a horizontal length of 6 yards. The vertical height is 18 inches. For that length, he realizes that he needs to put in braces as shown in the following diagram. Assuming that the braces are equally spaced, find the height of each brace. The measurements are to the nearest eighth of an inch.

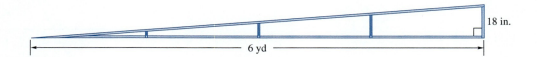

12. Framing the Shed Wall. Jay is building a storage shed. The two end walls are to be framed as shown in the following figure. The vertical studs are 16 inches apart measured from the centers of the studs. If the tallest of the studs in the triangular region is 30 inches, determine the length that Jay should cut each of the other studs in the triangular region. The measurements are to the nearest eighth of an inch.

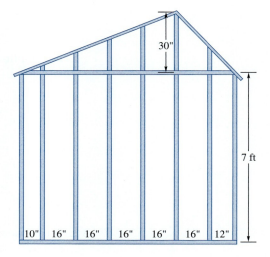

13. In the following figure m ∠GCH = 60°, line CH is parallel to JD, and lines BH, GC, and FD are vertical and perpendicular to line AE. Determine the measures of ∠CHB, ∠CDJ, and ∠DJC.

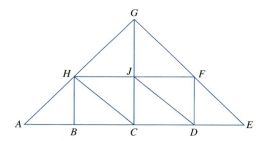

14. Lines PA and PB are tangent to the circle in the following figure, and m $\angle APB = 25°$. Find the measure of $\overset{\frown}{AB}$ and the measure of $\angle OBA$.

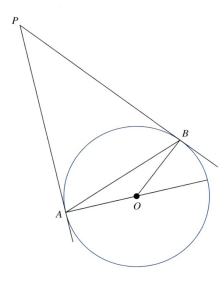

15. Line AC is tangent to the following circle. Angle AOC is a right angle, segment $AO = 16$ cm, and segment $OC = 20$ cm. What is the radius of the circle?

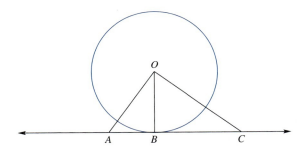

Activity Set 9.4

9.4 Scale Drawings

1. **Shortest Route.** The following is part of the street map of Fantasy City. Draw the shortest route a person can walk between *A* and *B*. Determine the actual length of this route. The map *is* drawn to scale.

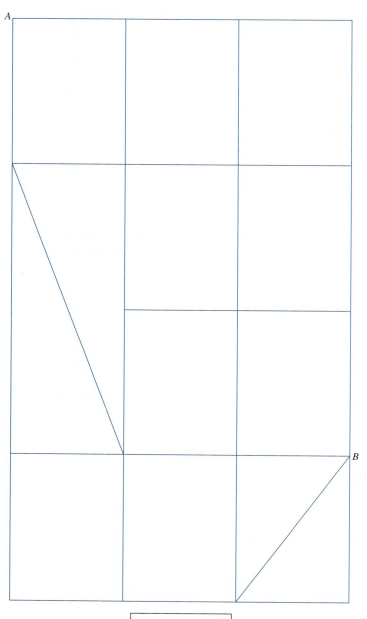

Scale: 3 cm = 44 yd

2. A forester is trying to estimate the height of a tree. Standing 43 feet from the base of the tree, the forester determines that the angle between the ground and the top of the tree is 52°. The following drawing is *not* to scale. Using a scale of 1 cm = 3 feet, draw the situation to scale. Use your drawing to approximate the actual height of the tree.

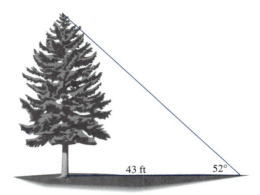

3. Corrine is trying to determine where to place her furniture in a room in her new home. A rough sketch of the room is shown here. Draw the room to scale on a piece of paper so that Corrine can try with different furniture arrangements. Make the scale drawing as large as possible. What scale did you use?

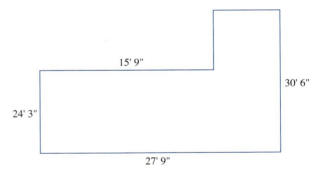

Discussion 9.4

In the activities, we used scale drawings or created scale drawings to answer questions. This is an extension of the idea of similar triangles. A scale drawing is a figure that is similar to the original. That is, a scale drawing has the same shape but not necessarily the same size as the original figure. We can make scale drawings of any type of figure, not just triangles. Scale drawings are often used to show objects on paper that would be either too large or too small to be shown at their actual size. Blueprints, maps, and detail pictures of microorganisms are examples of scale drawings.

As we saw in the activities, scale drawings can be used to solve problems in which making direct measurements would be difficult. When making a scale drawing, the angle measurements in the drawing are equal to the original angles, and the lengths of the sides are proportional to the original. The ratio of the drawing to the original figure is called the scale ratio or scale factor. When creating a scale drawing, we need to make the drawing a reasonable size, meaning that it needs to be small enough to fit on the size paper we are using but large enough to retain as much accuracy as necessary.

Example 1

Width of the River. A team of students is trying to determine the approximate width of a river. They have stretched a string between two stakes 20 feet apart along the bank of the river. One of the stakes is directly across the river from a tree stump. Siting from the other stake to the tree, they determine the angle between their string and the line of sight to the stump to be approximately 32°. How can they use this information to find the distance across the river?

Solution Let's first make a rough sketch to visualize the situation.

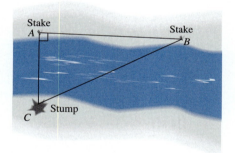

By accurately drawing a scaled-down version of the situation, we can determine the distance from A to C. On our scale triangle, we let each 2 ft be represented by 1 cm. This means that on our drawing, the distance between the stakes is $20 \text{ ft} * \frac{1 \text{ cm}}{2 \text{ ft}} = 10$ cm. We can draw a 10-cm line to correspond to AB in the sketch. A common technique in labeling scale drawings is to use A' to correspond to the original point A, B' to correspond to B, and so on. The symbol A' is read as "A prime." Using this notation, our 10-cm line is labeled $A'B'$.

Next, using a protractor, we can draw an accurate 32° angle at B'. Because the stump is directly across from A, we can draw an accurate 90° angle at A'.

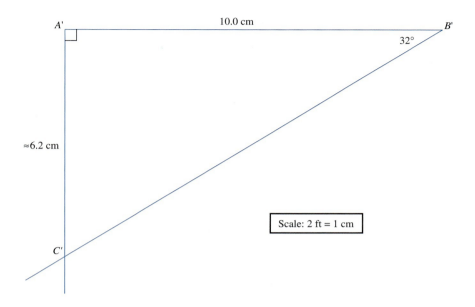

After carefully drawing the triangle, we can measure the length $A'C'$. This is found to be approximately 6.2 cm. We can then use unit fractions to convert back to the actual measure from the scale drawing measure.

$$6.2 \text{ cm} * \frac{2 \text{ ft}}{1 \text{ cm}}$$
$$= 12.4 \text{ ft}$$

We conclude that the distance across the river is approximately 12.4 ft.

Notice in the preceding example that the scale is labeled on the drawing. This is essential for the reader to interpret the drawing.

Angles in Context

In many problem situations, angles are given with reference to compass directions. The angles are measured with respect to the north–south line. If we are told to go N38°W, we first locate north and then measure an angle of 38° toward west from the north line. This direction is shown in the following figure. Directions given in this way are often referred to as **bearings** or **headings.**

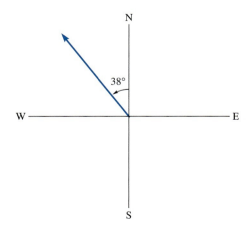

The following figure shows a bearing of S64°E.

Notice that in both examples, north or south is the direction given first. We start at the north–south line and indicate at what angle toward east or west to head. Bearing angles are often read starting with the angle measurement. For example N38°W can be read as 38° west of north.

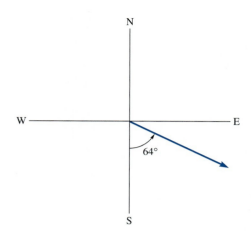

Example 2

Outdoor School. A sixth-grade outdoor school class has just learned how to read a compass. They are going to try their new skills on an outdoor course that was set up by one of their counselors. They are to follow these directions.

- Walk 15 feet due east.
- Head N20°E for 24 feet.
- Head N85°E for 10 feet.
- Walk 35 feet due south.
- Head S30°W for 12 feet.
- At this point you should be able to find a marker. Sign your name on the marker, and then head back to where you started for lunch.

Assuming that the sixth-graders make it to the marker successfully and return to the starting point in a straight course, how far will they need to walk back to the starting point?

Solution We start by making a rough sketch of the situation.

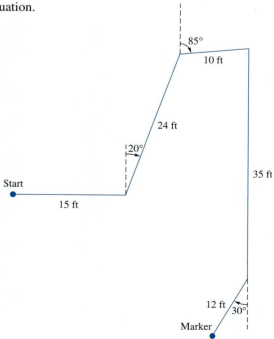

Next, we need to choose a scale. We have some fairly large lengths. The largest length is 35 ft. If we let 1 cm = 2 ft, then 35 ft is represented by 17.5 cm. It is not clear if this will all fit on a page with this scale. Therefore, let's chose our scale to be 1 cm = 3 ft. Then we can convert all of our lengths into centimeters.

$$15 \text{ ft} * \frac{1 \text{ cm}}{3 \text{ ft}} = 5 \text{ cm} \qquad 35 \text{ ft} * \frac{1 \text{ cm}}{3 \text{ ft}} \approx 11.7 \text{ cm}$$

$$24 \text{ ft} * \frac{1 \text{ cm}}{3 \text{ ft}} = 8 \text{ cm} \qquad 12 \text{ ft} * \frac{1 \text{ cm}}{3 \text{ ft}} = 4 \text{ cm}$$

$$10 \text{ ft} * \frac{1 \text{ cm}}{3 \text{ ft}} \approx 3.3 \text{ cm}$$

With these lengths we can now make our scale drawing. Notice that we must indicate our scale with our drawing.

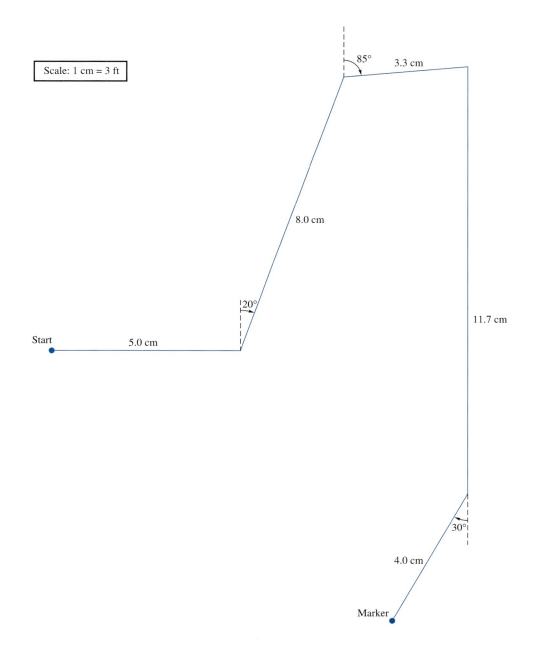

Finally, we can measure the distance between the marker and the start. Then, convert this distance to feet.

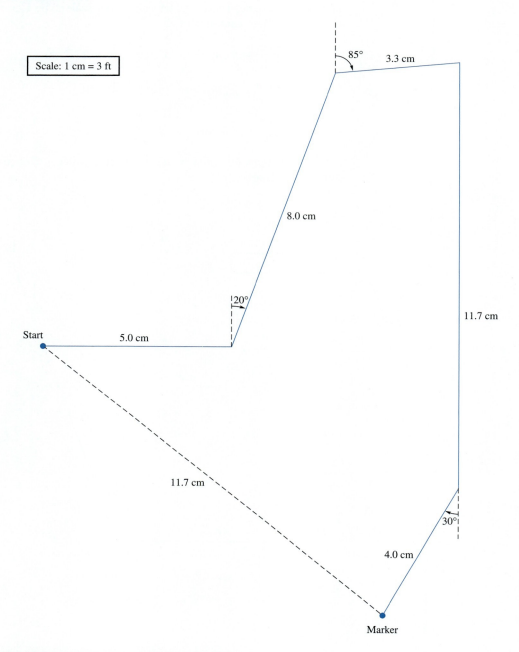

The distance measures approximately 11.7 cm.

$$11.7 \text{ cm} * \frac{3 \text{ ft}}{1 \text{ cm}} = 35.1 \text{ ft}$$

We conclude that they must walk about 35 feet back to the start.

In this section, we explored ways of using scale drawings to solve problems. As you worked on some of these problems with others in your class, you may have noticed that answers did not always match exactly. Any time we rely on our abilities to measure and draw accurately, we must expect some error.

9.4 Scale Drawings

Assume that all figures are *not* drawn to scale unless otherwise indicated or unless you are directed to measure the figure.

Problem Set 9.4

1. Create scale drawings to determine all of the missing lengths and all of the missing angles.

   a.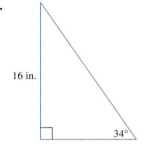

   b.

   c.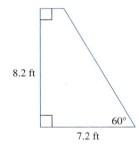

   d.

2. Use scale drawings and the following figure to approximate the height of the tree.

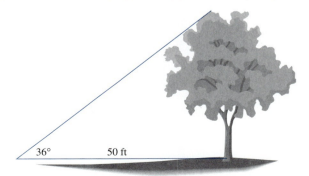

3. A flagpole that is 22.0 ft high is secured with a guy wire as shown in the following picture. The distance from the base of the pole to the spot where the guy wire is attached is 7.5 ft.
   a. Carefully draw the situation to scale. Use your scale drawing to determine the length of the guy wire.
   b. Use the Pythagorean theorem to determine the length of the guy wire. Compare your result with what you found in part a.
   c. Which of these methods is better in this situation? Why?

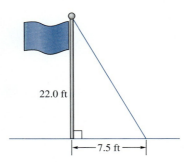

4. A flagpole is located on a hill. A guy wire is again attached to secure the pole as shown in the following picture.
   a. Carefully draw the situation to scale. Use your scale drawing to determine the length of the guy wire.
   b. Explain why we cannot use the Pythagorean theorem to determine the length of the guy wire for this situation.

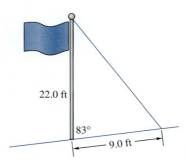

5. **Support Plate.** On a blueprint for a building, the dimensions of a rectangular support plate are $\frac{1}{8}$ inch by $1\frac{3}{8}$ inches. If the scale for the blueprint is 1 in = 4.0 meters, what is the actual area of the support plate?

6. **Tracy's Trip.** Tracy's plane leaves the Millstone Airport flying N42°W.
   a. If the plane is flying at an average rate of 280 mph, how far will it have traveled after 1 hour 45 minutes?
   b. Use a scale drawing to determine how much farther north the plane is at that time.

7. Use a scale drawing to answer the questions in the following problem.

    Scouting Challenge. In a scout orienteering challenge, a group was told to hike from point A to a marker at point D. Point D is located northwest of point A, but, due to the density of the forest, they cannot see the destination marker from point A.

    The scout group began their hike heading due west. After 45 minutes, they spotted the marker through the forest. They turned at a heading of N15°W and hiked 1 hour and 15 minutes through the forest in a straight path to point D.

    a. If the group averaged 2.5 miles per hour, how far did they hike?

    b. One of the group members left his backpack at point A. After receiving much harassment from his fellow scouts he decided to walk directly back to point A to retrieve his pack. In what direction should he head? How far does he need to walk?

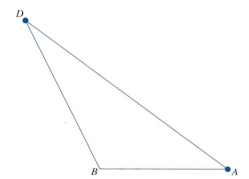

8. Use a scale drawing to answer the questions in the following problem.

    Dr. Martin. Dr. Martin left work at 3:00 P.M. on Friday. She drove due south for 32 miles. Then she turned east onto Moon St. She proceeded on Moon St. for 10.6 miles when Moon St. veered S30°E. After another 5.6 miles, Moon St. ended at a stop sign. Dr. Martin made a 90° right turn onto Half Moon Ave. At about 2.4 miles Half Moon Ave. has a hairpin turn and heads due east. An oncoming car crossed the centerline at this corner and struck Dr. Martin's car. Life Flight, which is based at Dr. Martin's work, was called to the scene. Life Flight is a helicopter used to transfer emergency medical patients to a medical facility.

    a. How far did Dr. Martin travel before she was hit?

    b. How far must Life Flight travel to reach the scene of the accident?

9. Use a scale drawing to answer the questions in the following problem.

    Captain Dan. Captain Dan is sailing his ship from Beach Bay to Portville. Portville lies 55 km directly north of Beach Bay, so Captain Dan heads out of the harbor heading due north. After traveling 11 km along this course, Captain Dan's first mate, Albert, notices a large rocky island on the map between Beach Bay and Portville. Captain Dan orders the ship to change course to N35°E. They sailed this course for 12 km while Albert studied the maps. Albert discovered that if they continue following their new heading they would still run into the island. Captain Dan quickly adjusts their course to due east, which they travel for 9 km. At that point they resume a due north course for 13 km. Albert points out that if they don't adjust this course, they will end up sailing to a point east of their destination, Portville. What heading adjustment should they make to sail directly to Portville from their present location? How much longer was their trip than expected?

10. Room Layout. Pick a living room, family room, or bedroom of your home or apartment. Measure the dimensions of the room. Make a scale drawing of the room with doors, windows, and any immovable objects marked. Measure the floor dimensions of the major pieces of furniture in the room, and make flat scale models of the furniture. Arrange the pieces in at least two different ways in the room. Explain the advantages of a scale drawing in this situation.

Activity Set 9.5

9.5 Geometric Solids and Surface Area

> **Definition**
> A **polyhedron** (polyhedra is the plural) is a three-dimensional figure in space whose sides are polygons. The sides of a polyhedron are called **faces,** and the faces intersect in **edges** and **vertices.**

Prisms and pyramids are special types of polyhedra.

> **Definition**
> A **prism** is a polyhedron with two parallel, congruent faces. These are called the bases of the prism. If the lateral sides of the prism are perpendicular to the bases, the prism is called a **right prism.** Prisms that are not right are called **oblique.**

> **Definition**
> A **pyramid** is a polyhedron that has a base consisting of any polygon and sides that are triangles.

NOTE: For the remainder of this text, assume that all prisms whose lateral sides appear to be perpendicular to the bases are right prisms. Prisms and pyramids are named for their bases. The following figure shows some examples of prisms and pyramids.

Triangular right prism

Quadrilateral right prism

Rectangular right prism (or rectangular solid)

Hexagonal right prism

Oblique triangular prism

Triangular pyramid (or tetrahedron)

Square pyramid

Hexagonal pyramid

1. For this activity your instructor will provide you with templates similar to the ones that follow. Cut out your templates. By folding the patterns on the lines, which of the following patterns can be folded into a cube with no open faces?

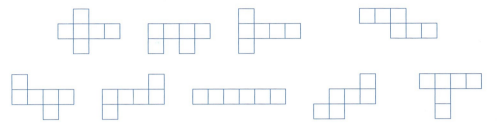

2. In Activity 1, you found one-piece patterns that can be used to build a cube. Draw a one-piece pattern to build a rectangular solid with a length of 3 inches, width of $1\frac{1}{2}$ inches, and height of $2\frac{1}{4}$ inches. Cut out and fold your pattern to confirm that you are correct.

3. Draw a one-piece pattern to build a triangular right prism with a height of 7.0 cm and bases that are right triangles with legs of 3.0 cm and 5.0 cm. The following is a picture of the base. Cut out and fold your pattern to confirm that you are correct.

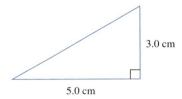

4. Draw and cut out two circles the size of the following. These represent the bases of a right circular cylinder (can). Draw a one-piece pattern of the lateral surface area of a right circular cylinder for which the top and bottom are the same size as the circle in the figure and the cylinder is $3\frac{1}{2}$ inches tall. Cut out your one-piece pattern and use it together with your circles to build your cylinder and confirm that you are correct.

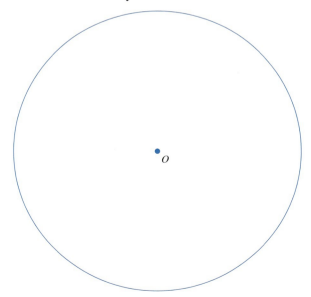

5. Find the surface area of the solids you built in Activities 2–4.

Discussion 9.5

Geometric Solids and Surface Area

In the activity set we defined prisms and pyramids, which are special kinds of polyhedra. In addition to these solids, three commonly used solids are cylinders, spheres, and cones.

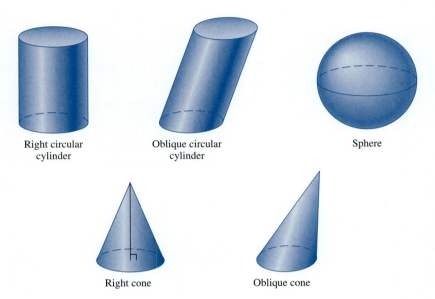

Right circular cylinder   Oblique circular cylinder   Sphere

Right cone   Oblique cone

For some applications we are interested in finding the surface area of a geometric solid. For example, we may want to paint the surface of a cylindrical water tank. If this tank is placed with one base on the ground, we only need to paint the top and the side surfaces. We might want to line a box with insulation material to make a cooler. In both of these cases we need to know the surface area of the given object.

In the activity set you made flat patterns to build a rectangular solid, a triangular prism, and a cylinder. These patterns can be useful in picturing the surface area of a figure.

Example 1

Find the surface area of the triangular right prism pictured here.

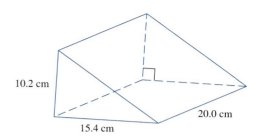

Solution To find the surface area, it can be helpful to draw a pattern that will fold into the given figure. The figure shows one possible pattern.

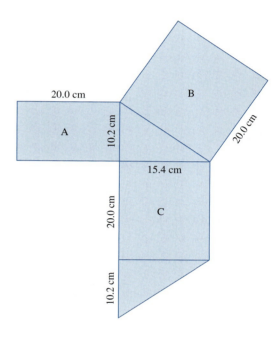

We can see that the pattern contains two congruent right triangles and three rectangles. The dimensions of rectangle A are 10.2 cm by 20.0 cm. The dimensions of rectangle C are 15.4 cm by 20.0 cm. To find the dimensions of rectangle B, we need to determine the length of the hypotenuse of the triangle. We can use the Pythagorean theorem.

$$\text{Hypotenuse}^2 = \text{Leg}^2 + \text{Leg}^2$$
$$\text{Hypotenuse}^2 = (10.2 \text{ cm})^2 + (15.4 \text{ cm})^2$$
$$\text{Hypotenuse} = \sqrt{(10.2 \text{ cm})^2 + (15.4 \text{ cm})^2}$$
$$\text{Hypotenuse} \approx 18.5 \text{ cm}$$

Therefore, the dimensions of rectangle B are 18.5 cm by 20.0 cm. Now, we are ready to determine the surface area of the figure.

Surface area
$= 2 * (\text{area of triangle}) + \text{area of rectangle } A + \text{area of rectangle } B + \text{area of rectangle } C$
$= 2\left(\frac{1}{2} * 10.2 \text{ cm} * 15.4 \text{ cm}\right) + 10.2 \text{ cm} * 20.0 \text{ cm} + 18.5 \text{ cm} * 20.0 \text{ cm} + 15.4 \text{ cm} * 20.0 \text{ cm}$
$\approx 1040 \text{ cm}^2$

The surface area is approximately 1040 square centimeters.

The following are special formulas for finding the surface area of spheres and cones.

Surface Area of a Sphere

The surface area of a sphere is given by the formula

$$SA = 4\pi r^2$$

where SA is the surface area and

r is the radius.

Sphere

Surface Area of a Cone

The surface area of a cone is given by the formula

$$SA = \text{area of base} + \text{lateral surface area}$$
$$SA = \pi r^2 + \pi r s$$

where SA is the surface area

r is the radius, and

s is the slant height.

Right cone

Example 2

Shorty's Corner delivers oil to homes that heat with oil-burning furnaces. The truck they use to deliver the oil carries a tank similar to the one pictured here. The ends of the tank are not exactly hemispheres, but we can use hemispheres to approximate any needed information. The tank is about 6 feet 4 inches high and 11 feet 11 inches long from end to end. The tank needs to be painted periodically. The owners have one gallon of paint left over from previous years. If the paint they use covers about 250 square feet per gallon, do they need to buy more paint to paint the tank this year?

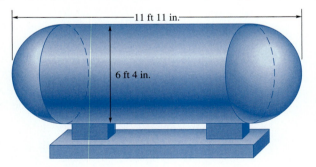

Solution

First, let's convert all of the measurements to inches.

11 ft 11 in.

$= 11 \text{ ft} * \dfrac{12 \text{ in.}}{1 \text{ ft}} + 11 \text{ in.}$

$= 143 \text{ in.}$

6 ft 4 in.

$= 6 \text{ ft} * \dfrac{12 \text{ in.}}{1 \text{ ft}} + 4 \text{ in.}$

$= 76 \text{ in.}$

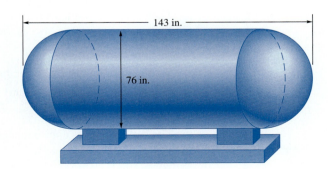

To determine the amount of paint, we need to find the surface area of the tank. The surface area of the tank is the surface area of the two hemispheres plus the surface area of the cylinder, not including the top and bottom of the cylinder. The radius of each hemisphere and the radius of the cylinder is one half the height, $\frac{76 \text{ in.}}{2} = 38$ in. The length of the cylinder is the total length of the tank minus two radii. That is,

Length of the cylinder = 143 in. − 2(38 in.)
$\qquad\qquad\qquad\qquad\;\;$ = 67 in.

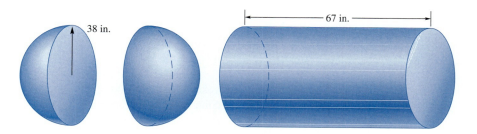

We have a formula for the surface area of a sphere, but we do not have a formula for the surface area of a cylinder. However, if we imagine slicing the cylinder along its length and unwrapping it, it will look like a rectangle. One dimension of this rectangle is the length of the cylinder and the other dimension is the circumference of a circle, $2\pi r$. Therefore, the surface area of the cylinder, not including the top and bottom, is the circumference of the circular base times the length of the cylinder.

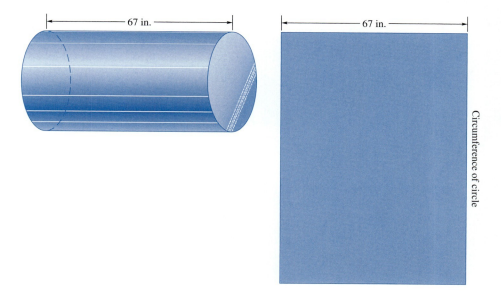

Now, we can find the surface area of the tank.

Surface area of tank

= 2(surface area of a hemisphere) + lateral surface area of cylinder

= surface area of sphere + lateral surface area of cylinder

= $4\pi * \text{radius}^2$ + (circumference of circle * length of cylinder)

= $4\pi * \text{radius}^2 + (2\pi * \text{radius} * \text{length of cylinder})$

= $4\pi * (38 \text{ in.})^2 + 2\pi * 38 \text{ in.} * 67 \text{ in.}$

$\approx 34{,}100 \text{ in.}^2$

Next, we can use the surface area to determine the number of gallons of paint.

$$34{,}100 \text{ in.}^2 * \left(\frac{1 \text{ ft}}{12 \text{ in.}}\right)^2 * \left(\frac{1 \text{ gal}}{250 \text{ ft}^2}\right)$$

$$= 34{,}100 \text{ in.}^2 * \left(\frac{1^2 \text{ ft}^2}{12^2 \text{ in.}^2}\right) * \left(\frac{1 \text{ gal}}{250 \text{ ft}^2}\right)$$

$$= \frac{34{,}100}{12^2 * 250} \text{ gal}$$

$$= 0.95 \text{ gal}$$

It will take about one gallon to paint the tank. Therefore, they do not need to buy any more paint this year.

In this section, we looked at ways of finding the surface area of solids. Often we can find the surface area of solids by finding the area of the sides of the solid using formulas we already know. In the case of a cylinder, we can imagine slicing the cylinder along its length and unwrapping it to form a rectangle. Then the surface area of the side of the cylinder is the area of the rectangle formed. If we want the total surface area of a cylinder, we then need to add the area of the circular top and bottom. To find the surface area of a sphere or cone we need to use the formulas that were given.

Problem Set 9.5

Assume that all figures are *not* drawn to scale unless otherwise indicated or unless you are directed to measure the figure.

1. Determine the surface area of each of the following figures.

   a.

   8.0 in.
   14.0 in.
   10.0 in.

   b.
   2.5 cm

   c.

   12.5 ft
   36.0 ft

   d.
   5.8 cm
   13.0 cm
   4.2 cm
   7.5 cm

2. Each of the patterns a through d can be folded into a solid shape. Match the patterns a–d with their corresponding solids I–IV.

a.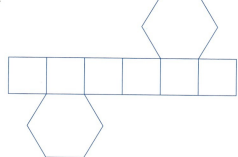

b.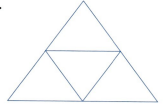

c.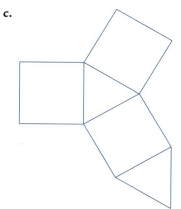

d.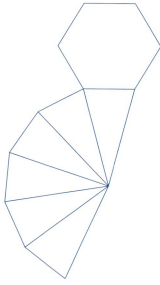

I  II  III  IV

3. Simplify each expression. Express your results using only positive exponents. Verify your results.

   a. $4R^3(R^2 + 9R)$

   b. $(3T^2)^3 - T^5(7T)$

   c. $\dfrac{10xy^2}{x^2y}$

   d. $\dfrac{x^4 * x^5}{x^3}$

   e. $\dfrac{x^4 + x^5}{x^3}$

   f. $(2\,m)^3 + 5\,m^2(3\,m)$

   g. $7x^{-3}(4x^7)$

   h. $10p^{-2}$

   i. $\dfrac{R^4 * R}{R^{-3}}$

4. The following expression arose from a geometry application. Determine the numerical value and units of the result.

   $(3\text{ cm})^3 + 2\text{ cm}^2(5\text{ cm})$

5. a. Name the following figure.
   b. Determine the surface area of the figure.

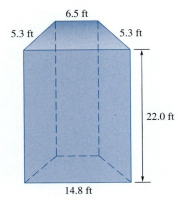

6. **Lead Box.** A rectangular box is to be used to store radioactive material. The box needs to be lined with lead. The inside of the box is 11.0 inches long, 8.0 inches wide, and 6.5 inches deep. Determine the area that the lead needs to cover to line the box, including the lid.

7. **Terry's Living Room.** Terry is painting his living room. The living room is a rectangle which is 17′6″ long by 14′9″ wide by 8′ tall. In one of the short walls is an opening 5′6″ wide by 8′ tall. If the paint that Terry plans to buy covers 500 square feet per gallon, how much paint does Terry need to buy to paint the ceiling and the walls?

8. Write an expression for the surface area of the box in terms of L, W, and H.

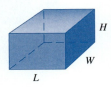

9. Write an expression for the surface area of the cylinder in terms of r and h. Be sure to include the area of the top and bottom.

10. The two following formulas are used for circles. Explain how you can use units to identify which formula is used to determine the circumference of a circle and which is used to determine the area.

   $\boxed{\pi r^2}$ $\boxed{2\pi r}$

11. **80-Foot Pool.** Aqua Pool charges 30¢ per square foot plus a service fee of $200 to refinish the inside of a pool. How much will Aqua Pool charge to refinish the pool pictured in the figure?

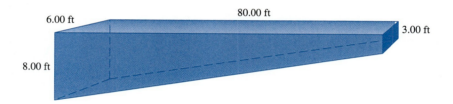

12. **Can Labels.** A regional cannery needs to order paper to put labels on their canned foods. The cans have a $2\frac{1}{4}$-inch diameter and are 5 inches tall. The labels on the cans overlap by about $\frac{1}{2}$ inch. The paper they use comes in rolls that are 5 inches wide and 200 feet long. If they need to label 2000 cans, how many rolls of paper do they need to order?

13. **a.** Write a formula for the surface area of the triangular right prism shown in the figure.
    b. Solve your formula in part a for P. Verify your solution.

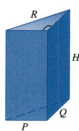

14. In the following figure, side RS = 2.11 cm, side ST = 3.06 cm, and side VT = 3.47 cm. Determine the lengths of sides SV and RW.

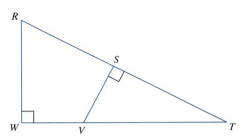

15. **Mirror.** The following figure shows the path of light from a source at point S to the point P. The light reflects off the mirror at point M so that indicated angles are equal. Determine the length of the path the light travels from S to P.

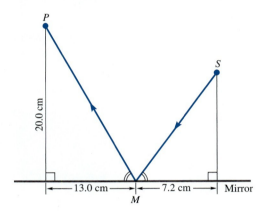

9.6 Volumes of Prisms and Cylinders

Activity Set 9.6

1. For this activity your instructor will provide you with a template similar to the one shown in the following figure. Cut out your template, and fold on the heavy lines to form a box. Tape the sides and bottom of your box, but do not tape the lid shut.

   a. What are the dimensions of your box?
   b. How many cubes fit in your box?
   c. Does your box look the same as your teammate's? As those in other teams?
   d. Does your box hold the same number of cubes as your teammate's? As other team's?

Discussion 9.6

In the activities you constructed a rectangular box similar to the one in the following picture. To determine the volume of the solid we fill the rectangular box with cubic units. You found that this box can hold 24 cubic units (units^3), which can be determined by finding the product of the length, width, and height of the rectangular box.

$$4 \text{ units} * 3 \text{ units} * 2 \text{ units} = 24 \text{ units}^3$$

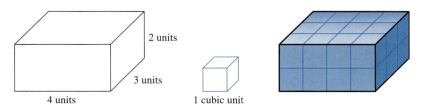

However, we need a formula that can be used more generally. To do this let's analyze the rectangular solid in more detail. If we look at the bottom layer of cubes, we notice that the number of cubes in this layer is equal to the area of the bottom (base) of the box.

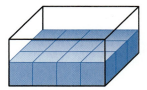

Two layers represent the height of the box. Therefore, the volume of this box is equal to the area of the base times the height.

*volume of box = area of base * height*

This formula can be generalized to any prism.

Let's look at the triangular right prism pictured here. The base of this prism is a right triangle whose legs have lengths 7.5 units and 8 units. The height of the prism is 16 units. To determine the volume we begin filling the prism with cubes. The number of cubes in the bottom layer is equal to the area of the triangular base

$$\frac{1}{2} * 7.5 * 8 = 30 \text{ cubes}$$

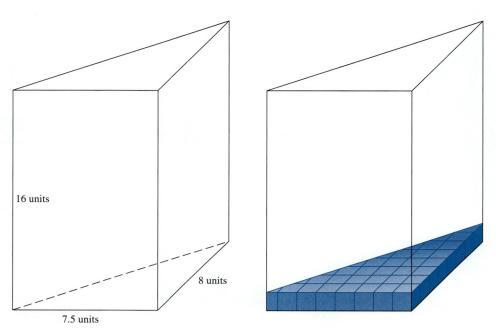

Next, we begin adding layers of cubes. The number of layers of cubes is equal to the height of the prism. Because 30 cubes are in the bottom layer and there are 16 layers of cubes, the total number of cubes is $30 * 16 = 480$ cubes in the triangular prism. Notice that this calculation is equivalent to multiplying the area of the base times the height. That is,

*volume of a prism = area of the base * height*

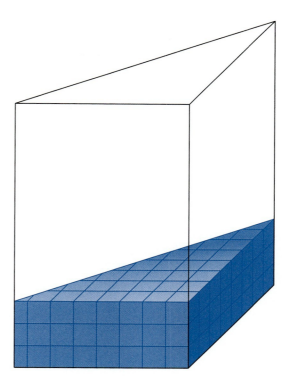

$$\text{volume} = \left(\tfrac{1}{2} * 7.5 \text{ units} * 8 \text{ units}\right) * 16 \text{ units}$$
$$= 480 \text{ units}^3$$

The volume of this prism is 480 cubic units.

In the triangular right prism in the previous figure, the triangular base has a height and the prism has a height. We needed to use both of these heights to find the volume of the prism. This can be confusing. Therefore, when working with three-dimensional figures we need to be as clear as possible in labeling and identifying the heights of the various pieces.

Example 1

Determine the volume of both of the following figures.

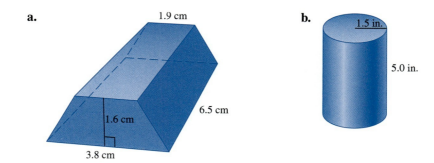

Solution   a. The figure in part a represents a prism with a trapezoidal base. The trapezoidal base of the prism is facing us, and the height of the prism is 6.5 cm. The volume of a prism is equal to the *area of the base* times the *height*. In this case the base is a trapezoid. Therefore,

$$\text{volume} = \text{area of trapezoidal base} * \text{height of the prism}$$
$$= \left(\tfrac{1}{2} * \text{height of trapezoid} * \text{sum of the bases}\right) * \text{height of the prism}$$
$$= \tfrac{1}{2} * 1.6 \text{ cm} * (3.8 \text{ cm} + 1.9 \text{ cm}) * 6.5 \text{ cm}$$
$$\approx 30. \text{ cm}^3$$

The volume of this trapezoidal prism is about 30. cm^3.

b. Although a cylinder is not a prism, we can visualize the volume of a cylinder in the same way we did for a prism. That is, the number of cubes in the bottom level of a cylinder is equal to the area of the circular base, and the number of layers of cubes is equal to the height. Therefore, the volume of a cylinder is equal to the area of the base times the height. The base of this cylinder is a circle with radius 1.5 inches. Therefore,

$$\text{volume} = \text{area of circular base} * \text{height}$$
$$= (\pi * \text{radius}^2) * \text{height}$$
$$= \pi * (1.5 \text{ in.})^2 * 5.0 \text{ in.}$$
$$\approx 35 \text{ in}^3.$$

The volume of this cylinder is approximately 35 in^3.

So far we looked at examples of right prisms and right cylinders. Does our formula work for oblique prisms or oblique cylinders? Consider the following oblique rectangular solid.

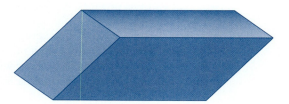

Now imagine filling this solid with sheets of plywood. This solid takes seven sheets of plywood.

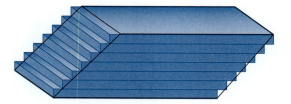

The volume of these seven sheets of plywood is the same whether we stack the sheets in the shape of an oblique rectangular solid or in the shape of a right rectangular solid.

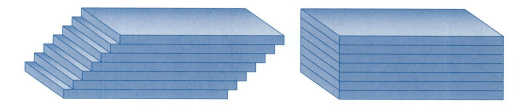

Therefore, the volume of an oblique rectangular solid is the same as the volume of the right rectangular solid that has the same base and the same height as the original. The height is the perpendicular distance between the bases.

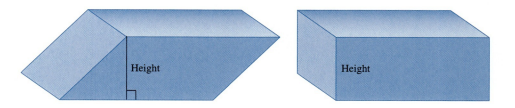

Volumes of Pyramids and Cones

The volume of a pyramid and its corresponding prism with the same base and height are related. To see this, consider a cube. How many pyramids with square bases fit in a cube?

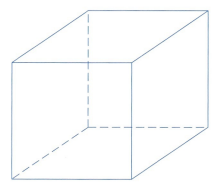

One pyramid fits with its base on the bottom of the cube and its vertex opposite the base at the top front right-hand corner.

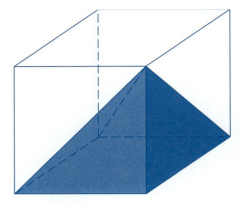

A second pyramid fits with its base on the left-hand side of the cube and its vertex opposite the base at the top front right-hand corner.

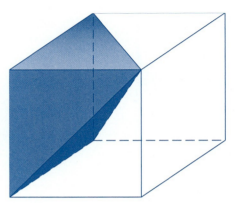

A third pyramid fits with its base at the back of the cube and its vertex opposite the base at the top front right-hand corner.

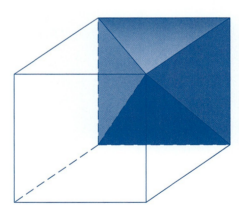

These three pyramids together fill the entire cube. Therefore, each pyramid must be one third the volume of the cube. In general, the volume of a pyramid is equal to one third the volume of the prism with the same base and height. Similarly, the volume of a cone is equal to one third the volume of the cylinder with the same base and height. Because the volume of a prism is equal to the area of the base times the height, we obtain the following formula for the volume of a pyramid.

> **Volume of Pyramids and Cones**
>
> $$\text{volume of a pyramid (or cone)} = \tfrac{1}{3} * \text{area of base} * \text{height}$$

Example 2

A grain silo consists of a cylinder and a cone. Determine the number of tons of grain that will fit in the silo pictured, if grain weighs approximately 15 pounds per cubic foot.

Solution

We need to find the volume of the silo.

volume of silo = volume of cylinder + volume of cone

$$= \text{area of base} * \text{height of cylinder} + \frac{1}{3} * \text{area of base} * \text{height of cone}$$

$$= \pi * (4.5 \text{ ft})^2 * 35 \text{ ft} + \frac{1}{3}\pi * (4.5 \text{ ft})^2 * 2.5 \text{ ft}$$

$$\approx 2300 \text{ ft}^3$$

The volume of the silo is approximately 2300 ft³. Next, we need to determine the number of tons of grain that fit in the silo. We can use unit fractions to accomplish this.

$$2300 \text{ ft}^3 = 2300 \text{ ft}^3 * \frac{15 \text{ lb}}{1 \text{ ft}^3} * \frac{1 \text{ ton}}{2000 \text{ lb}}$$

$$= \frac{2300 * 15}{2000} \text{ tons}$$

$$\approx 17 \text{ tons}$$

The grain silo can hold about 17 tons of grain.

At this time, we will not develop the formula for the volume of a sphere. The formula is provided in the box.

Volume of a Sphere

The volume of a sphere is given by the formula

$$V = \frac{4}{3}\pi r^3$$

where V is the volume and
r is the radius.

Sphere

In this section, we looked at volumes of solids. If the solid is a prism, then the volume of the solid is equal to the *area of the base times the height*. This general principle is true for right prisms, oblique prisms, right cylinders, and oblique cylinders; however, if the solid is oblique, we need to be sure that our height is perpendicular to the base. Because three pyramids fit into a rectangular solid, we found that the volume of a pyramid is equal to one third the area of the base times the height. This also works for the volume of a cone.

Problem Set 9.6

Assume that all figures are *not* drawn to scale unless otherwise indicated or unless you are directed to measure the figure.

1. Determine volume of each figure.

a.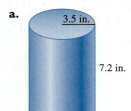

b.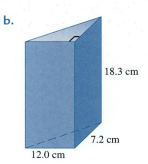

c.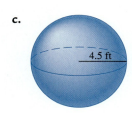

d.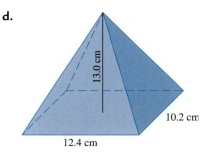

e.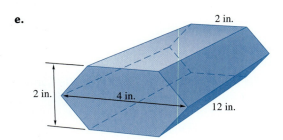

The hexagonal base in this solid is both vertically and horizontally symmetrical.

2. Find the surface area of the figures in Problem 1.

3. **80-Foot Pool.** Determine the volume of the pool depicted in the following figure. How many gallons of water will it take to fill the pool?

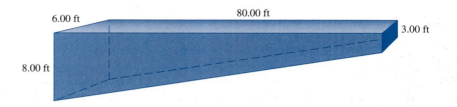

4. Each of the following formulas is used to determine a length, an area, or a volume. Each variable represents a length. Identify which kind of measurement the formula is used for (length, area, or volume). Explain how you know, without using a table of formulas.

   a. $d + \pi d$

   b. $2\pi rh + 4\pi r^2 + \pi r^2$

   c. $s^2h + \dfrac{\pi r^2}{2} * h$

   d. $(s_1 s_2 + s_1 s_3) s_4$

   e. $\dfrac{w s_2 (b_1 + b_2)}{2}$

   f. $\dfrac{2R + r}{2}$

5. Simplify each expression. Express your results using only positive exponents. Verify your results.

   a. $\dfrac{-(5n^2)^3}{n^2}$

   b. $\left(\dfrac{-5n^2}{n^2}\right)^3$

   c. $(-5x)^2 - 3x^{-4}(2x^6 + 10x^4)$

   d. $\dfrac{12k^{-2} - 7k^3}{k^{-3}}$

   e. $\dfrac{12k^{-2} * 7k^3}{k^{-3}}$

   f. $\pi(12\ m^2) + 2\pi(4\ m)(7\ m)$

6. **The Long Driveway.** Russ and Maria have a gravel driveway that is 0.7 mile long and 6 ft 8 in. wide. If they are going to resurface the driveway with a 3-in. layer of gravel, how many cubic yards of gravel do they need?

7. **Concrete Walkway.** How many cubic yards of cement are needed to make a concrete walkway around a rectangular garden that is 20 feet by 35 feet, if the walkway is to be 3 feet wide and 4 inches thick?

8. **Cords of Wood.** A cord of wood is the amount of wood that when stacked measures 4 feet by 4 feet by 8 feet. Approximately how many cords of wood fit in a woodshed measuring 12 feet long by 5 feet wide by 6 feet high?

9. The following figure shows a right prism and a right pyramid. The height of each figure is 9.2 cm. The bases of the prism and the base of the pyramid are the exact same seven-sided polygon. The area of the 7-gon is 23 cm². Determine the volume of the prism and the volume of the pyramid.

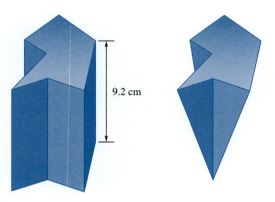

10. The following expression arose from a geometry application. Determine the numerical value and units of the result.

$$4.19(13.4 \text{ ft}^2)(5.5 \text{ ft}) + [(3.2 \text{ ft})(6.8 \text{ ft}) + (3.2 \text{ ft})^2](5.5 \text{ ft})$$

11. The base of the following triangular right prism is an equilateral triangle whose side length is 4.0 inches. The prism is 8.5 inches tall. Determine the surface area and volume of the prism.

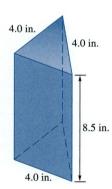

12. What is the diameter of a circle whose area is 1 square foot?

13. What is the length of the side of a cube whose surface area is 150 square centimeters?

14. A cylindrical can holds one quart and is $8\frac{1}{2}$ inches tall. What is the diameter of the can?

15. The Pool. A new swimming pool needs to be sealed with a protective paint and then filled with water. The paint used to seal the pool covers approximately 400 square feet per gallon. How many gallons of paint are needed to seal the pool, and how many gallons of water are needed to fill the pool? All distances are measured to the nearest centimeter.

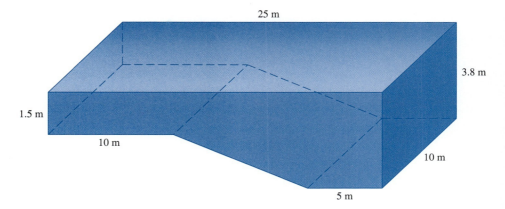

16. Gasket. A rubber gasket is 9.83 cm in diameter and 0.48 cm thick. It has a hole that is 2.54 cm in diameter. Determine the volume of material needed to fill an order for 1000 gaskets.

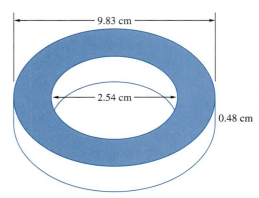

17. Water Tank. A water storage tank is in the shape of a cylinder with a hemisphere on top. We measure the diameter of the tank to be about 6.0 feet, but we are unable to measure the height. We know that the tank holds about 6500 gallons of water when filled to the top of the cylinder. Determine the height of the tank.

18. **The Beaker.**

a. The beaker depicted in the figure has a diameter of 15 centimeters and a height of 7 centimeters. What volume of liquid would be in the beaker if you filled it to a depth of 1 centimeter? 2 centimeters?

b. Make a table for the volume in the beaker at a given height. What values for the height should you include in your table?

c. Graph the volume in terms of the height. If your graph is linear, find the slope of this graph.

d. Write an equation for the volume in terms of the height of the liquid.

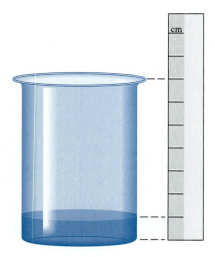

19. **The Flooded Basement.** Reino has just discovered that his unfinished basement has flooded. With a ruler he determines that the water is 18 inches deep. A sketch of the basement room is shown here. He rushed out and rented a pump that can pump 150 gallons an hour. Reino begins to pump the water out at 6:00 P.M.

a. How many gallons of water were in the basement before Reino begins to pump it out?

b. Assuming that no additional water is flooding into the basement, how many gallons are in the basement at 7:00 P.M? At 9:00 P.M?

c. Write an equation for the amount of water in the basement (in gallons) in terms of the number of hours past 6:00 P.M.

d. Draw a graph of the amount of water in the basement (in gallons) in terms of the number of hours past 6:00 P.M.

e. How much water will be in the basement at 6:00 A.M. the next morning?

f. Will Reino be able to return the pump within 24 hours, or will he have to rent it for more than one day? Explain how you decided.

g. At what time will all of the water be pumped out of the basement?

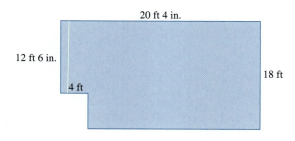

20. Determine the volume of the figure.

Chapter Nine Summary

In Chapter 9 we studied a variety of topics in geometry. We began the chapter by looking at angle relationships.

Vertical angles are angles on opposite sides of the intersection point of two lines. The measures of vertical angles are always equal. Vertical angles are sometimes called **opposite angles.**

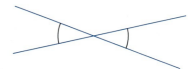

Supplementary angles are angles whose measures add up to 180°. The sum of the measures of **complementary angles** is 90°.

We discussed the following theorems. We most often use these theorems to determine the measures of angles if we have two parallel lines. However, the theorems can also be used to show that two lines are parallel if we have certain congruent angles.

> *Alternate Interior Angles Theorem*
>
> If two lines are cut by a transversal, then the two lines are parallel if and only if **alternate interior angles** are congruent.

> *Corresponding Angles Theorem*
>
> If two lines are cut by a transversal, then the two lines are parallel if and only if **corresponding angles** are congruent.

The sum of the measures of the angles in a triangle is 180°. In any convex polygon, we can create two fewer triangles than the number of sides in the polygon by starting at one vertex and drawing lines to each of the other nonadjacent vertices. This means that the sum of the measures of the interior angles in a convex polygon is 180° times two less than the number of sides.

In a circle we can create a **central angle** whose vertex is at the center of the circle and an **inscribed angle** whose vertex is on the circle. By definition the measure of the central angle is equal to the measure of its intercepted arc. We showed that the measure of an inscribed angle is equal to one half the measure of its intercepted arc, or equivalently, one half the measure of the corresponding central angle.

After looking at angle relationships, we went on to **similar triangles.** In general, two figures are similar if they have the same shape but not necessarily the same size. In Section 9.3 we found that similar triangles have special properties.

> *Properties of Similar Triangles*
>
> 1. Similar triangles have equal angles. Conversely, triangles with equal angles are similar.
> 2. The ratios of corresponding sides in similar triangles are a constant. Conversely, if the ratios of the corresponding sides are constant, the triangles are similar.

We used these properties to determine whether two triangles are similar, to determine lengths of sides in a triangle, and to determine distances that could not be measured directly. It is important to write the ratios of corresponding sides in a consistent fashion when using these properties. Then, we extended the idea of similar triangles to **scale drawings.**

The **surface area** of a three-dimensional figure is the total area of its surface. In the case of a prism, pyramid, cylinder, or cone, the surface area is the area of the lateral sides plus the area of the base(s). In many instances it is helpful to "unwrap" the figure to find its surface area. The units on surface area are always square units.

The formula used most often in finding volumes is the volume of a prism.

$$\text{\textbf{volume of a prism (or cylinder)}} = \text{area of the base} * \text{height}$$

Then, we found that the volume of a pyramid is one third the volume of the corresponding prism.

$$\text{\textbf{volume of a pyramid (or cone)}} = \tfrac{1}{3} * \text{area of the base} * \text{height}$$

These two formulas incorporate a huge collection of geometric figures. Other geometric figures can be broken down into these. The units on volume are always cubic units.

The formulas used to find area, circumference, surface area, or volume can often be confused with one another. Analyzing the units on the formulas can help eliminate this confusion. For example, two formulas associated with circles are πr^2 and $2\pi r$. In the first formula πr^2, the units are squared, so this must give the area of a circle. In the second formula $2\pi r$, the units are not squared, so this gives the circumference of a circle.

Many problems in daily life can be solved using geometric concepts. To solve these problems we often have to combine a collection of many of the geometric ideas we learned.

Chapter Ten

Statistics and Probability

Chapter 10 will focus on the study of statistical methods for describing data, probability, and chance. We will learn how to look critically at the construction of graphs in order to make informed decisions about the data being represented. Descriptive statistics will allow us to report about and summarize large sets of data.

We will also look at probability. Performing probability experiments can help us in understanding the mathematics of chance. This will facilitate our ability to reason about uncertainty and risk. The study of probability and statistics can better prepare us to make decisions about complex issues and to understand arguments presented numerically or graphically.

10.1 Graphical Displays of Data

1. Both of the following graphs represent the same information.
   a. Which of the two graphs appears to indicate that interest rates for home improvements are lower?
   b. Which of the two graphs makes it seem that interest rates have been going down significantly since 1981?
   c. Are both of the graphs correctly drawn? Explain.

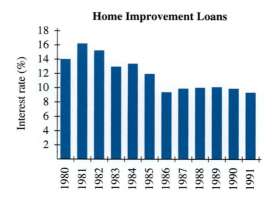

2. A manager at the local credit union used the same information as in the two graphs in Activity 1. How is the following graph different from the graphs in Activity 1? Do you think that this graph is misleading? Explain.

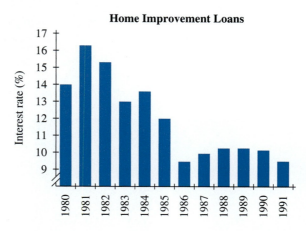

3. A graphic artist added some details to the graph that was produced by the credit union manager. What effect does this have on someone reading the graph?

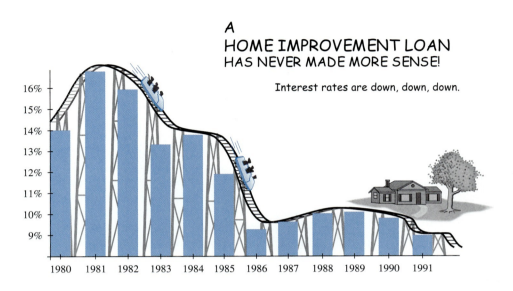

4. a. Choose any number for the length of the side of a square. Determine the area of your square. Enter this information in the first row of the following table.
   b. Double the side length of your square. Determine the area of this new square. Enter this information in row 2 of the table.
   c. Compute the ratio of the area of the second square to the area of the first square. What happens to the area of a square when you double the length of each side?

	Side Length of Square	Area of Square
1.		
2.		

5. a. Choose any two numbers for the length and width of a rectangle. Determine the area of your rectangle. Enter this information in row 1 of the following table.
   b. Now double the length and the width of your rectangle. Determine the area of this new rectangle. Enter this information in the table.
   c. Double the length and width again and compute the new area. Enter this information in the table.
   d. Compute the ratio of the consecutive areas. What happens to the area of a rectangle when you double the length and the width?

	Length of Rectangle	Width of Rectangle	Area of Rectangle
1.			
2.			
3.			

6. Use a strategy similar to Activity 4 to determine what happens to the area of a square when the side length is tripled.

7. a. Use a strategy similar to Activity 4 to determine what happens to the volume of a cube when the side lengths are doubled.

   b. Use a strategy similar to Activity 4 to determine what happens to the volume of a cube when the side lengths are tripled.

Discussion 10.1

More and more we find that information is presented to us in the form of graphs. Graphs have always been used to represent numerical information in technical and scientific fields. Now, with the abundant use of computers and software that can produce professional graphs, we frequently see graphical information in daily news and advertising. These graphs are being generated to inform us and to persuade us in marketing and advertising. It is important that those responsible for creating the graphs portray the data honestly, clearly, and accurately, but because this is not always the case, it is equally important that each of us learn to read graphs critically.

We saw in the activities the effect of changing the vertical scale of the graph. This can emphasize the amount of change in the data by increasing the slope of the graphs, making slight increases or decreases seem more dramatic. The information in a graph can also be distorted by breaking the horizontal or vertical scale. None of the graphs in the activities is truly "wrong." However, changing the scale on the vertical axis and breaking the axes can affect the impression the graph gives. We also saw in the activities how enhancing graphs with suggestive titles and pictures can help readers reach a desired conclusion. When reading this type of information, we must read critically.

The following graph of math achievement scores was prepared by an urban school district. The horizontal scale jumps from 1981 to 1988. Extra space falls between 1981 and 1988, but not enough to represent seven years. This makes the improvement in math scores over the years between 1981 and 1988 appear dramatic. Creating a scale with unequal increments (either horizontally or vertically) is incorrect. The creator of the following graph also begins the graph at 20% and ends it at 80%. Visually, it appears that the lowest displayed score is close to zero instead of over 20%. The top score appears to be near 100% but is instead less than 80%. The purpose of this graph was to give the illusion of both higher scores at the top and greater gains overall. Careful reading of the scale, however, helps us to interpret the information correctly.

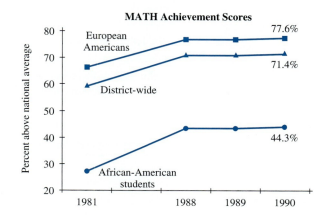

In comparison, the next graph does not condense the horizontal axis, and the vertical axis extends from 0 to 100%. The increments of one year on the horizontal axis are about the same size as 10% increments on the vertical axis. This graph gives a straightforward representation of the data.

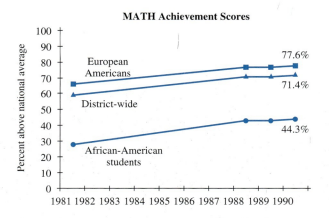

In the third graph of this situation, the vertical axis is condensed. Each increment is now 20% instead of 10%. In this graph, the school district's rise in test scores appears even less impressive because the slope on this graph appears less steep. Additionally, the difference between the results of black students appears closer to the district-wide results. This graph appears different from the previous one; however, they are both correct.

All graphs have an effect on the reader. Adjusting the scale and using breaks on the axes influence how graphs are perceived. The reader must carefully read the information given on the graph so as to make an informed decision about the data that are presented.

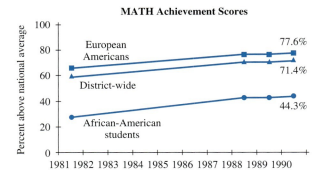

Another method of distorting data visually is seen in the following **picture graph.** How much larger does the graduate in 1987 appear to be compared with the graduate in 1979? How much larger does the graduate appear to be in 1995 compared with 1987? In each case, from 1979 to 1987, and from 1987 to 1995, the tuition approximately doubled. We saw in Activity 5 that when we double both the length and width of a rectangle, the area is actually four times larger. In the following graph, the height of the graduates is doubled; however, their widths also doubled. What we see is the area of each of the pictures and, because the length and the width both doubled, the graduate in 1987 appears four times larger than the graduate in 1979. The same is true for the figures between 1987 and 1995. The graduate in 1995 is about four times taller than the graduate in 1979 because the tuition quadrupled in this time span. How much larger does the 1995 graduate appear to be compared with the 1979 graduate?

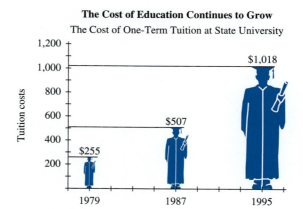

Not all picture graphs are visually distorted. Look at the next picture graph. This graph also uses pictures of graduates to display the information. Each graduate in the graph is the same size and represents about $250 in tuition. By stacking the graduates on top of one another, we change the height of each represented year's graph, but not the width. The overall area of the pictures representing tuition in 1987 is twice as large as those representing tuition in 1979, and the area of the pictures representing tuition in 1995 is four times larger than that for 1979. From this, we see that this graph is an accurate representation of the data.

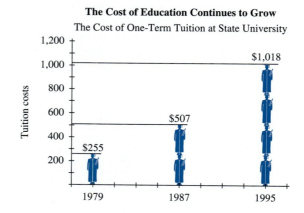

In this section, we saw how graphs can be used to misrepresent results by inappropriate scales or breaking axes. We saw that the way the data are displayed can make a difference in how it is interpreted. We, as consumers, must read graphs critically.

Problem Set 10.1

1. **Public School Dollars.** The three following graphs (I–III) all represent the same data. Use them to answer the following questions.

   a. If you were writing an advertisement in support of a local bond measure to provide more money to schools, which graph would you use? Explain.

   b. If you are an author of a ballot measure to reduce local taxes and you want to show that schools have not been hurt by cutbacks in government spending and are getting a large share of the tax dollars, which graph would you use? Explain.

   c. If you are an editor of the local newspaper and wish to display the data in a fair and unbiased graph, which graph would you use? Explain.

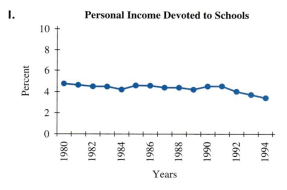

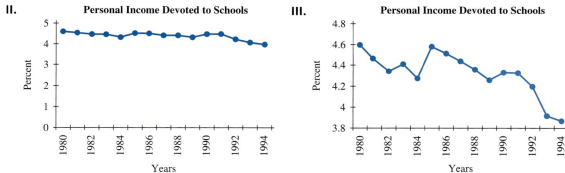

2. **Higher Enrollments in Higher Education.** The following graph represents the 1995 and 1996 fall term headcount in public four-year colleges in the state of Oregon. Beginning in 1998, the figures are predictions. The graph was included in an article about the "baby boom echo" (the bulge in population resulting from the children of people born between 1945 and 1960) reaching college age. Will the number of students enrolling in higher education double between 1995 and 2004? Assuming the current data and the predictions are correct, do you feel that this graph is either misleading or incorrect? Explain.

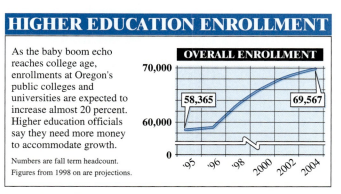

Source: Oregon University System.

3. **Selling a Home.** The following picture graph represents the average sales price of a home in the Portland, Oregon, metropolitan area. Discuss whether you think this graph is misleading. Give specific reasons for your decision.

Comparisons represent the average sales price of residential listings as reported to RMLS™, the Portland, Oregon, metro area multiple listing service, for the period ending Sept. 30, 1996.

Source: Realtors Multiple Listing Service.

4. **Revenue Report.** The following table gives the revenue for a local manufacturing company.
   a. Prepare a bar graph of the data for the stock investors. You will want to make the increases in revenue seem high.
   b. Prepare a second bar graph for the workers to convince them that asking for a pay increase at this time is not warranted. You will want to minimize the amount of the revenue increases.

Year	1990	1991	1992	1993	1994	1995
Revenue in $1000	8.7	9.3	9.6	9.2	10.4	11.0

5. **U.S. Economic Activity.** The following graph displays the annual growth rate of the economy by quarters for the last four years. Is the graph either misleading or incorrect? If yes, explain in what way. Be specific.

Copyright 1996 Time, Inc. Reprinted by permission.

6. **Last-Minute Santas.** On the night before Christmas everyone rushes to the mall, or so it seems. How many of us really wait until the last minute to finish our Christmas shopping?

   a. The following graph is compiled from data from the International Council of Shopping Centers. Is the graph either misleading or incorrect? If yes, explain in what way. Be specific.

   b. The following stockings-and-chimney graph was drawn to display the same data given in part a. Is the graph either misleading or incorrect? If yes, explain in what way. Be specific.

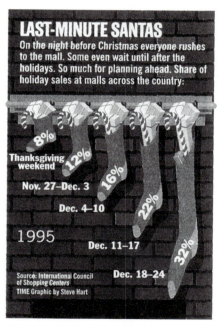

   Copyright 1996 Time, Inc. Reprinted by permission.

7. Cancer Research. Explain what is being presented in the following graph. Is the graph either misleading or incorrect? If yes, explain. (*Note:* The top row for each cancer type represents data from 1962 and the bottom row represents 1998 data.)

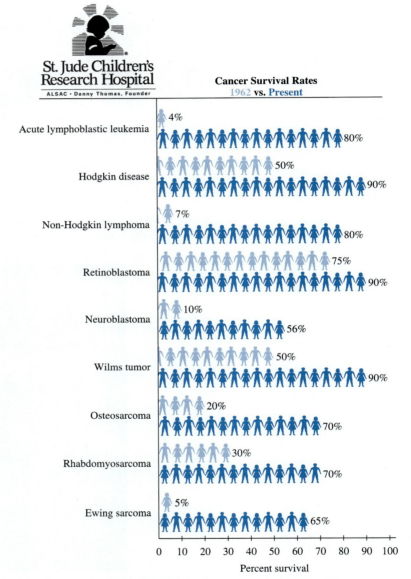

Source: Data provided by St. Jude Public Relations, ALSAC-St. Jude Children's Research Hospital, Memphis, TN.

10.2 Descriptive Statistics

Discussion 10.2

Stem-and-Leaf Diagrams

The two following tables list the fourth-week enrollment in several mathematics classes at Mt. Hood Community College for winter 1995 and winter 1996.

NUMBER OF STUDENTS ENROLLED IN MATH CLASSES FOURTH WEEK WINTER 1995

38	34	36	35	35	38
34	29	37	19	27	39
39	38	21	13	21	20
29	21	39	21	30	

NUMBER OF STUDENTS ENROLLED IN MATH CLASSES FOURTH WEEK WINTER 1996

39	38	43	40	39	40
39	37	30	33	32	31
43	36	40	36	14	34
14	36	18	29	32	28

As presented, the data are difficult to analyze or compare. **Descriptive statistics** are used to summarize and organize lists of data to make them more useful. One method of organizing and sorting a list is to use a **stem-and-leaf diagram.** To create a stem-and-leaf diagram of the data listed in the previous tables the "stems" represent the tens digits and the "leafs" represent the units. We can first sort the data by the stems. Let's first look at the 1995 data. The data include items with tens digits of 1, 2, and 3. These become the stems. The first item in the 1995 data is 38. The stem is then 3 and the leaf is 8. The number 38 is shown in the start of the stem-and-leaf diagram.

Stems	Leaves
1	
2	
3	8

The remaining data from winter 1995 have been added to the next stem-and-leaf diagram.

Stems	Leaves
1	9 3
2	9 9 1 1 7 1 0 1
3	8 4 9 4 6 7 9 5 8 5 0 8 9

The 1996 data need stems of 1, 2, 3, and 4. The following stem-and-leaf diagram represents the data from the winter 1996 table.

Stems	Leaves
1	4 8 4
2	9 8
3	9 9 8 7 6 6 0 3 6 9 2 2 1 4
4	3 3 0 0 0

The data are now somewhat easier to compare, but we can make them more usable. We next sort the leaves in numerical order. If the diagrams are to be used to present the data, we need to add titles and to provide a key to indicate what the stems and leaves represent. The key can be a statement like "1 | 3 represents 13 students," which indicates that the stem of 1 is the tens digit and the leaf of 3 is the ones digit. A key of "5 | 7 represents 570 units" indicates that the stem is the hundreds digit and the leaf the tens digit.

NUMBER OF STUDENTS ENROLLED FOURTH WEEK 1995

1 | 3 represents 13 students

Stem	Leaves
1	3 9
2	0 1 1 1 1 7 9 9
3	0 4 4 5 5 6 7 8 8 8 9 9 9

NUMBER OF STUDENTS ENROLLED FOURTH WEEK 1996

1 | 4 represents 14 students

Stem	Leaves
1	4 4 8
2	8 9
3	0 1 2 2 3 4 6 6 6 7 8 9 9 9
4	0 0 0 3 3

The data are now organized in such a way that they are easier to read and compare. Now we can make a general statement such as "The classes were bigger in winter 1996 than in winter 1995." It still is difficult to make any detailed comparisons.

Measures of Central Tendency

In addition to using stem-and-leaf diagrams to organize data, numerical values can be used to summarize the data. Often a single number is used to represent the "center" of the data. We look at three different ways to determine a measure that represents the "center" of a set of data. These numbers or measures are often referred to as **measures of central tendency.**

> **Definition**
>
> The **mode** is the value that occurs most often in a set of data.

Once the data are organized, as we did with the stem-and-leaf diagrams, it is very easy to identify the mode. In the enrollment data for the fourth week from winter 1995, we can see that the mode is a class size of 21 because the number 21 occurs four times, and no other number occurs more than three times. Sometimes more than one data entry occurs most frequently. In the enrollment data for fourth week from winter 1996, we can see that class sizes of 36, 39, and 40 all appear three times. We say that the winter 1996 data have three modes: class sizes of 36 students, 39 students, and 40 students. In this case the mode is not useful for comparing the data. In fact the mode of 21 students gives the impression that classes were much smaller than they really were. The mode might be useful if we were ordering clothing stock in different sizes. Knowing what size had sold most in the past would be useful information.

The mean is what we usually think of as the average. The mean is easily computed and takes into consideration all of the data values. The mean is often represented by the symbol $\bar{x}$ and is read as "x bar." It is the number we get by adding together all of the data and then dividing the sum by the number of data items.

> **Definition**
>
> The **mean** of a set of data is the average found by adding all of the data entries and then dividing by the number of data entries.

$$\text{mean} = \frac{\text{sum of the data entries}}{\text{number of data entries}}, \text{ or symbolically as}$$

$$\bar{x} = \frac{x_1 + x_2 + x_3 + \cdots + x_n}{n}$$

where $x_1, x_2, x_3, \ldots, x_n$ represents the data entries and n represents the number of data entries.

> **Definition**
>
> The **median** is the value that falls in the center of a data set sorted in ascending or descending order. The median splits the data into approximately the upper 50% and the lower 50%.

> **How to Find the Median**
>
> To find the median, first sort the data. Then,
>
> - If the number of data entries is odd, the value that falls in the middle is the median.
> - If the number of data entries is even, the average of the two middle values is the median.

Example 1

Find the mean and the median for each of the following two data lists.

List A: 3, 5, 8, 6, 4
List B: 23, 79, 84, 109, 96, 79

Solution

The mean of list A is found by adding all of the data items and dividing by the number of data items.

$$\bar{x} = \frac{3 + 5 + 8 + 6 + 4}{5} = 5.2$$

The mean of list A is 5.2.

To find the median of list A, we must first sort the data, that is, arrange the data in descending or ascending order. Then we must locate the value that falls in the middle.

List A

3
4
5 ← the middle of the sorted data
6
8

The median of list A is 5 because it is the middle of the sorted data.

We can compute the mean of list B.

$$\bar{x} = \frac{23 + 79 + 84 + 109 + 96 + 79}{6} \approx 78.3$$

The mean of list B is approximately 78.3.

To find the median of list B, we first sort the data. Because list B has an even number of data entries, no data value falls directly in the middle. To determine the median we locate the middle two values and compute the average of these two values.

List B

23
79
79 ⎫
84 ⎬ Median is $\frac{79 + 84}{2} = 81.5$
96
109

The median of list B is 81.5.

The mean of a data set is often reported, probably because it is familiar to most people and is also fairly easy to compute. It does not require us to sort the data as the mode and median do. Although the mean is often a useful measure of central tendency, it sometimes gives us a distorted view. The next example illustrates this point.

Example 2

Western-Net Cellular Phone Company. Being a relatively new company, Western-Net has only eight employees. Their annual salaries are listed in the following table.

Employee	Salary ($)
President	225,000
Financial officer	95,000
Office manager	32,000
Sales rep 1	35,000
Sales rep 2	30,000
Sales rep 3	15,000
Sales rep 4	18,000
Sales rep 5	24,000

a. Determine the mean and median salaries for the company.

b. Which measure would be of most interest to you if you were considering going to work as a sales representative for the company? What other questions might you ask?

Solution

a. The mean can be found by using the formula from the definition.

$$\bar{x} = \frac{x_1 + x_2 + x_3 + \cdots + x_n}{n}$$

$$= \frac{225{,}000 + 95{,}000 + 32{,}000 + 35{,}000 + 30{,}000 + 15{,}000 + 18{,}000 + 24{,}000}{8}$$

$$= \$59{,}250$$

The mean salary at Western-Net is $59,250.

To find the median, we must first put the salaries in order.

$$225{,}000 \quad 95{,}000 \quad 35{,}000 \quad 32{,}000 \quad 30{,}000 \quad 24{,}000 \quad 18{,}000 \quad 15{,}000$$

The middle numbers are 32,000 and 30,000. The average of 32,000 and 30,000 is 31,000.

We conclude that the median salary at Western-Net is $31,000.

b. The mean salary in this example is quite a bit larger than any of the salaries of the sales representatives. The large salaries of the president and financial officer pull the mean up. If you were considering working for the company as a sales representative, this number is not representative of the salary you might receive.

The median lets you know that half of the salaries are above $31,000 and half are below $31,000. This gives you a more realistic view of what you might expect.

Asking for the mean and median salaries of just the sales representatives gives you an even better idea of your potential salary.

Example 3

Fourth-Week Enrollments Revisited.

a. Determine the mean of each of the data sets giving fourth-week enrollments for winter 1995 and winter 1996 at Mt. Hood Community College.

b. Determine the median of the data sets giving fourth-week enrollments for winter 1995 and winter 1996 at Mt. Hood Community College.

Solution

a. To determine the mean, we know we must find the sum of all of the data entries and divide by the number of data items. For the winter 1995 data, the sum of the enrollments for the 23 sections listed is 693 students. The mean is then $\frac{693}{23}$, or approximately 30.1 students per section.

For the winter 1996 data, the sum of the enrollments for the 24 sections is 801 students. The mean is $\frac{801}{24}$, or approximately 33.4 students per section.

b. We know that the median is the middle number of the *sorted* data. We can use the stem-and-leaf diagram to determine the median. In winter 1995 because there were 23 sections, the one in the middle is the 12th section in an ordered list. That leaves 11 numbers above and 11 numbers below the middle number. Looking at the ordered stem-and-leaf diagram we see that the number in the 12th position is 34.

NUMBER OF STUDENTS ENROLLED FOURTH WEEK 1995

1 | 3 represents 13 students

Stem	Leaves
1	3 9
2	0 1 1 1 1 7 9 9
3	0 4 4 5 5 6 7 8 8 8 9 9 9

We conclude that 34 students was the median class size in winter 1995.

In winter 1996, there were 24 sections. Therefore, the 12th and 13th sections share the middle position. We average their values to determine the median. Because the 12th and 13th entries are both 36, their average is 36.

NUMBER OF STUDENTS ENROLLED FOURTH WEEK 1996

1 | 4 represents 14 students

Stem	Leaves
1	4 4 8
2	8 9
3	0 1 2 2 3 4 6 6 6 7 8 9 9 9
4	0 0 0 3 3

We conclude that the median class size for winter 1996 was 36 students.

When reporting the central value of a data set, we usually use only one of the three measures. The one used depends on which gives the best measure of center for the given data set.

- The **mean** works well if the data set has no excessively high or low values.
- The **median** is a good choice for data sets that contain only a few numbers that are excessively high or low.
- The **mode** is the appropriate measure when we are interested in the number that occurs the most often.

Quartiles and Box Plots

We saw how to find the measure of the center of a data set. This is often used to help describe a set of numbers. But the center only gives limited information about a set of numbers. Suppose we want to compare the level of anxiety of a group of third-graders. We want to know if they are more anxious at recess or in the classroom. Do third-graders already experience performance anxiety in the classroom or are they more likely to be anxious about their ability to play sports well or fit in socially with their peers? Let's look at some data collected in a study of 14 randomly selected third-graders. A scale from 0 to 10, where 0 represented low anxiety and 10 represented high anxiety, was created to quantify observed behaviors that indicated when students were feeling anxious. The results are recorded in the next two tables.

ANXIETY LEVEL OF STUDENTS AT RECESS			
8.7	6.1	5.8	4.2
4.0	3.6	6.0	6.7
6.7	6.9	2.5	2.4
5.0	4.0		
Median = 5.4			

ANXIETY LEVEL OF STUDENTS IN THE CLASSROOM			
6.2	6.8	6.0	5.1
3.9	4.7	8.2	4.1
5.7	4.1	3.8	6.7
7.8			
Median = 5.7			

In the data presented, we see that the median anxiety level for the group was higher in the classroom than at recess. Do we have enough information to say that children are more anxious in the classroom? Is the range of levels of anxiety larger in the classroom or at recess? Are the scores of the most anxious child the same in both settings? What about the least anxious child? Does the range of the anxiety levels of the middle group of students differ? Clearly, knowing the median is not enough to answer all of these questions. We will now see how to determine and report the spread of a data set.

> Definition
>
> The **range** of a data set is found by finding the difference between the maximum value and the minimum value.

In the data set in the Anxiety Level of Students at Recess table the maximum value is 8.7 and the minimum value is 2.4. We can compute the range of the data set by finding the difference between these values.

range = maximum value − minimum value
range = 8.7 − 2.4
range = 6.3 units on the anxiety scale

The range of the data set in the Anxiety Level of Students in the Classroom table is

$range = $ maximum value $-$ minimum value

$range = 8.2 - 3.8$

$range = 4.4$ units on the anxiety scale

The range of the data is greater during recess than in the classroom.

You may have noticed that in computing the range for recess, one child measured 8.7 on the anxiety scale. This value was much higher than the rest of the entries. Using just the two extreme numbers places a lot of emphasis on the highest or the lowest values in a data set. We need to find other numbers to help us describe the data set. We already know how to determine the median, which divides the data into approximately the top and bottom 50%. Two other numbers, called quartiles, divide the data further.

Definition

If the numbers are arranged in order, the **lower quartile** (Q_1) is the median of the lower half of the data and the **upper quartile** (Q_3) is the median of the upper half of the data. These values along with the median divide the data into four groups of about equal size.

We say that about 25% of the data lies below Q_1, about 50% lies below the median, and about 75% lies below Q_3.

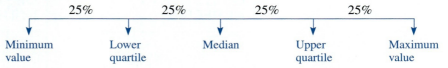

To find the quartiles for the Anxiety Level of Students at Recess, first sort the data. Next, determine the median of the data set. Then find Q_1 by determining the median of the lower half and Q_3 by determining the median of the upper set of numbers.

ANXIETY LEVEL OF STUDENTS AT RECESS

8.7	6.1	5.8	4.2
4.0	3.6	6.0	6.7
6.7	6.9	2.5	2.4
5.0	4.0		
Median = 5.4			

$\left.\begin{array}{l} 2.4 \\ 2.5 \\ 3.6 \\ 4.0 \\ 4.0 \\ 4.2 \\ 5.0 \end{array}\right\} Q_1 = 4.0$

$\text{Median} = \dfrac{5.0 + 5.8}{2} = 5.4$

$\left.\begin{array}{l} 5.8 \\ 6.0 \\ 6.1 \\ 6.7 \\ 6.7 \\ 6.9 \\ 8.7 \end{array}\right\} Q_3 = 6.7$

When the median falls between two numbers in the data set, that is, the number of data entries is even, we divide the data into two equal sections. Q_1 is the median of the lower half of the data, and Q_3 is the median of the upper half of the data.

We can now further describe the data set by saying that the median anxiety level for these children at recess was 5.4 and that 25% of the children measured at or below 4.0 on the anxiety scale and 25% measured at or above 6.7. Another way to say this is that the middle 50% of the children measured between 4.0 and 6.7 on the anxiety scale.

> **Definition**
>
> The difference between the upper quartile and lower quartile is called the **interquartile range.**

$$\text{interquartile range} = Q_3 - Q_1$$
$$= 6.7 - 4.0$$
$$= 2.7 \text{ units on the anxiety scale}$$

We now look at how to determine the quartiles if the number of data entries is odd.

ANXIETY LEVEL OF STUDENTS IN THE CLASSROOM

6.2	6.8	6.0	5.1
3.9	4.7	8.2	4.1
5.7	4.1	3.8	6.7
7.8			

Median = 5.7

3.8
3.9
4.1
4.1 $Q_1 = \dfrac{4.1 + 4.1}{2} = 4.1$
4.7
5.1

5.7   Median = 5.7

6.0
6.2
6.7
6.8 $Q_3 = \dfrac{6.7 + 6.8}{2} = 6.75$
7.8
8.2

When the median is an actual number in the data set, that is, the number of data entries is odd, we group the data below the median and group the data above the median. Q_1 is the median of the lower group of data, and Q_3 is the median of the upper group of data.

In our example, the minimum anxiety reading for the students in class was 3.8 and the maximum was 8.2. The median anxiety level for the 13 students was 5.7. About 25% of the students recorded levels of anxiety at or below 4.1 and about 25% recorded levels above 6.75. Using the minimum, maximum, median, and quartiles is referred to as the **five number summary of the data.** These five numbers can be helpful in comparing two sets of numerical data, especially when we are comparing large data sets. The five number summary gives an indication of the middle data and the amount of spread around the middle.

> **Definition**
>
> A **box plot** or **box-and-whisker plot** is a graphical representation of the five number summary. A horizontal scale is used to determine where to plot the minimum, maximum, quartiles, and median.

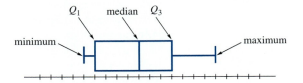

To create a box-and-whisker plot, we draw a rectangle with Q_1 and Q_3 as the ends of the rectangle above an appropriate horizontal scale. This rectangle is called the "box." The median is shown by drawing a vertical line in the box. The minimum and maximum are located, and the "whiskers" are drawn as horizontal lines from these extreme values to the edge of the box. Recall that the box represents approximately the middle 50% of the data set. The whisker from the minimum to the box represents approximately the lower 25% of the data, and the whisker from the maximum to the box represents approximately the upper 25% of the data.

Comparing the box plots of the data sets measuring classroom anxiety can help us visualize the relationship between anxiety ratings of the students at recess and in the classroom. From the graph we can see that there is more variability in how students react at recess than in the classroom. The highest and the lowest anxiety levels were recorded at recess. The median anxiety level is lower at recess than in the classroom, but this is not enough evidence to say that children are more anxious in the classroom. When we compare the boxes we see that the middle 50% of the students appear to have about the same levels of anxiety in these two situations. By analyzing all five values in the box plot it is possible to see a more complete picture and make better comparisons.

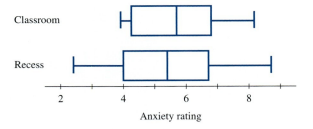

Example 4

A sociologist prepared a questionnaire to evaluate the amount of stress workers felt on the job. After interviewing workers in 54 different occupations the sociologist summarized the results in the following box-and-whisker diagram. Use this diagram to answer the following questions.

a. What percent of the workers experience a stress level of 5.3 (the median) or higher?
b. What percent of the workers experience a stress level of 7.2 (the upper quartile) or higher?
c. The middle 50% of workers experience stress between what levels?
d. Determine the levels of stress for the lowest 25% of all of the workers.

Solution a. We know that the median of the data is 5.3. Therefore, we can say that about 50% of the workers experience a stress level of 5.3 or higher.

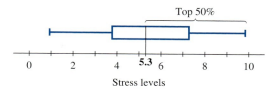

b. We know that 7.2 is equal to the upper quartile; about 25% of the data is at or above this number. Therefore about 25% of workers experience a stress level of 7.2 or higher.

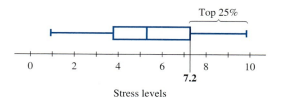

c. We know that the interquartile range from Q_1 to Q_3 gives us approximately the middle 50% of a set of data. So we want to locate the values for the ends of the box in the box-and-whisker plot.

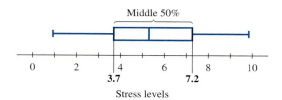

We can estimate the ends of the box to be about 3.7 and 7.2. This means that 50% of workers experience stress levels between 3.7 and 7.2.

d. Approximately 25% of workers with the lowest stress levels fall between the minimum value and the first quartile.

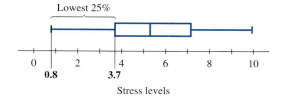

We can estimate the minimum to be about 0.8 and the first quartile to be about 3.7. The levels of stress for the lowest 25% of the workers fall between 0.8 and 3.7.

In this section, we were introduced to the **stem-and-leaf diagram** as a way to organize information. We then looked at three **measures of central tendency** for data sets. The first was the **mode,** the item that occurred the most times. The second measure was the **mean,** or average, which we find by dividing the sum of the data items by the number of data items. The third measure was the **median,** the item in the middle of a sorted set of data. Although all of these can be useful in analyzing or comparing sets of data, usually only one of the measures is used. Which one to use depends on which gives the best measure of center for the given data set.

Problem Set 10.2

1. **Coffee-Makers.** Nineteen coffee-makers were rated by a consumer's bureau. The prices (rounded to the nearest dollar) of the 19 models varied significantly and are shown in the following stem-and-leaf diagram. Use the diagram to answer the questions.

   a. What is the cost of the most expensive coffee-maker?

   b. What is the cost of the least expensive coffee-maker?

   c. What is the mode?

   d. Determine the median cost of the 19 coffee-makers.

PRICE FOR THE 19 COFFEE-MAKERS IN THE STUDY						
2	0	0	0	0	2	5
3	0	0	7			
4	0	0				
5	0	0	2			
6	0	5				
7						
8	5					
9	0	0				

   Key: 2 | 0 represents $20.

2. **Skyscrapers.** A double stem and leaf compares two sets of data. The stem is in the center, and the leaves extend to the left and right. The following diagram shows the heights of buildings over 700 feet, not including TV towers and antennas, in Chicago and New York.

   a. Explain what the following line from the diagram means.

       6 5 | 8 | 1 1 5

   b. How tall is the tallest building in these two cities, and where is it located?

   c. Which city has more buildings over 700 feet? Over 1000 feet?

HEIGHTS OF BUILDINGS OVER 700 FEET

Chicago			New York														
Leaves		Stem	Leaves														
		7	0	1	1	2	2	3	3	4	4	5	5	5	8	9	
6	5	8	1	1	1	5											
	6	9	2	3	5												
	1	10	0	5													
4	3	11															
		12	5														
		13	6	7													
	5	14															

Key: 6 | 9 | means about 960 feet;
      | 9 | 2 means about 920 feet.

3. Each data set in this problem must have at least five items.
   a. Make up two different sets of data with the same mean.
   b. Make up two different sets of data with the same median.
   c. Make up two different sets of data with a mean of 7 and a median of 6.
   d. Make up a data set with a mean of 25 and a median of 6.

4. **TV Show Audience Ratings.** During the 1996–1997 television season, NBC had 18 shows in the top 50 rated TV shows, and ABC had 18. The following double stem-and-leaf diagram displays the average audience rating for these two television networks. The audience rating is the percentage of all U.S. households with TV sets that were tuned into a certain show. For example, in the first row of the table, 7 | 8 | means that 8.7% of all TVs were tuned into a particular show on ABC.

   a. Explain what the following line from the diagram means.

   $\quad$ 0 | 14 | 1

   b. Which network had the top rated show in 1996–1997? What percent of the TV audience watched this show?

   c. How many shows on ABC had an average audience rating over 15%? How many shows on NBC had an average audience rating over 15%?

AVERAGE AUDIENCE RATING FOR ABC AND NBC TV SHOWS, 1996–1997

ABC							NBC					
Leaves						Stem	Leaves					
					7	8						
5	5	4	3	2	1	9	5					
				6	1	10	5	5	8	8		
		9	7	5	2	11	0	0	0	4	5	8
				8	5	12						
						13						
					0	14	1					
						15						
					2	16	5	8	8			
						17	0					
						18						
						19						
					5	20	5					
						21	2					
Key: 2	16	means 16.2%;	16	5 means 16.5%.								

5. **The Cafeteria.** The cafeteria is trying to decide whether to continue to stay open from 4:00 P.M. to 6:00 P.M. During the last five weeks they gathered the data listed in the table.

   a. Make a stem-and-leaf diagram of the data. (Because there are no Friday evening classes the cafeteria is closed at 3:30 P.M. on Fridays.) (*Hint:* Use the ones digits as the leaves and the two-digit hundreds-tens as the stems.)

   b. What was the median number of people per day visiting the cafeteria between 4 P.M. and 6 P.M. during the five weeks?

   c. Would you recommend that the cafeteria stay open from 4 P.M. to 6 P.M.? Do you need other information before you can make the recommendation? If so, what information?

NUMBER OF PEOPLE VISITING THE CAFETERIA BETWEEN 4 P.M. AND 6 P.M.				
	M	T	W	Th
Week 1	345	352	300	324
Week 2	340	360	310	322
Week 3	329	342	350	342
Week 4	349	345	354	324
Week 5	333	339	342	324

6. **Worksheet Scores.** The following table shows the scores received by a math class on a recent worksheet.

   a. How many students turned in the worksheet?
   b. How many students received a score of 22?
   c. What was the mean of the worksheet scores?
   d. What was the median of the worksheet scores?

Worksheet Scores	Number of Students
25	5
24	6
22	7
20	8
19	4
18	3
15	1
12	1

7. **Amazing Puzzle Company.** The Amazing Puzzle Company has 58 employees who earn a salary of $35,000, 42 employees who earn a salary of $27,500, 16 employees who earn a salary of $18,000, 5 employees who earn a salary of $52,000, and 2 employees who earn a salary of $73,000. What is the average annual salary at the company?

8. **The PGA Tour.** In 1998, 26 golfers made more than one million dollars on the PGA (Professional Golfers Association) Tour. The following table shows the amounts earned, rounded to the nearest $100,000. What was the average income for these top golfers?

Amount Earned in $1,000,000's on the 1998 PGA Tour	Number of Golfers
2.6	1
2.2	1
2.0	1
1.8	4
1.7	2
1.5	2
1.4	1
1.3	4
1.2	4
1.1	4
1.0	2

9. **Mathematics Study Time.** The following bar graph depicts the number of hours students spend studying mathematics outside of class per week. Use the graph to answer the following questions.

   a. How many students study at least 7 hours a week, but less than 8?

   b. Two hundred forty students were in this study. Approximately what percent study mathematics 9 hours or more outside of class per week?

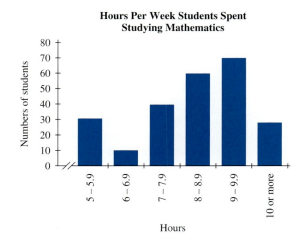

10. **Credit Card Costs.** The following box plot shows the annual cost in interest and fees for several credit cards assuming a $2000 debt balance for the year. Use the box plot to answer the questions.

    a. Approximately how much did the most expensive credit card cost?
    b. Approximately how much did the least expensive credit card cost?
    c. What was the median cost for carrying a $2000 balance on a credit card?
    d. What percentage of the credit cards charge more than $228 in a year?
    e. What percentage of the credit cards charge more than $190 in a year?
    f. If your bank charged you $240 for credit last year, assuming you carried a balance that averaged $2000, would you change card companies? Explain.

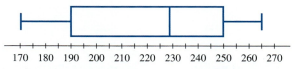

Annual Cost for Credit Cards Assuming a $2000 Debt Balance

11. **Basketball.** The number of points scored per game by two different college basketball players are summarized in the following box-and-whisker plot. Use this to answer the following questions.

    a. What is the median number of points per game scored by player A? Player B?
    b. What is the maximum number of points per game scored by player A? Player B?
    c. Which player was more consistent during the season? Explain.
    d. In what percentage of games did player B score over 12 points? Over 10 points?
    e. An NBA team wants to pick the player who has the potential to score the most points. Which player would you recommend they choose? Explain why.

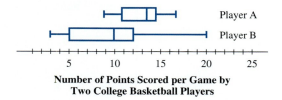

Number of Points Scored per Game by Two College Basketball Players

12. **Standardized Mathematics Test.** The following box plot displays the national results of four different groups of students on a standardized mathematics test. Notice that only the interquartile range for each group of results is displayed. Toni scored 44 on this test. Compare Toni's results to each of the national groups.

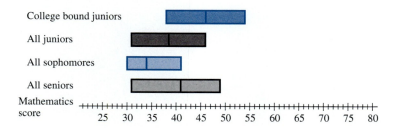

13. Professor C's Class. The following pie graph is missing its percent labels. Using a protractor, carefully measure the central angles, and complete a table similar to the following.

Majors	Percent
Sciences/Math	
Engineering	
Technical	
Health	
Art	
Humanities	
General studies	
Undecided	

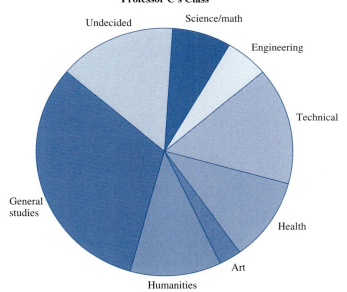

Professor C's Class

14. Population Statistics. The following table shows the population, area, and population density of the 36 counties in Oregon. The bottom row gives the data for the entire state. Use the table to answer parts a–h. (These data are from the 1990 U.S. Census.)

a. Verify that the areas of Baker County given in square kilometers and square miles are equivalent (1.609 km ≈ 1 mile).

b. Using the data in the second, third, and fourth columns of the table, verify that the population density given for Marion County is correct in columns five and six.

c. What is the population of Lane County?

d. What is the population density of Lane County?

e. What is the mean population density for the 36 counties of Oregon?

f. What county has a population density closest to the mean?

g. What counties would you guess are largely urban?

h. What countries would you guess are largely rural?

i. Using the technology available to you, sort the data by population per square mile. Determine the median of population densities in the 36 counties in Oregon.

County	Population	Area (km²)	Area (miles²)	Population (per km²)	Population (per mile²)
Baker	15,317	7,946.9	3,068.3	1.9	5.0
Benton	70,811	1,752.1	676.5	40.4	104.7
Clackamas	278,850	4,838.9	1,868.3	57.6	149.3
Clatsop	33,301	2,142.7	827.3	15.5	40.3
Columbia	37,557	1,701.0	656.8	22.1	57.2
Coos	60,273	4,145.4	1,600.5	14.5	37.7
Crook	14,111	7,717.0	2,979.5	1.8	4.7
Curry	19,327	4,215.0	1,627.4	4.6	11.9
Deschutes	74,958	7,817.4	3,018.3	9.6	24.8
Douglas	94,649	13,045.4	5,036.8	7.3	18.8
Gilliam	1,717	3,118.7	1,204.1	0.6	1.4
Grant	7,853	11,729.6	4,528.8	0.7	1.7
Harney	7,060	26,249.4	10,134.9	0.3	0.7
Hood River	16,903	1,353.0	522.4	12.5	32.4
Jackson	146,389	7,214.1	2,785.4	20.3	52.6
Jefferson	13,676	4,612.5	1,780.9	3.0	7.7
Josephine	62,649	4,246.4	1,639.5	14.8	38.2
Klamath	57,702	15,396.5	5,944.6	3.7	9.7
Lake	7,186	21,073.1	8,136.3	0.3	0.9
Lane	282,912	11,795.3	4,554.2	24.0	62.1
Lincoln	38,889	2,537.3	979.7	15.3	39.7
Linn	91,227	5,934.8	2,291.4	15.4	39.8
Malheur	26,038	25,609.3	9,887.7	1.0	2.6
Marion	228,483	3,069.1	1,185.0	74.4	192.8
Morrow	7,625	5,265.1	2,032.8	1.4	3.8
Multnomah	583,887	1,127.3	435.3	517.9	1,341.4
Polk	49,541	1,919.4	741.1	25.8	66.8
Sherman	1,918	2,132.3	823.3	0.9	2.3
Tillamook	21,570	2,854.7	1,102.2	7.6	19.6
Umatilla	59,249	8,327.7	3,215.3	7.1	18.4
Union	23,598	5,275.0	2,036.7	4.5	11.6
Wallowa	6,911	8,146.6	3,145.4	0.8	2.2
Wasco	21,683	6,167.3	2,381.2	3.5	9.1
Washington	311,554	1,874.6	723.8	166.2	430.4
Wheeler	1,396	4441.9	1715.0	0.3	0.8
Yamhill	65,551	1,853.4	715.6	35.4	91.6
Oregon	2,842,321	248,646.4	96,002.5	11.4	29.6

15. 1996 Presidential Election. The following table lists the election results in terms of percent of votes, state by state for the 1996 presidential election. President Clinton won this election with 379 electoral votes to Senator Dole's 158. Clinton won the election in 30 states plus Washington, D.C. Dole won in 20 states.

   a. Determine the minimum, maximum, median, and quartiles for the state-by-state percentages given in the table for both Dole and Clinton.

   b. On a single graph, draw two box plots to compare the percentages of each candidate.

   c. Write a paragraph comparing Dole's versus Clinton's percentages state by state. Include in your paragraph observations using the five number summary.

States	Clinton (%)	Dole (%)	States	Clinton (%)	Dole (%)
Alabama	43.3	50.7	Montana	41.6	44.6
Alaska	35.1	53.5	Nebraska	35.0	53.5
Arizona	47.4	44.5	Nevada	46.1	44.0
Arkansas	54.5	37.3	N. Hampshire	50.0	40.2
California	53.2	39.1	New Jersey	54.1	36.9
Colorado	45.9	47.3	New Mexico	51.0	42.7
Connecticut	53.3	36.4	New York	60.0	31.8
D.C.	52.2	36.9	North Carolina	44.3	49.0
Delaware	88.4	9.9	North Dakota	40.5	47.3
Florida	48.3	42.5	Ohio	47.8	41.5
Georgia	46.1	47.5	Oklahoma	40.6	48.5
Hawaii	59.2	30.3	Oregon	49.8	39.1
Idaho	34.2	52.9	Pennsylvania	49.1	39.9
Illinois	54.5	37.3	Rhode Island	61.6	27.3
Indiana	41.5	48.1	South Carolina	44.1	50.3
Iowa	51.0	40.5	South Dakota	43.3	47.0
Kansas	36.5	54.8	Tennessee	48.4	46.0
Kentucky	46.2	45.2	Texas	44.1	49.1
Louisiana	52.8	40.3	Utah	34.1	55.7
Maine	53.4	31.9	Vermont	55.4	32.1
Maryland	54.7	38.6	Virginia	45.5	47.7
Massachusetts	62.4	28.6	Washington	52.9	37.6
Michigan	52.2	38.9	West Virginia	51.6	37.1
Minnesota	52.1	35.8	Wisconsin	50.0	39.4
Mississippi	44.2	49.9	Wyoming	37.3	50.2
Missouri	44.2	48.1			

16. Speeding Costs. Exceeding the posted speed limit or driving too fast for conditions is one of the most prevalent factors contributing to motor vehicle accidents, according to the National Highway Traffic Safety Administration (NHTSA). The economic cost to society is estimated by NHTSA to be $28.9 billion per year. The following table shows the estimated costs of speeding-related crashes for noninterstate roads. Each column represents the costs of accidents in the state that occurred on roads posted with a specific speed limit. Use the table to answer the following questions.

a. For roads posted with a speed limit of 55 mph, determine the minimum, maximum, median, and quartiles of the costs of speeding-related accidents.

b. Repeat part a for roads posted 45 mph.

c. Repeat part a for roads posted 35 mph.

d. On a single graph, draw three box plots to compare the costs of speed-related crashes for speed limits of 55 mph, 45 mph, and 35 mph.

e. What conclusions can you make *based on your graph*?

f. Iowa shows an estimated cost of $0 for roads posted at 40 mph. Do you think that it is reasonable that, in 1997, no costs are related to excessive speeding crashes on roads with 40 mph speed limits in Iowa? If this is unreasonable, explain what you think the zero in this column means.

ESTIMATED COSTS OF SPEEDING-RELATED MOTOR VEHICLE ACCIDENTS BY SPEED LIMIT ON NONINTERSTATE ROADS FOR 1997 (MILLIONS OF DOLLARS)

State	55 mph	50 mph	45 mph	40 mph	35 mph	<35 mph
AL	118	9	133	24	40	25
AK	5	5	3	4	1	3
AZ	80	24	51	47	26	38
AR	114	1	19	4	16	18
CA	328	47	127	83	167	127
CO	58	8	23	25	29	38
CT	4	3	18	16	16	58
DE	5	25	1	4	4	6
FL	97	28	113	55	85	91
GA	158	4	58	20	48	24
HI	4	4	8	1	9	8
ID	18	11	4	6	11	12
IL	206	1	28	12	56	75
IN	69	18	28	31	28	31
IA	29	2	4	0	6	15

17. Crime Rates. The crime rate (per 1000 residents) is displayed in the box plots shown below. The data have been divided into crime rates in suburban, urban, and rural areas of a state. Use the box plots to answer the following questions.

  a. Estimate the median, quartiles and extremes for the crime rate in urban counties.
  b. Which area has the largest interquartile range?
  c. If you say that the most dangerous area is the one with the highest maximum crime rate, which area is most dangerous? Which is second? Which is third?
  d. If you say that the most dangerous area is the one with the highest median crime rate, which area is most dangerous? Which is second? Which is third?

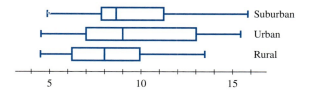

10.3 Probability

Activity Set 10.3

1. **a.** If you flip a coin 30 times, how many heads would you expect to get? Tails?
    b. Flip a coin 30 times. Record your results in the following table.

	Tally
Heads	
Tails	

    c. Determine the ratio of the number of heads to the total number of tosses.
    d. Determine the ratio of the number of tails to the total number of tosses.

    In Activities 1c and 1d you computed what is called **experimental probability,** which is the ratio of the number of times an event occurs during an experiment and the total number of trials. That is,

    $$\text{experimental probability of an event} = \frac{\text{number of occurrences of the event}}{\text{total number of trials}}$$

2. On the board is a diagram similar to the following. On the board, initial the region that applies to you. For example, if you went to bed after 11:00 P.M. and ate breakfast, initial the region where the circles overlap. If you went to bed after 11:00 P.M. but did not eat breakfast, initial the circle on the left where it does not overlap the other circle. If you did not go to bed after 11:00 P.M. and did not eat breakfast, initial the portion of the picture outside of both circles but inside the rectangle.

    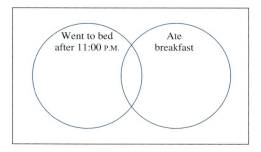

    a. What is the probability that a person from this class went to bed after 11:00 P.M.?
    b. What is the probability that a person from this class ate breakfast?
    c. What is the probability that a person from this class went to bed after 11:00 P.M. and ate breakfast?
    d. What is the probability that a person from this class went to bed before 11:00 P.M. and ate breakfast?
    e. What is the probability that a person from this class went to bed before 11:00 P.M. and did not eat breakfast?

3. In this activity, you will be playing a game and deciding whether the game is fair. You will need a partner to play this game. In the game, two coins are flipped. Player A gets 1 point if a match occurs and because there are two ways to get a match, player B gets 2 points if there is no match.

   a. Do you think this is a fair game? Why or why not? If you are not sure, guess.

   b. Next you will play the game to help you determine if your decision was correct or to help you make your decision. You and your partner must first decide which player will be player A and which will be B. Flip two coins. Record your results in the following table. Continue playing the game for about 3 minutes.

	Tally	Total Points
Player A: Match		
Player B: No match		

   c. Based on your results from playing the game, do you think the game is fair? Why or why not? If you believe the game is unfair, how would you change it so it was fair?

   d. Recall,

   $$\text{experimental probability of an event} = \frac{\text{number of occurrences of the event}}{\text{total number of trials}}$$

   Determine the experimental probability of getting a match and the experimental probability of getting no match. Do these probabilities support your answer from part c? Explain.

4. In this activity, you will be playing a game and deciding whether the game is fair. You will need a partner to play this game. In this game, you spin a paper clip on the spinner that follows. Player A gets 1 point if the paper clip lands in the shaded area, and player B gets 1 point if the paper clip lands in the unshaded area.

   a. Do you think this is a fair game? Why or why not? If you are not sure, guess.

   b. Next you will play the game to help you determine if your decision was correct or to help you make your decision. You and your partner must first decide which player will be player A and which will be B. Begin playing the game. Record your results in the following table. Continue playing the game for about 3 minutes.

	Tally	Total Points
Player A: Shaded area		
Player B: Unshaded area		

   c. Based on your results from playing the game, do you think the game is fair? Why or why not? If you believe the game is unfair, how would you change it so it was fair?

   d. Determine the experimental probability of landing in the shaded area and the experimental probability of landing in the unshaded area. Do these probabilities support your answer from part c? Explain.

   e. Use your protractor to determine the percent of the circle that is shaded and the percent of the circle that is unshaded. How do these percentages compare with your experimental probabilities in part d?

450 Chapter 10 Statistics and Probability

Spinner for Activity 4

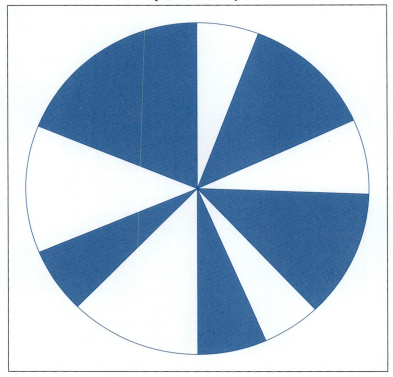

Discussion 10.3

Probability

In this section we look at probabilities. Knowing the probabilities of different events allows us to make predictions and decide whether situations are fair.

You were asked to determine whether the game in Activity 3 was fair. One way to decide whether a game is fair is to determine the probability of each event in the game. There are two types of probabilities: experimental and theoretical.

> **Definition**
>
> The **experimental probability** of an event is the ratio of the number of observed occurrences of the event and the total number of trials.

$$\text{experimental probability of an event} = \frac{\text{number of occurrences of the event}}{\text{total number of trials}}$$

> **Definition**
>
> The **theoretical probability** of an event is the ratio of the number of ways to obtain the event and the total number of possible outcomes, assuming all outcomes are equally likely to occur.

$$\text{theoretical probability of an event} = \frac{\text{number of ways to obtain the event}}{\text{total number of possible outcomes}}$$

For example, if a coin is tossed 100 times and 60 heads occur, then the experimental probability of obtaining a head is $\frac{60}{100} = \frac{3}{5}$. However, exactly two outcomes are possible when a coin is flipped and only one way to obtain a head. This means that the theoretical probability is $\frac{1}{2}$. So, you can think of the theoretical probability as the probability of an ideal experiment. In this scenario, ideally we would have obtained 50 heads after tossing the coin 100 times.

Example 1

A six-sided die was tossed 20 times, and the following results were recorded.

Number	1	2	3	4	5	6
Tally	|||	????	||	|||	|||	||||

a. Determine the experimental and theoretical probability of obtaining a 6.
b. Determine the experimental and theoretical probability of obtaining a number less than 5.

Solution   a. In the table we observed that a 6 occurred four times out of a total of 20 tosses. Therefore, the experimental probability of obtaining a 6 is $\frac{4}{20}$ or $\frac{1}{5}$.

Probabilities are often expressed as decimals or percents. As a decimal $\frac{1}{5}$ would be 0.2; as a percent it would be 20%.

Because there is one way to obtain a 6 out of six possible outcomes, the theoretical probability of obtaining a 6 is $\frac{1}{6}$. As a decimal this is approximately 0.167; as a percent this is $\approx 16.7\%$.

b. In the table we observed that a number less than 5 occurred 13 times out of a total of 20 tosses. Therefore, the experimental probability of obtaining a number less than 5 is $\frac{13}{20}$. As a decimal this is 0.65; as a percent it is 65%.

Because a number less than 5 (1, 2, 3, or 4) can be obtained four ways out of six possible outcomes, the theoretical probability of obtaining a number less than 5 is $\frac{4}{6}$ or $\frac{2}{3}$. As a decimal this is approximately 0.667; as a percent this is $\approx 66.7\%$.

Example 2

If a bag contains three red balls, five blue balls, and two white balls, what is the probability of drawing one red ball from the bag? One blue ball? One white ball?

Solution When we ask for the probability of an event we are referring to the theoretical probability. The notation P(red) represents the probability of drawing a red ball. Because 10 balls are in the bag,

$$P(\text{red}) = \frac{3}{10}$$

$$P(\text{blue}) = \frac{5}{10} = \frac{1}{2}$$

$$P(\text{white}) = \frac{2}{10} = \frac{1}{5}$$

Notice, in Example 2,

$$P(\text{red}) + P(\text{blue}) + P(\text{white}) = \frac{3}{10} + \frac{5}{10} + \frac{2}{10} = 1$$

That is, the sum of the probabilities of all possible events is equal to 1.

Example 3

a. In the previous example, suppose we obtain a red ball on the first draw and then replace the ball. What's the probability of obtaining a red ball on the second draw?

b. If we obtain a red ball on the first draw and do not replace the ball, what is the probability of obtaining a red ball on the second draw? Obtaining a blue ball on the second draw?

Solution a. If we replace the red ball, then 10 balls are still in the bag when we draw the second ball. Therefore, the probability of obtaining a red on the second draw is $\frac{3}{10}$.

b. If we do not replace the red ball, then two red balls, five blue balls, and two white balls are in the bag when we draw the second ball. Therefore,

$$P(\text{red on second draw}) = \frac{2}{9}$$

$$P(\text{blue on second draw}) = \frac{5}{9}$$

In Activity 3, you were asked to determine the experimental probability of obtaining a match when two coins are tossed and of getting no match. You probably found that both probabilities were close to one half. Now, let's determine the theoretical probabilities for each of these events.

NOTE: From now on, the word *probability* will mean *theoretical probability* unless otherwise stated.

Example 4

Determine the probability of obtaining a match and the probability of obtaining no match when two coins are tossed.

Solution

To determine the indicated probabilities we must determine all of the possible outcomes of tossing two coins. This is referred to as the sample space.

> **Definition**
> The set of all possible outcomes of an experiment is called the **sample space.**

To help us determine the sample space, imagine tossing two coins, one of which is a penny and one a dime. The outcome in which the penny lands heads and the dime lands tails is different from the outcome when the penny lands tails and the dime lands heads.

This means that the sample space for tossing two coins is

HH HT TH TT

where H stands for heads and T stands for tails.

We can visualize this in the following tree diagram. A **tree diagram** is a visual way to list the sample space and determine probabilities. Two coins are used in this experiment, and we will be making two observations. First we will observe the outcome of the first coin, then we will observe the outcome of the second coin. We say that this experiment has two **stages.** In the first stage we want to display all of the possible outcomes for the first coin. Two possible outcomes are: heads or tails. So, we draw the first stage of the tree diagram with two **branches.**

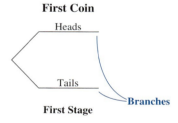

If the first coin lands heads up, the second coin could be a head or a tail. Therefore, for the second stage we need to draw two branches off of the first heads' branch. Similarly, if the first coin lands tails up, the second coin could be a head or a tail. So the first stage tails' branch has two branches from it.

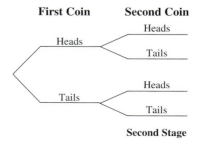

To read the outcomes from a tree diagram, we begin at the point on the left and trace along each path. In the following figure, one path is marked and the outcome is recorded to the right of the path. Note that HT means you get heads on the first coin and tails on the second.

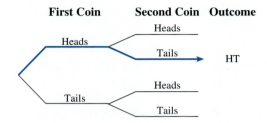

Finally, we complete the tree diagram by recording the outcome of each of the four branches.

```
First Coin     Second Coin   Outcome
                 Heads         HH
     Heads
                 Tails         HT

                 Heads         TH
     Tails
                 Tails         TT
```

Next, we assign probabilities to each of the four outcomes. The chance of heads or tails is equally likely on each toss of the coins, therefore the probability of each outcome is equal. Therefore the probability for each of the outcomes is $\frac{1}{4}$.

Using this information, there are two ways to obtain a match out of four possible outcomes and two ways to obtain no match out of four possible outcomes. Therefore,

$$P(\text{match}) = \frac{1}{2}$$

$$P(\text{no match}) = \frac{1}{2}$$

Many people believe that the probability of a match when tossing two coins is $\frac{2}{3}$. This common misconception comes from thinking that a match can be achieved in two ways, HH or TT, but "no match" can be obtained only one way, with one H and one T. This is the reason why the game in Activity 3 awarded 1 point for a match and 2 points for a no match. We can see from both our experimental and theoretical probabilities that the probabilities of obtaining a match or a no match are equal. Therefore, the game in Activity 3 is unfair. How can the game be made fair?

Example 5

Consider the situation in which a bag contains two white balls and one black ball. We draw one ball from the bag, note the color, replace the ball, and draw a second one. Draw a tree diagram for this situation.

a. What is the probability that we receive a match?
b. What is the probability we receive no match?

Solution In this experiment, we draw two balls and record their color. Because two observations are recorded, this experiment has two stages. Three colors are possible for each ball, so our tree diagram needs three branches for the first stage, and for each branch of the first stage we need three branches for the second stage. Each of the outcomes is equally likely.

First Ball	Second Ball	Outcome
White	White	WW
	White	WW
	Black	WB
White	White	WW
	White	WW
	Black	WB
Black	White	BW
	White	BW
	Black	BB

a. From the diagram, five outcomes are possible in which the balls match out of a total of nine. Therefore, P(match) = $\frac{5}{9}$.

b. Similarly, there are 4 outcomes where the balls do not match. So, P(no match) = $\frac{4}{9}$.

As you can probably imagine, drawing this tree diagram when there are many balls would be a burden. However, we can accomplish the same goal by drawing a simpler tree diagram in which the branches contain the probabilities for each outcome. For example, in the previous situation, rather than drawing two branches containing the white outcome, we can draw one branch containing white and indicate the probability of a white is $\frac{2}{3}$. A tree diagram containing the probabilities is called a **probability tree**. The complete probability tree is shown here.

First Ball	Second Ball	Outcome and Probability
White 2/3	White 2/3	WW (2/3)*(2/3) = 4/9
	Black 1/3	WB (2/3)*(1/3) = 2/9
Black 1/3	White 2/3	BW (1/3)*(2/3) = 2/9
	Black 1/3	BB (1/3)*(1/3) = 1/9

Notice, to determine the probability that this first ball is white *and* that the second ball is black we *multiply* the probabilities on the tree, that is, P(WB) = $\frac{2}{3} * \frac{1}{3} = \frac{2}{9}$. Similarly, the probability that this first ball is black *and* the second ball is black is P(BB) = $\frac{1}{3} * \frac{1}{3} = \frac{1}{9}$.

Because the sum of the probabilities computed in the probability tree is one $\left(\frac{4}{9} + \frac{2}{9} + \frac{2}{9} + \frac{1}{9} = 1\right)$, we can feel somewhat confident they are correct. Now we can use this diagram to answer the questions in Example 5.

$$P(\text{match}) = P(WW \text{ or } BB) = P(WW) + P(BB) = \frac{4}{9} + \frac{1}{9} = \frac{5}{9}$$

$$P(\text{no match}) = P(WB \text{ or } BW) = P(WB) + P(BW) = \frac{2}{9} + \frac{2}{9} = \frac{4}{9}$$

To determine the probability that the outcome is BB *or* WW we *add* the probabilities for these two different outcomes.

Example 6

A bag contains three red balls, five blue balls, and two white balls. We draw one ball from the bag, note its color, and draw a second ball *without* replacing the first. Draw a probability tree for this situation.

a. Let R represent red, B blue, and W white. What is the probability of the outcome RR? RW? BW? RB?

b. What is the probability of drawing two balls that match?

Solution As we saw in Example 3, P(red on the first draw) = $\frac{3}{10}$. Therefore, the first branch containing red is labeled $\frac{3}{10}$. If the first ball drawn is red and not replaced then only two red balls are in the bag out of nine on the second draw, so P(red on the second draw if red was drawn on the first draw) = $\frac{2}{9}$.

The following is a diagram of the entire tree.

First Ball	Second Ball		Outcome
Red 3/10	Red 2/9	RR	(3/10)*(2/9) = 6/90
	Blue 5/9	RB	(3/10)*(5/9) = 15/90
	White 2/9	RW	(3/10)*(2/9) = 6/90
Blue 5/10	Red 3/9	BR	(5/10)*(3/9) = 15/90
	Blue 4/9	BB	(5/10)*(4/9) = 20/90
	White 2/9	BW	(5/10)*(2/9) = 10/90
White 2/10	Red 3/9	WR	(2/10)*(3/9) = 6/90
	Blue 5/9	WB	(2/10)*(5/9) = 10/90
	White 1/9	WW	(2/10)*(1/9) = 2/90

a. $P(RR) = \frac{6}{90} = \frac{1}{15}$

$P(RW) = \frac{6}{90} = \frac{1}{15}$

$P(BW) = \frac{10}{90} = \frac{1}{9}$

$P(RB) = \frac{15}{90} = \frac{1}{6}$

b. $P(\text{match}) = P(RR \text{ or } BB \text{ or } WW) = \frac{6}{90} + \frac{20}{90} + \frac{2}{90}$

$= \frac{28}{90}$

The probability of a match is $\frac{28}{90}$, which is about 31%.

Example 7

Hit and Run. A hit-and-run accident occurs. A witness claims that the car that fled the scene is green. It is known that people may think they see one thing when they are seeing something else. To determine the reliability of the witness's statement, the witness is shown clips of 100 cars, 30 of which are green and 70 of which are blue. After each clip the witness is to identify the color of the car seen in the clip. The following chart shows the results.

The first entry in the table, 24, indicates that in 24 instances the car in the clip was green and the witness identified the car as green. The 6 in the second row represents the instances when the car in the clip is green and the person identified the car as blue.

	The car in the clip is **green**.	The car in the clip is **blue**.
The witness identifies the car in the clip as **green**.	24	14
The witness identifies the car in the clip as **blue**.	6	56

a. If the car in the clip was green, what is the probability that the witness identifies the car as green?

b. If the car in the clip was blue, what is the probability that the witness identifies the car as blue?

c. If the person identifies the car in the clip as green, what is the probability that the car was green?

d. If the person identifies the car in the clip as blue, what is the probability that the car was blue?

e. What is the probability that the car in the clip is green and the witness identifies the car as green?

Solution It will be helpful to determine the sum of the rows and columns to answer the questions.

	The car in the clip is **green**.	The car in the clip is **blue**.	
The witness identifies the car in the clip as **green**.	24	14	38
The witness identifies the car in the clip as **blue**.	6	56	62
	30	70	

a. If the car in the clip is green, we are restricted to the first column. A total of 30 green cars are shown, and the witness identifies 24 of them as green. Therefore,

$$P(\text{witness identifies green when the car shown is green}) = \frac{24}{30} = 80\%$$

b. The second column of the table represents all of the cases in which the car in the clip is blue. Seventy blue cars are shown, and the witness identifies 56 of these as blue. Therefore,

$$P(\text{witness identifies blue when the car shown is blue}) = \frac{56}{70} = 80\%$$

c. If the person identifies the car in the clip as green, what is the probability that the car was green? We often phrase this question as "What is the probability that the car is green, given that the person identifies the car as green?" The first row of the table represents all of the cases when the person identifies the car as green. In 38 cases the witness identifies the car as green. Therefore,

$$P(\text{the car is green, given the witness identifies green}) = \frac{24}{38} \approx 63\%$$

d. The second row of the table represents all of the cases when the person identifies the car as blue. In 62 cases the witness identifies the car as blue. Therefore,

$$P(\text{the car is blue, given the witness identifies blue}) = \frac{56}{62} \approx 90\%$$

e. A total of 100 cars are shown. In 24 cases the car shown is green, and the witness identifies the car as green. Therefore, the probability that the car is green and the witness identifies the car is green is 24%.

$$P(\text{the car is green and the witness identifies green}) = \frac{24}{100} = 24\%$$

What does Example 7 tell you about the reliability of eyewitness accounts?

The probabilities in Example 7 parts a, b, c, and d are called **conditional probabilities.** In part a, we were looking for the probability that the witness identifies a green car given the car in the clip is green. The notation used for a conditional probability is $P(A \mid B)$, read probability of A given B. In part a we use the notation P(witness identifies green $\mid$ car is green).

Example 8

If a coin is flipped ten times and all of the outcomes are heads, what is the probability of a tail on the 11th flip?

Solution Some people believe that the probability of a tail is much greater on the 11th flip than a head because the first ten coins landed heads up. But remember that each time the coin is flipped the probability of a head or a tail is still 50%. Therefore, the probability of the coin landing tails up on the 11th flip is still 50%.

Let's look at this probability another way. We know that the first ten tosses resulted in heads. What are the possible outcomes when the coin is flipped the 11th time? The result is either ten heads followed by a head or ten heads followed by a tail.

   H H H H H H H H H H H

or

   H H H H H H H H H H T

Because two outcomes are possible: one of which results in a tail on the 11th flip, the probability of a tail on the 11th flip given heads on the first ten is $\frac{1}{2}$ = 50%. In shorthand, P(tail on the 11th $\mid$ ten heads) = 50%.

In this section we looked at **probabilities,** both experimental and theoretical. The **experimental probability** of an event is the ratio of the number of times that an event occurs during an experiment and the total number of trials. The **theoretical probability** of an event is the ratio of the number of ways to obtain the event and the total number of possible outcomes, assuming all outcomes are equally likely to occur. You can think of the theoretical probability as the probability associated with an ideal experiment. In everyday language, the word probability refers to theoretical probability. We used **tree diagrams** and **probability trees** to determine various probabilities.

Problem Set 10.3

1. In a standard bridge deck, which contains 52 cards, determine the following probabilities.
    a. What is the probability of drawing one card from the deck and obtaining an ace?
    b. What is the probability of drawing one card from the deck and obtaining a face card?
    c. Draw one card from the deck and replace the card. What is the probability that on the second draw the card is of the same suit?
    d. Draw one card from the deck and *do not replace* the card. What is the probability that on the second draw the card is of the same suit?
    e. Draw one card from the deck and replace the card. What is the probability that on the second draw the card is the same card?

2. A box contains seven green and three orange marbles. Determine the following probabilities.
    a. What is the probability of drawing a green marble? An orange marble?
    b. Suppose we obtain a green marble on the first draw and replace it. What is the probability of drawing a green marble on the second draw? An orange marble?
    c. Suppose we obtain a green marble on the first draw and *do not replace* it. What is the probability of drawing a green marble on the second draw? An orange marble?

3. A box contains four black and two purple marbles. Consider the two-stage experiment of randomly selecting a marble from the box, replacing the marble, and randomly selecting a marble a second time.
    a. Draw a probability tree.
    b. What is the probability of selecting two black marbles?
    c. What is the probability of selecting a black marble first and then selecting a purple marble?
    d. What is the probability of selecting *at least* one black marble?

4. Redo Problem 3 assuming the first marble is *not replaced.*

5. A Family of Four
    a. What is the probability that a family with four children has three girls and one boy, with the youngest being the boy?
    b. What is the probability that a family with four children has four girls?
    c. If a family has three children, which are all girls, what is the probability that the fourth child is a boy?

6. Often probabilities are expressed as *odds*. For example, if you go to a race track to bet on a horse, the odds for each horse are given. Let's say that the odds for the horse Midnight Star to win are 2 to 3, written 2:3. This means that in five races Midnight Star is expected to win two times and lose three times. Therefore, the probability that Midnight Star will win is $\frac{2}{5} = 40\%$ and the probability that Midnight Star will lose is $\frac{3}{5} = 60\%$.

   a. If the odds of living to age 65 are 7:3, what is the probability of living to age 65?
   b. If the odds that the Blazers will beat the Sonics are 2:5, what is the probability that the Blazers will beat the Sonics?
   c. If the odds of Carl beating Joan in poker are 3:5, what is the probability that Joan will win?
   d. If the probability of rain today is 80%, what are the odds in favor of rain?

7. Smoking. At a local hospital the cause of death was recorded along with whether the patient was a smoker. The data are recorded in the following table. Use the table to answer the following questions.

   a. How many total patient deaths are recorded in the table?
   b. How many of the patients were smokers? How many patients died of cancer?
   c. What percent of the patients recorded in the table died of cancer? What percent died of heart disease?
   d. From the table, what percent of the patients who smoked died of cancer? What percent of the nonsmokers died of cancer?
   e. From the table, determine the probability of a patient dying of heart disease given that he or she smoked.

	Cancer	Heart Disease	Other
Smoker	135	310	205
Nonsmoker	55	155	140

8. Drug Testing. It has been shown that a test for illicit drug use can fail in two ways. The test can give a false-positive reading or a false-negative reading. Suppose a manufacturer of a drug test claims its test is 98% reliable (this means it fails 2% of the time). We assume this means that 2% of drug users receive a false-negative reading and 2% of nonusers receive a false-positive reading. One thousand people apply to work for a company that drug tests its applicants. Of these, 5% are drug users.

   a. Complete the following table.

	Drug Users	Nonusers
Drug test is positive.		
Drug test is negative.		

   b. How many of the 1000 people undergoing testing should be eliminated because of drug use?
   c. How many *are* eliminated by drug testing?
   d. How many drug users get the job in spite of testing?
   e. Should the company use the test? Why or why not?

9. **Advertisements.** Is it likely that television viewers believe an advertisement that includes a hidden camera interview? A random sample of 500 adults was taken. The opinions of these adults regarding the ads as well as their educational background is displayed in the following table. Use the table to answer the following questions.

   a. What percent of the sample found the ads to be believable?

   b. What percent of the sample found the ads to be believable, given that they had less than a high school education? Some college? Were college graduates?

   c. Based on your results from part b, are people more likely or less likely to believe an advertisement that includes a hidden camera interview if they are more educated?

	Less than High School	High School Graduate	Some College	College Graduate
Believable	11	42	61	24
Unbelievable	19	77	140	96
Not sure	2	13	11	4

10. **HIV/AIDS Knowledge Study.** A study to determine HIV/AIDS knowledge was completed in 1990. A random-digit-dialed interview of persons at least 18 years old was conducted. In Oregon, 3308 people participated in the study. The following table shows some of the responses of those surveyed in Oregon by demographic characteristics.

Characteristics	A person who is infected with the AIDS virus can look and feel healthy. (Yes; %)	How effective do you think a condom is in preventing getting the AIDS virus through sexual activity? (Very/Somewhat; %)	Would you be willing to work with a person who is infected with the AIDS virus? (Yes; %)
Age			
<65 years	81	87	76
≥65 years	48	64	46
Sex			
Male	73	83	69
Female	75	82	70
Education			
Less than high school	53	68	53
High school graduate	70	80	64
Some college	83	89	78
Income			
<$20,000/year	67	79	66
≥$20,000/year	81	89	76

    a. What percent of those surveyed who are under the age of 65 answered yes to the question, "Can a person inflicted with AIDS look and feel healthy?" Over the age of 65?

    b. Did the level of education seem to make a difference in the percentage of yes responses to the question "Can a person inflicted with AIDS look and feel healthy?" Explain.

    c. If we assume the *correct* answers are those in parentheses, are people who make less than $20,000 a year more or less likely to know about HIV/AIDS than people who make more than $20,000 a year?

    d. Which demographic category(ies) (age, sex, education, or income) seemed to make no difference in the number of correct responses?

11. Sum of Two Dice

  a. Read the following experiment, and guess which column will contain the most X's.
  b. With a partner perform the experiment.
  c. Determine the experimental probability of each sum.
  d. Determine all of the possible outcomes when the red and white dice are thrown.
  e. Determine the theoretical probability of each sum.

The Experiment

Let one person roll the two dice (one white and one red) and announce the *sum* of the roll to his or her partner. The partner makes an X in the appropriate box on the following chart starting at the bottom of each column. Stop when the X's reach the top in any one column.

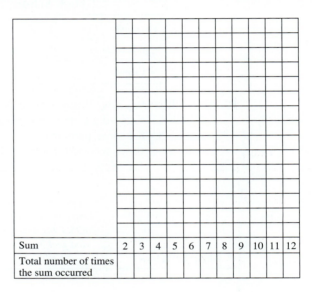

12. Product of Two Dice

   a. Read the following experiment, and guess which column will contain the most X's.

   b. With a partner perform the experiment.

   c. Determine the experimental probability of each product.

   d. Determine all of the possible outcomes when the red and white dice are thrown.

   e. Determine the theoretical probability of each product.

   The Experiment

   Let one person roll the two dice (one white and one red) and announce the *product* of the roll to his or her partner. The partner makes an X in the appropriate box on the following chart starting at the bottom of each column. Stop when the X's reach the top in any one column.

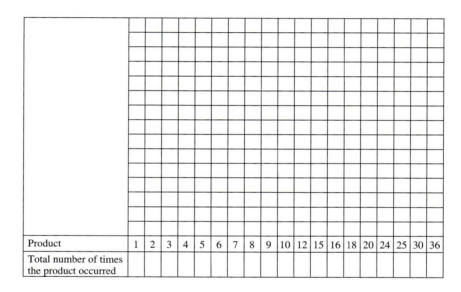

Product	1	2	3	4	5	6	7	8	9	10	12	15	16	18	20	24	25	30	36
Total number of times the product occurred																			

13. **Circle Area Estimation.** The following figure consists of a square, each side length of which is 4.00 inches, and a circle with a diameter of 2.95 inches.

   a. Take a handful of rice and drop the rice onto the following figure.

   b. Count the number of grains of rice that land in the circle, and count the total number of grains that land in the square and circle combined.

   c. Of all the grains that landed inside the figure, what proportion landed in the circle?

   d. Multiply your proportion in part c by the area of the square. Your result is an estimate of the area of the circle.

   e. Use the diameter of the circle to compute its area. Compare this result with your estimate from part d.

14. Area Estimation. Use the method in Problem 13 to determine the area of the blob inside the following square.

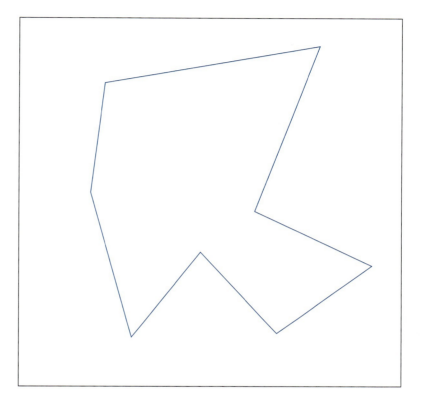

15. The Dart Board. A dart board is represented in the following figure. The diameters of the circles from the smallest to the largest are $\frac{9}{16}$ in., $1\frac{3}{8}$ in., $7\frac{3}{8}$ in., $8\frac{3}{8}$ in., $12\frac{3}{8}$ in., and $13\frac{3}{8}$ in.

a. The innermost circle is called the bull's-eye. Assuming that a dart is thrown and lands in the board, what is the probability of hitting the bull's-eye?

b. The narrow outermost ring is called the doubles ring, and the narrow middle ring is called the triples ring. If your dart hits the bull's-eye, you receive a score of 50 points. If your dart lands in the next bigger circle, you receive a score of 25 points. If the dart lands in the doubles ring, your score is twice the number shown on the outside of the wedge. Similarly, if your dart lands in the triples ring, your score is three times the number of the wedge. If your dart lands in any other location on the board, your score is equal to the number of the wedge.

Where on the dart board should you aim? Explain your reasons and any assumptions you are making.

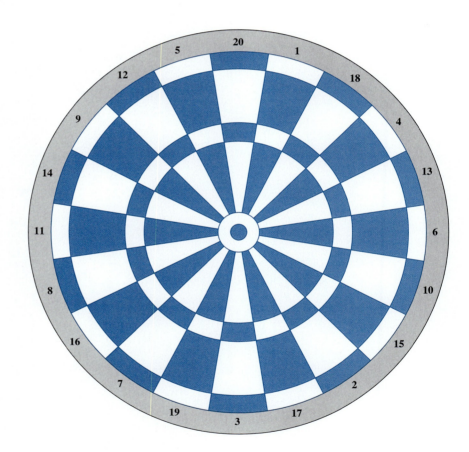

Chapter 10 Summary

Graphical and numerical displays of data are often used to summarize large amounts of data. Use of graphical software on computers has advanced the techniques available for producing innovative tables and graphs. Today, more than ever before, we need to be aware of how to produce honest representations of data, and we need to know how to read graphs so that we can properly interpret what they mean. In this chapter we saw how the way a graph is constructed can affect our perception of the data. When reading information from a graph, remember to

- Look carefully at the scale that is used for both axes.
- Look for breaks in the axes.
- In a picture graph, decide if the artist created pictures in which the area (or volume) of the picture is sized correctly.

A **stem-and-leaf** diagram allows us to easily organize large sets of data and gives a quick way to check how many numbers fall into a certain range. Unlike other displays, you can still see every piece of the data.

We looked at different methods of summarizing data with statistics. Summaries of data typically tell us about the central value and how much variation occurs in the data. Three values can be used to describe the center or typical value for a set of data: the mean, the median, and the mode.

The **mean** is an average of a set of numbers. It is found by adding up all of the numbers and dividing by the number of values.

The **median** is the number that falls in the middle of a set of data when the data are arranged from least to greatest.

The **mode** is the value that occurs the most often in a set of numbers.

When reporting the central value of a data set, we must select the appropriate measure to use.

- The **mean** works well if the data set has no excessively high or low values.
- The **median** is a good choice for data sets that contain only a few numbers that are excessively high or low.
- The **mode** is the appropriate measure when we are interested in the number that occurs the most often.

We looked at different ways to report the variability of a data set. The range is one measure of variability. The **range** is equal to the difference between the maximum value and the minimum value. The range can give us an idea of how spread out the data are.

Quartiles are another indicator of variability. **Quartiles** group the data into four groups of approximately equal size. There are three quartiles: Q_1 (lower), Q_2 (median), and Q_3 (upper). Roughly speaking, these three quartiles split the data into quarters. The difference between the third quartile and the first quartile is called the **interquartile range** and represents approximately the middle 50% of the data.

A **box-and-whisker plot** displays the median, the quartiles, and the extreme values of a set of data. A box-and-whisker plot can be used to compare two or more sets of data.

The probability of an event gives us an indication of how likely the event is to occur. We can determine the probability of an event by performing the event many times and recording the results. This gives us the experimental probability of an event. The **experimental probability** of an event is the ratio of the number of times the event occurred during the experiment and the total number of trials.

If we know that each outcome of an event is equally likely to occur, we can compute the theoretical probability of the event. The **theoretical probability** of an event is the ratio of the number of ways to obtain the event and the total number of possible outcomes, assuming all outcomes are equally likely to occur.

To calculate the theoretical probability of an event, you need to know all the different outcomes possible. A **sample space** is a list of all of the possible outcomes. We can determine the sample space for an experiment by making an organized list or a **tree.** In general, theoretical probability can predict the results of a probability experiment. As the number of attempts increases, the experimental probability gets closer to the theoretical probability.

Chapter Review 7-10

1. Solve each equation. Check your solution.

   a. $6k + 71 = 47 - 2k$

   b. $2n - 3(n + 5) = 7n$

   c. $\dfrac{24.2}{10.5} = \dfrac{k}{16.4}$

   d. $4 - 3(4 - x) = 12x + (x - 5)$

   e. $12.1m - 20.29 = 3.7(m + 1.1)$

   f. $\dfrac{2p + 1}{p} + \dfrac{5}{3p} = 6$

   g. $\pi\left(\dfrac{D}{2}\right)^2 = 1.0$

   h. $\dfrac{x^2 + 9}{x} = \dfrac{2x - 3}{2}$

2. Solve the following inequalities, graph the solutions on a number line, and check the solutions.

   a. $\tfrac{2}{3}x - 13 > {-25}$ b. $23 \le {-4}(3w + 7)$ c. $2.1x - 17 < 4.7x - 10$

3. List three solutions to each equation.

   a. $y = x$ b. $y = \sqrt{4x}$

4. List three solutions to each equation.

   a. $y = 3x^2 - 1$ b. $y = 24 - 10x$

5. Graph the following equations.

   a. $y = 5x - 4$

   b. $y = x$

   c. $8x - 5y = 10$

   d. $y = \tfrac{-2}{5}x + 12$

   e. $y = 2x^3$

   f. $y = \dfrac{x}{5}$

   g. $y = 35 - 8.50x$

   h. $y = 2x(x - 1)$

6. Solve the equation $23.00 + 13.25x = 14(20.50 - 2x)$ graphically. Check your solution.

7. a. *Use the graph* (do not graph your own) to solve the following equation.

   $$3(x + 3) - (x + 14) = 0.5x - 2.5 - x$$

   b. *Use the graph* (do not graph your own) to solve the following system of equations.

   $$\begin{cases} y = 3(x + 3) - (x + 14) \\ y = 0.5x - 2.5 - x \end{cases}$$

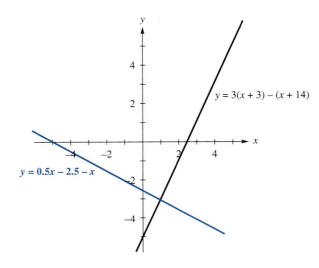

8. Solve the following equation graphically. Check your solution.

$$\frac{5(2x - 6)}{4} - \frac{x - 11}{2} = 13 - x$$

9. Suppose you are asked to solve each of the following equations either algebraically or graphically. Without actually solving them, decide which equations you would solve algebraically and which graphically. Explain how you made your decision.

   a. $-0.1x + 3.2 = \frac{x}{4} - 1$

   b. $30 - 5k = 50$

   c. $4(18 - 3x) + 20 = 5 - 3(1 - x)$

   d. $\frac{5m}{3} + 6 = \frac{-3}{5} + \frac{m}{3}$

10. *Use the graph* to solve the equation $5(18 + x) - 115 = 2(42 - 3.25x) + 6$. Do not graph this on your own.

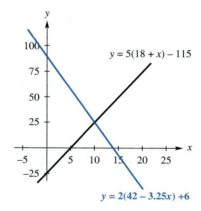

11. *Use the graph* to solve the equation $0.08x - 4(1 - 0.07x) = 14$. Do not graph this on your own.

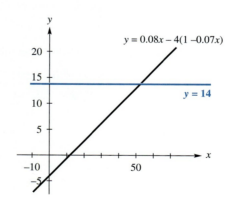

12. Solve each two-by-two linear system. Numerically check all solutions.

   a. $\begin{cases} 2x - 3y = 5 \\ x + y = 13 \end{cases}$

   b. $\begin{cases} 4x - 6 = y \\ 0.5x + y = 18 \end{cases}$

   c. $\begin{cases} 0.65x - 0.2y = 13 \\ y = 4.25 - x \end{cases}$

   d. $\begin{cases} 5y + 83 = 3x \\ x + y = 105 \end{cases}$

13. **Wobbles.** A toy company developed a new toy called Wobbles. It wants to produce 12,000 Wobbles in time for Christmas store deliveries. It will cost $4500 over this time period plus $8 for each Wobble produced. The problem is that the toy company cannot possibly produce enough Wobbles in time. Another factory promises to produce and deliver the additional Wobbles at a cost of $11.50 per Wobble. If both companies work to produce a total of 12,000 Wobbles, the costs will total $116,075. How many Wobbles will be made at each factory?

14. Without using a calculator, evaluate the following expressions.
    a. 8^{-2} b. $(-8)^2$ c. $(-8)^{-2}$ d. -8^{-2}

15. Rewrite each expression so that the exponent is positive.
    a. x^{-3}
    b. $-x^{-1}$
    c. $\dfrac{5}{x^{-4}}$
    d. $8x^{-3}$
    e. $\left(\dfrac{4}{x}\right)^{-2}$
    f. $(2x)^{-3}$

16. Rewrite each expression in the form ax^n, where a is the numerical coefficient and n is either a negative or positive integer. For example, we can rewrite $\dfrac{2}{x^4}$ as $2x^{-4}$.
    a. $\dfrac{5}{x^2}$ b. $\dfrac{2}{x^{-5}}$ c. $\dfrac{24}{(2x)^3}$ d. $\left(\dfrac{1}{x}\right)^5$

17. Rewrite each numerical expression in the form x^n, where n is either a negative or positive integer and x is an integer.
    a. $\left(\dfrac{1}{2}\right)^5$ b. $\left(\dfrac{1}{4}\right)^3$ c. $\dfrac{1}{5^3}$ d. $\dfrac{1}{10^{-3}}$

18. Without plotting the points, identify which of the following tables graph as straight lines. For those that would graph as lines, determine the slope.

a.
Input	Output
1	21
2	17
3	13
4	9

b.
Input	Output
1	2
2	5
3	9
4	14

c.
Input	Output
0	108
5	124
10	140
15	156

d.
Input	Output
0	200
5	150
10	100
20	0
40	-200

e.
Input	Output
0	24
2	28
4	32
8	36
12	40

19. Write the equation of each of the following lines.

a.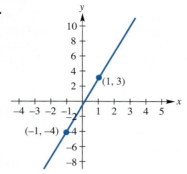

b.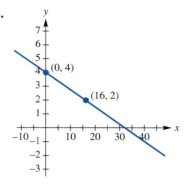

20. For each part, write the equation of the line.
  a. Passing through the points $(-10, -1)$ and $(5, -6)$
  b. Passing through the point $(-2, 1)$ with a slope of $\frac{3}{4}$
  c. Passing through the points $(2, 112)$ and $(10, 168)$
  d. Parallel to the line $y = -5x + 3$ and passing through the point $(4, 8)$
  e. Perpendicular to the line $y = -5x + 3$ and passing through the point $(4, 8)$

21. For each part, write the equation of the line.
  a. Passing through the point $(0, 2068)$ with a slope of -150
  b. Parallel to the line $y = \frac{3}{4}x - 1$ and passing through the point $(4, 0)$
  c. Perpendicular to the line $y = \frac{3}{4}x - 1$ and passing through the point $(4, 0)$
  d. Parallel to the line $y = 15$ and passing through the point $(4, 7)$

22. Geewhiz Computers grossed $158,000 in sales in 1997 and $192,000 in 1999.
  a. Assuming that the relationship between the gross sales and the year is linear, write an equation that models this relationship.
  b. Use your equation to predict the gross sales for the year 2004.
  c. Do you think that it's reasonable to assume that this relationship is linear? Explain.

23. The Boiling Pot. The temperature of a pot of water on a stove is being monitored. After 5 minutes the temperature is 101°F, after 7 minutes it is 113.8°F, and after 9 minutes it is 126.6°F.
  a. Is the temperature rising linearly? Explain how you decided.
  b. Write an equation for the temperature of the water in terms of the number of minutes.
  c. What was the temperature of the water when it was placed on the stove?
  d. When will the water begin to boil?

24. a. A driveway requires 26,000 pounds of gravel. How many tons is this?
  b. The speed of sound is about 525 feet per second. Convert this speed to miles per hour.
  c. The surface area of the moon is about $1.52 * 10^8$ square kilometers. Convert this to acres.
  d. The moon travels about 2,415,000 kilometers in 28 days in one rotation around the earth. Convert the moon's velocity to miles per hour.
  e. Water weighs about 62.4 pounds per cubic foot. Convert this to grams per cubic centimeter.

25. Solve and check the following equations.

    a. $\dfrac{3}{5} = \dfrac{135}{t}$

    b. $\dfrac{20}{x} = \dfrac{10}{x-5}$

    c. $\dfrac{3}{2m} + \dfrac{5}{3m} = \dfrac{1}{12}$

    d. $\dfrac{5}{x+3} = \dfrac{7}{x-3}$

    e. $\dfrac{5}{2R} - 2 = \dfrac{4}{5R}$

26. Solve for the indicated variable. Verify your solutions.

    a. $G = \left(\dfrac{V}{H}\right) * 100$ for H

    b. $\dfrac{V_c}{M_c} = \dfrac{V_d}{M_d}$ for M_d

    c. $KE = \dfrac{1}{2}mv^2$ for m

    d. $KE = \dfrac{1}{2}mv^2$ for v

    e. $1 = \left(\dfrac{c}{a}\right) - \left(\dfrac{b}{2a}\right)$ for a

    f. $\dfrac{4}{h} = \dfrac{L+1}{h} - \dfrac{L}{L+h}$ for h

27. Line L is parallel to side AC of the following triangle. Find the measures of $\angle 1$, $\angle 2$, and $\angle 3$.

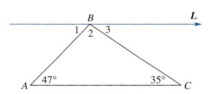

28. In the following figure, line PO is parallel to line AB, m $\angle 1 = 15°$, and m $\angle 2 = 25°$. Find the measures of angles 3 and 4.

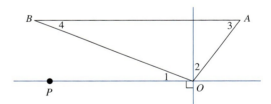

29. Simplify each expression. Verify your results.

    a. $x^3(x^4)$

    b. $\dfrac{A^6}{A^{-4}}$

    c. $\dfrac{24m^{-2}}{8m^{-5}}$

    d. $20t^3 * 5t^{-8}$

    e. $4x^{-3}(5x^5 - x) + (3x)^2$

    f. $\left(\dfrac{2}{x^2}\right)^{-1}$

    g. $n^0(n^{-3})(n^5)$

    h. $(4r^{-2})^3 + 3r(5r^{-7})$

30. Simplify each expression. Verify your results.

    a. $6p^3 * p^{-2}$

    b. $\dfrac{y^2 y^5 y^{-3}}{y^4}$

    c. $(18.2\ m^2)(4\ m) + (3.5\ m)^3$

    d. $0.5k^2 * k^{-4} - (2k^3)^{-2}$

    e. $\dfrac{4P^2 + P^6}{P^2}$

31. a. Without using a calculator, estimate the value of the expression

$$\frac{(3.76 * 10^{49})(8.19 * 10^{30})}{6.12 * 10^{-27}}$$

b. Evaluate the expression in one step using your calculator. Round your result to three significant digits. Compare your results from parts a and b.

$$\frac{(3.76 * 10^{49})(8.19 * 10^{30})}{6.12 * 10^{-27}}$$

32. In the figures below $\triangle ABC \sim \triangle XYZ$. If you know the measurements that are shown on the figures, determine the lengths of the missing sides. The triangles are not drawn to scale.

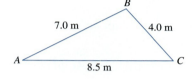 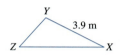

33. In the following triangle, m $\angle AED = 70°$, m $\angle ACB = 70°$, $AE = 4.8$ cm, $AD = 7.9$ cm, $EB = 12.8$ cm, $ED = 8.1$ cm, and $DC = 2.8$ cm. Determine the length of side BC.

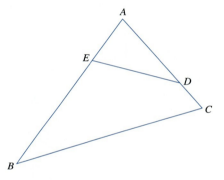

34. Find the value of x in the following figure. Angles that are equal are marked with same symbol.

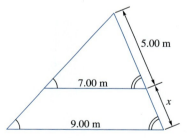

35. Identify the three similar triangles in the following figure. Indicate corresponding sides by the way you name the triangles. Explain how you know the triangles are similar. Angle ABC is a right angle.

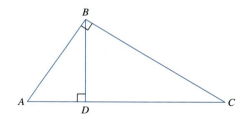

36. a. Write a formula for the volume of the following figure.
  b. Solve the formula you wrote in part a for h.

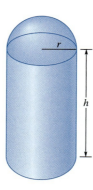

37. a. The pattern shown here could be folded into a solid. Name the solid.
  b. Write a formula for the volume V of the solid from part a.
  c. Solve the formula from part b for C.
  d. Write a formula for the surface area S of the solid from part a.
  e. Solve the formula from part d for C.

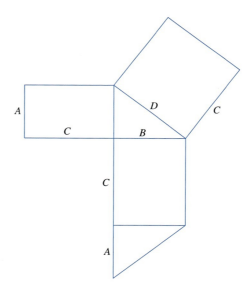

38. For each expression, determine the numerical value and units of the result.

  a. $\dfrac{14 \text{ cm}^3}{56 \text{ cm}^2} * \dfrac{20 \text{ g}}{1 \text{ cm}^2}$

  b. $\dfrac{320 \text{ mL}}{\dfrac{10 \text{ mL}}{27 \text{ sec}}}$

39. For the following expression, determine the numerical value and units of the result. The units are not real units.

$$14 \text{ noot} * \dfrac{3.5 \text{ arps}^2}{1 \text{ noot}} * \dfrac{1 \text{ bmr}}{5 \text{ arps}} * \dfrac{1 \text{ bmr}}{5 \text{ arps}} * \dfrac{8 \text{ wuf}}{1 \text{ bmr}^2}$$

40. Each of the following formulas is used to determine a length, an area, or a volume. Each variable represents a length. Identify which kind of measurement the formula is used for (length, area, or volume). Explain how you know, without using a table of formulas.

a. $\dfrac{\pi\left(\dfrac{d}{3}\right)^3}{2}$ b. $w_1(w_2^2 + w_1 w_2)$ c. $\dfrac{D + \dfrac{d}{2}}{2}$ d. $s^2 \pi + rs$

41. Use a scale drawing to answer the questions in the following problem.

Shawn's Visit.   Shawn just flew into the Millstone airport. He rented a car to drive to his friend's house. He left the airport heading N45°W for 15 miles. He took exit 12, which heads N20°E. He continued on this road for another 9 miles. He then turned left onto Country Road. Country Road heads due west for 24.6 miles. At this point Shawn headed N45°E for 37.2 miles, at which point he reached his friend's house. His friend has a helicopter, so when Shawn was ready to leave he flew Shawn back to the Millstone airport.

a. How far did Shawn drive to reach his friend's house?

b. How far did they fly back to the airport?

42. a. A triangle has sides of 72 cm and 36 cm. The angle between the two given sides is 90°. Determine the length of the third side.

b. A triangle has sides of 72 cm and 36 cm. The angle between the two given sides is 135°. Determine the length of the third side.

43. Determine the surface area and the volume of the following figure. Assume that sides and faces that appear to be parallel are parallel. This figure is not a pool, so be sure to include all sides in the surface area.

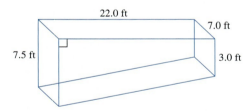

44. In the following figure, AB is the diameter of the circle, CB is tangent to the circle at B, and the measure of $\angle ACB$ is 68°.

a. Find the measure of $\angle CAB$, $\angle ADB$, and $\angle DBA$.

b. Identify a pair of similar triangles in the figure.

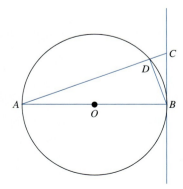

45. Waterbed. A king-sized waterbed mattress measures 5.5 feet by 6.5 feet by 8 inches.
   a. How many gallons of water does it take to fill the mattress?
   b. How much does this water weigh? (*Hint:* Water weighs about 62.4 pounds per cubic foot.)
   c. When all the water from the mattress has leaked onto the floor of the bedroom, it is about $1\frac{1}{2}$ inches deep. What is the approximate square footage of this bedroom?

46. An aerial photograph is taken that includes three trees. The distances in the photograph between the three trees are given in the following rough sketch at the left. By driving along the road, Joan was able to measure the actual distance between Trees A and B to be 0.70 miles. This is shown in the sketch at the right below. Determine the actual distance between Trees A and C and between Trees B and C (assuming the ground to be fairly flat).

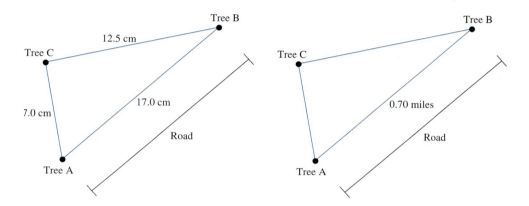

47. Drinking Cup. The drinking cup depicted in the figure is a portion of a cone.
   a. Determine the surface area of the cup.
   b. Determine the number of ounces of liquid the cup will hold.

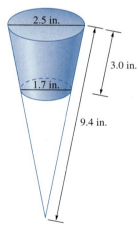

48. a. Determine the volume of the following solid.
   b. Determine the surface area of the solid.

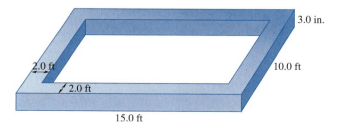

49. **The Garage.** George and Harriet are building a new garage. They are in the process of making the forms for the concrete foundation. However, they need to be sure that their foundation will be square (that is, each corner forms a right angle). They have measured the lengths indicated in the following figure.

   a. Explain how they can use these lengths to determine whether the foundation will be square.

   b. Based on the measurements given, is their foundation square?

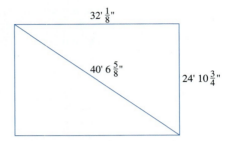

50. **Teenage Drunk Driving.** The number of people killed in crashes involving drunk teenage drivers decreased from about 3600 in 1982 to 1200 in 1995. During this period, the percentage of teenagers who report that they have had a drink in the last 30 days has also dropped.

   a. From the following table, plot the number of fatalities on the vertical axis and the percent drinking on the horizontal axis.

   b. Draw an "eyeball fit" line through the data.

   c. Write an equation for the line you have drawn.

Year	Percent of Teenagers Who Report Having a Drink in the Last 30 Days	Number of Fatalities in Crashes Involving Drunk Teenage Drivers
1982	35	3600
1985	41	2600
1988	34	2450
1990	34	2200
1991	27	1800
1992	20	1450
1994	20	1200
1995	19	1190

51. Determine the mean, the median, and the mode for the following data set.

   Number of Miles Driven to Students' Vacation Destination

   125   100   40   250   290   80   350   100
   60   15   70   50   50   150 50   370

52. **Rodeo Prize Money.** Many people think that cowboys and rodeos are a part of our history, but rodeos are becoming big business and the prizes for the top participants are increasing as seen in the following graph. Is the graph misleading or incorrect? If yes, explain in what way.

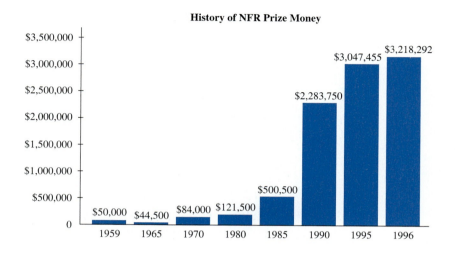

53. **Promotions.** Two women and three men in State Bank's financial office are all equally qualified to be promoted.

   a. What is the probability, if the position is chosen randomly, that a woman gets the next promotion?

   b. If two openings for promotion just arose, what is the probability that both of the positions are filled by a woman, if the positions are chosen randomly?

   c. If two openings for promotion just arose, what is the probability that one woman and one man will receive the promotions, if the positions are chosen randomly?

54. **New York Knicks.** During the 1999 regular National Basketball Association season the New York Knicks played 50 games. They won 27 games and lost 23. The number of points they scored in these games is listed in the table. Create a stem-and-leaf display of the data, and use it to answer the following questions.

   a. What is the least number of points scored by the Knicks in a single game? The most?

   b. What is the median for the number of points scored in a game?

   c. What is the mode?

POINTS SCORED BY NEW YORK KNICKS 1999 NBA SEASON

85	79	101	73	78	95	74	78	79	82
115	80	68	85	84	87	97	86	98	63
94	108	113	78	96	81	71	91	94	91
94	78	95	72	82	86	93	91	89	71
90	67	110	82	91	73	85	71	95	101

55. **Bigger Fish?** Fish have been collected from two lakes that were stocked at the same time. The lengths of the fish are given in the following table.

   a. Determine the minimum, maximum, median, and quartiles for the lengths of the fish in the samples from each lake. Use these to make a box-and-whisker plot for the lengths of the fish in each lake.

   b. What conclusions can you make based on your box-and-whisker plot?

   LENGTHS OF FISH IN CENTIMETERS

Duck Lake					Glacier Lake				
16	26	30	24	30	44	40	38	37	45
30	25	17	26	19	46	30	47	50	47
18	22	23	28	33	35	50	25	47	52
34	22	18	21	36	52	31	40	56	39
20	28	21	27	27	64	42	44		

56. **Top-Rated Network Television Shows.** Each week network television programs are rated and ranked. Television programs can be divided into two categories: those that are weekly series and those that are specials or sporting events. Use the following table to answer the questions.

   a. Which network had the most programs in the top-rated shows? What percent of the top-rated shows are on this network?

   b. What percent of the top rated shows were specials or sporting events?

   c. What percent of the top-rated shows on CBS were specials or sporting events? On FOX?

   d. What is the probability that a program is a weekly series given it was one of the top-rated shows shown on ABC?

   TOP-RATED NETWORK TELEVISION SHOWS

Networks	Number of Weekly Series	Number of Specials or Sporting Events
NBC	15	3
ABC	12	3
CBS	10	5
FOX	2	0

Appendices

The topics covered in the appendices should be familiar to you. We include them here for a refresher and as a reference. Depending on your background, any piece of this material may be new to you. It is up to you to read it and to get help as needed. We assume a working knowledge of these topics in the text.

Appendix A The Real Number System

It is often important to know what part or subset of our number system we are working in. We will review the real number system and its subsets in this section. Historically, numbers were developed to help people count things. We will start with this first set of numbers and then expand the set to include the numbers that will be used in this course.

> **Definition**
>
> The set of **natural numbers** is also known as the counting numbers. It consists of the numbers we use to count objects: 1, 2, 3, 4, . . .

One of the first additions made to this set was the number zero. This may seem like an unimportant distinction, but historically, the addition of zero was an important step in mathematical sophistication.

> **Definition**
>
> The set of **whole numbers** includes all of the natural numbers and zero. It consists of the numbers 0, 1, 2, 3, . . .

If we are subtracting whole numbers, the result is not always a whole number. For example, we need negative numbers to evaluate $6-10$, so we expand our numbers to include these negatives.

> **Definition**
>
> The set of **integers** includes the natural numbers, zero, and the opposites of all of the natural numbers. The integers are the numbers . . . , $-3, -2, -1, 0, 1, 2, 3, $ . . .

We can now add, subtract, or multiply any two integers and the result will be an integer. However, if we divide any two integers, the result is not always an integer. We need to add fractions to our set of numbers to be able to evaluate $5 \div 3$.

> **Definition**
>
> The set of **rational numbers** consists of all numbers that can be written as ratios of two integers, provided the divisor is not zero. That is, all numbers that can be written in the form $\frac{a}{b}$, where a and b are integers and $b \neq 0$, are rational numbers.

The rational numbers include all of the integers, because any integer can be written as a ratio, for example, $4 = \frac{4}{1}$. The rational numbers include numbers represented by common fractions, improper fractions, and mixed numbers because these can be written as ratios also. It can be shown that any ratio of integers can be written as either a terminating or a repeating decimal. So the rational numbers include all terminating and repeating decimals as well. Because integers, common fractions, improper fractions, and mixed numbers can all be expressed as terminating or repeating decimals, the rational numbers can be defined as those numbers that can be represented as terminating or repeating decimals.

Within the set of rational numbers, we can now do any of the four basic operations—addition, subtraction, multiplication, or division (except by zero)—and get an answer still in the set. Some additional numbers are not rational numbers. One type of number is the result we get when taking the square root of a number, if the number is not a perfect square. The decimal for $\sqrt{6}$ does not terminate or repeat; therefore, it is not a rational number. You probably know that π, the ratio of the circumference to the diameter of a circle, is also a nonterminating, nonrepeating decimal. We need a new set to include these numbers.

Notice that the sets mentioned so far are all nested; each one contains all of the numbers from the previous set. The following set does not include any of the previous sets.

> Definition
>
> An **irrational number** is any number that can be written as a nonterminating, nonrepeating decimal. In other words, any real number that *cannot* be written exactly in the form of a fraction is an irrational number.

This set includes such numbers as π and $\sqrt{2}$. When you enter $\sqrt{2}$ in your calculator, you get a decimal that may appear to be a rational number. What you can see on your calculator display is, in fact, a rational number but it is only an *approximation* of the irrational number you entered. The approximation is sufficient for most applications.

The number 2.020020002 . . . is a nonrepeating nonterminating decimal; therefore, it is irrational. Although we see a pattern to the decimal 2.020020002 . . . , it is not a repeating decimal. In a repeating decimal the exact same digits must repeat.

Notice that the irrational numbers are defined as those numbers that cannot be written in the form of a fraction, that is, they cannot be written as a rational number. We can think of irrational as meaning not rational.

> Definition
>
> The **real numbers** consist of all the rational numbers together with all the irrational numbers.

The following figure gives us a visual representation of the real number system. The picture can help us see which sets are contained in other sets.

A Visual Model of the Real Number System and Its Subsets

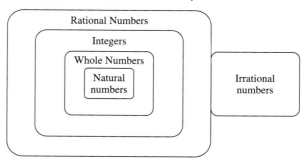

The following table lists the real number system and its subsets along with some examples of numbers belonging to the subsets. Because the entire table represents the real number system, any number in the table is also a real number.

THE REAL NUMBER SYSTEM	
Natural Numbers	1, 9, 27, 1080
Whole Numbers	0, 4, 56, 200
Integers	9, −10, −100, 0, 54
Rational Numbers	0.124, 2.3333..., −0.5, 17, $\frac{1}{2}$, −3$\frac{5}{7}$, −250
Irrational Numbers	π, $\sqrt{10}$, 0.010010001...

Example 1

Which of the following are whole numbers?

$\sqrt{8}$ 0   0.5 $\frac{8}{4}$ $2\frac{1}{3}$ −10   5 $\sqrt{36}$

Solution

The whole numbers consist of the numbers 0, 1, 2, 3, ... We can easily see that 0 and 5 are whole numbers. Fractions are not part of the set, so 0.5 and $2\frac{1}{3}$ are not whole numbers. But what about the fraction $\frac{8}{4}$? Because this fraction is equal to 2, it is a whole number. Negative numbers are not included in the set, so −10 is not a whole number. We need to be careful when deciding whether $\sqrt{8}$ and $\sqrt{36}$ are whole numbers. Because 8 is not a perfect square, $\sqrt{8}$ is an irrational number. Because $\sqrt{36}$ equals 6, $\sqrt{36}$ is a whole number. We conclude that 0, 5, $\frac{8}{4}$, and $\sqrt{36}$ are all whole numbers.

Notice from the example that when we are deciding what set a number belongs to, we need to think about the value of the number, not the form it is written in.

Problem Set A

1. For each part, a–f, identify all of the numbers from the list on the right that belong to the set.

   a. Natural numbers
   b. Whole numbers
   c. Integers
   d. Rational numbers
   e. Irrational numbers
   f. Real numbers

   $2\frac{3}{5}$
   0
   $\sqrt{7}$
   2.5
   $-\frac{1}{4}$
   15
   0.667
   −10
   0.151515...
   $-\sqrt{9}$
   0.010120123...

2. For each of the following numbers, list *all* of the number systems to which it belongs.
   a. 0.5
   b. $-6\frac{1}{2}$
   c. $\sqrt{43}$
   d. 100
   e. $\frac{2}{7}$
   f. $\sqrt{64}$
   g. -17
   h. π
   i. $\frac{23}{5}$
   j. -2.25

3. Decide if each of the following statements is true or false.
   a. Irrational numbers are not real numbers.
   b. All integers are also rational numbers.
   c. Every rational number is a real number.
   d. Every real number is a rational number.
   e. Zero is both a rational number and an irrational number.
   f. Every repeating decimal number is a rational number.
   g. $\sqrt{89}$ lies between the integers 8 and 9.
   h. $\sqrt{16}$ is a rational number.
   i. $\frac{6}{2}$ is a natural number.

4. Without using a calculator, locate the following numbers on a number line.
   a. 2.25   b. $-\frac{5}{4}$ c. $\frac{1}{3}$ d. $\sqrt{7}$ e. $-2\frac{7}{8}$ f. -3

5. How many natural numbers are there between $-\frac{3}{2}$ and $\frac{5}{2}$?
   a. None   b. 2   c. 4   d. Infinitely many

6. The numbers $\frac{22}{7}$, 3.14, and π are all
   a. integers   b. rationals   c. irrationals   d. reals

7. Which of the following is not a rational number?
   a. $\frac{22}{7}$ b. $\sqrt{11}$ c. 3.14   d. 0.7

Appendix B Arithmetic with Signed Numbers

Discussion B

In this section, we will review arithmetic operations on expressions involving both positive and negative numbers (signed numbers). There are many different ways to think about and discuss arithmetic on signed numbers. In your previous experiences you may have visualized them on a number line or thought about them as they relate to a checking account. For our purposes, we will visualize them using a helium balloon model.

Consider the following balloon. The top is filled with ten helium balloons and the bottom is filled with ten sandbags. Each helium balloon will make the basket rise one unit and each sandbag will make the basket fall one unit. For this reason, we can think of the helium balloons as representations of positive numbers and the sandbags as representations of negative numbers. Because we are starting with an equal number of helium balloons and sandbags, the basket is at zero on a number line.

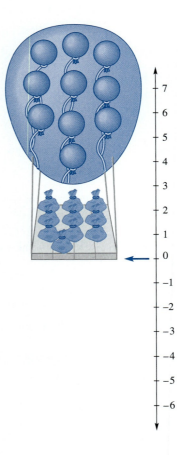

Addition and Subtraction of Signed Numbers

Now let's see how we can use this figure to model arithmetic. Suppose we want to determine the sum 5 + ⁻2. Using this model, we want to add 5 balloons and 2 sandbags. When we add five balloons the basket goes up five units. When we add 2 sandbags the balloon goes down two units. The basket ends up at 3. Therefore, 5 + ⁻2 = 3.

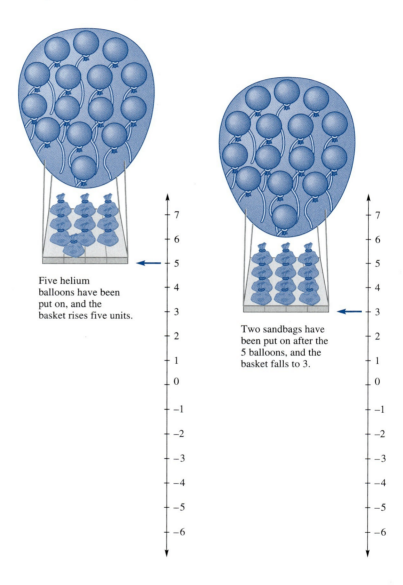

Five helium balloons have been put on, and the basket rises five units.

Two sandbags have been put on after the 5 balloons, and the basket falls to 3.

Subtraction works the same as addition except that we need to take off either helium balloons or sandbags. Let's try $^-2 - 4$. In terms of our model, we need to put on two sandbags because we are starting with $^-2$. Then we need to subtract (or take off) 4 balloons. When we put on 2 sandbags the balloon goes down two units. When we take off 4 balloons the balloon goes down four units. The basket will end up at $^-6$. Therefore, $^-2 - 4 = {^-6}$.

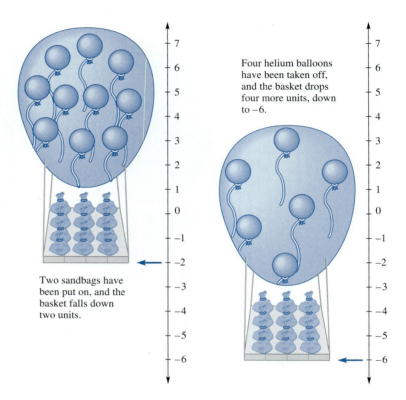

Example 1

Determine the following sums or differences.

a. $3 + {^-6}$ b. $4 - {^-1}$ c. $2 - 5$ d. $^-10 + 8$

Solution

a. Using the balloon model, we can think of $3 + {^-6}$ as adding 3 helium balloons and then adding 6 sandbags. When we add 3 helium balloons the basket rises up three units. Then, when we add 6 sandbags, the basket drops six units. This takes the basket down to $^-3$. Therefore, $3 + {^-6} = {^-3}$.

b. The expression $4 - {^-1}$ can be thought of as adding 4 helium balloons and then taking away one sandbag. By adding 4 helium balloons, the basket rises four units. Taking away one sandbag causes the basket to rise an additional one unit. The basket ends up at 5. Therefore, $4 - {^-1} = 5$.

c. The expression $2 - 5$, can be thought of as adding 2 helium balloons and then taking away 5 helium balloons. Adding two balloons cause the basket to rise two units. Taking away 5 balloons causes the basket to fall five units. The basket will end up at $^-3$. Therefore, $2 - 5 = {^-3}$.

d. The expression $^-10 + 8$ can be thought of as adding 10 sandbags and then adding 8 helium balloons. Adding 10 sandbags causes the balloon to fall 10 units. Adding 8 balloons will cause the balloon to rise 8 units. The balloon will end up at $^-2$. Therefore, $^-10 + 8 = {^-2}$.

We can imagine that taking away one sandbag produces the same effect as adding one balloon. Similarly, taking away one balloon produces the same effect as adding one sandbag. Mathematically, this means that we can rewrite any subtraction problem as an addition problem. Let's look at parts b and c from Example 1.

In part b we were looking at the expression $4 - {}^-1$. In this expression we are taking away one sandbag. This is equivalent to adding one balloon. Therefore, we can rewrite $4 - {}^-1$ as $4 + 1$.

In part c we were looking at the expression $2 - 5$. Taking 5 balloons away produces the same effect as adding 5 sandbags. Therefore, we can rewrite the expression $2 - 5$ and $2 + {}^-5$.

> **Subtraction as Addition of the Opposite**
>
> We can rewrite any subtraction problem as an addition problem. To do this, we change the subtraction to an addition and change the sign of the number that was originally being subtracted. We often say that we are "rewriting the subtraction by adding the opposite."

Multiplication of Signed Numbers

Multiplication is a shorthand notation for repeated addition. The product of $2 * 3$ means $3 + 3$, or $2 + 2 + 2$. For the purposes of this discussion, we will consider the first factor of the product as the number of times to add the second factor. In this example, we think of $2 * 3$ as $3 + 3$ or, in words, we are adding 3 twice. In terms of our balloon model this means that we need to add 3 helium balloons twice. This causes the basket to rise six units. Therefore, $2 * 3 = 6$ as we already know.

The product $2 * {}^-3$ then means to add 3 sandbags twice. This causes the basket to fall six units, ending up at $^-6$. Therefore, $2 * {}^-3 = {}^-6$.

Next, if the first factor is positive, we know that we are going to repeatedly add the second factor. If the first factor is negative we repeatedly subtract because a negative number is the opposite of a positive number.

Consider the product $^-2 * 3$. Because the first factor is negative, we repeatedly subtract the second factor. Because the second factor is positive, we subtract helium balloons. Putting this together, we need to subtract 3 helium balloons twice. This causes the basket to fall six units, ending up at $^-6$. Therefore, $^-2 * 3 = {}^-6$.

The product $^-2 * {}^-3$ can be thought of as subtracting 3 sandbags twice. This causes the basket to rise six units, ending up at 6. Therefore, $^-2 * {}^-3 = 6$.

Example 2

Determine the following products.

a. $5 * {}^-2$ b. ${}^-3 * 10$ c. ${}^-2 * 6$ d. ${}^-5 * {}^-10$

Solution

a. Using the balloon model, we can think of $5 * {}^-2$ as adding 2 sandbags five times. This will cause the basket to drop ten units, ending up at ${}^-10$. Therefore, $5 * {}^-2 = {}^-10$.

b. We can think of ${}^-3 * 10$ as subtracting 10 helium balloons three times. This causes the basket to drop 30 units, ending up at ${}^-30$. Therefore, ${}^-3 * 10 = {}^-30$.

c. We can think of ${}^-2 * 6$ as subtracting 6 helium balloons two times. This causes the basket to drop 12 units, ending up at ${}^-12$. Therefore, ${}^-2 * 6 = {}^-12$.

d. We can think of ${}^-5 * {}^-10$ as subtracting 10 sandbags five times. This causes the basket to raise 50 units, ending up at 50. Therefore, ${}^-5 * {}^-10 = 50$.

By looking at the previous examples, it appears that when we multiply two numbers, one of which number is positive and the other is negative, the result is a negative number. When we multiply two negative numbers the result is positive. To understand why this always works, let's consider the different combinations of numbers.

Consider the product where the first factor is positive and the second factor is negative, that is positive * negative. In this situation, we repeatedly add sandbags. This causes the basket to fall and the result of the product is a negative number.

If the first factor in the product is negative and the second factor is positive (that is, negative * positive), then we repeatedly subtract helium balloons. This causes the basket to fall and the result of the product is negative.

If both factors are negative (that is, negative * negative), then we repeatedly subtract sandbags. The basket rises and the result of the product is positive.

To summarize:

Products of Signed Numbers

1. The product of two positive numbers is a positive number.

   Positive * Positive = Positive

2. The product of a negative number and a positive number is a negative number.

   Negative * Positive = Negative
   Positive * Negative = Negative

3. The product of two negative numbers is a positive number.

   Negative * Negative = Positive

Let's see what happens when we multiply more than two numbers. Consider the product $2 * {}^-3 * {}^-4$. We multiply two numbers at a time and progress from left to right.

$2 * {}^-3 * {}^-4$

$= {}^-6 * {}^-4$ Because a positive times a negative is negative, $2 * {}^-3 = {}^-6$.

$= 24$ Because a negative times a negative is positive, ${}^-6 * {}^-4 = 24$.

In this example, we multiplied the numbers from left to right. However, because multiplication is associative and commutative, we can multiply the numbers in any order as long as we multiply two at a time. For example, if we had multiplied the second two factors first we would have $2 * 12 = 24$. Similarly, we can add a series of numbers in any order. However, subtraction and division are not associative, and the order that we perform either of these operations can affect the result.

Similarly, we compute the product $^-2 * {}^-3 * {}^-4$.

${}^-2 * {}^-3 * {}^-4$
$= 6 * {}^-4$ Because a negative times a negative is positive, $^-2 * {}^-3 = 6$.
$= {}^-24$ Because a positive times a negative is negative, $6 * {}^-4 = {}^-24$.

In general, when we multiply many factors, an odd number of negative numbers produces a negative product, and an even number of negative numbers produces a positive product.

Division of Signed Numbers

Suppose we want to compute $\frac{24}{-4}$. We need to determine the number whose product with $^-4$ produces 24. Because 4 is negative, we need to multiply by another negative number if we want to end up with positive 24. Therefore, $\frac{24}{-4} = {}^-6$. Similarly, $\frac{-24}{4} = {}^-6$ and $\frac{-24}{-4} = 6$. You can check these by multiplying the divisor by the quotient. For example in the case of $\frac{24}{-4} = {}^-6$, the divisor is $^-4$ and the quotient is $^-6$. The product $^-4 * {}^-6 = 24$ and the answer checks.

From these examples, we can see that the rules for dividing signed numbers are similar to those for multiplying them.

Quotients of Signed Numbers

1. The quotient of two positive numbers is a positive number.

   $$\frac{\text{positive}}{\text{positive}} = \text{positive}$$

2. The quotient of a negative number and a positive number is a negative number.

   $$\frac{\text{positive}}{\text{negative}} = \text{negative}$$

   $$\frac{\text{negative}}{\text{positive}} = \text{negative}$$

3. The quotient of two negative numbers is a positive number.

   $$\frac{\text{negative}}{\text{negative}} = \text{positive}$$

Example 3

Determine the following quotients.

a. $\dfrac{-150}{5}$ b. $\dfrac{36}{-9}$ c. $\dfrac{-60}{-3}$ d. $\dfrac{-2 * -24}{-4}$

Solution

a. Because there is one negative number in this quotient the result will be negative. Therefore, $\dfrac{-150}{5} = -30$.

b. Again there is one negative number, so the result is negative. Therefore, $\dfrac{36}{-9} = -4$.

c. In this quotient, both the numerator and denominator are negative, so the result is positive. Therefore, $\dfrac{-60}{-3} = 20$.

d. In the expression $\dfrac{-2 * -24}{-4}$, the product in the numerator is positive. Then we need to divide by a negative number. This produces a negative result. In steps,

$$\dfrac{-2 * -24}{-4}$$
$$= \dfrac{48}{-4}$$
$$= -12$$

Notice that in part d, an odd number of negative numbers occurred, and the result was a negative number.

We often see negative numbers in fractions. Consider the fraction $\dfrac{1}{-2}$. Because a positive number divided by a negative number produces a negative result, we often rewrite this fraction as $-\dfrac{1}{2}$. Similarly, $\dfrac{-1}{2}$ can be rewritten as $-\dfrac{1}{2}$, because a negative number divided by a positive number also produces a negative result. Therefore, $\dfrac{1}{-2}, \dfrac{-1}{2}$, and $-\dfrac{1}{2}$ are all equivalent fractions. We usually record a negative fraction with the negative sign in the numerator or in front of the fraction.

If both the numerator and denominator are negative numbers, then the fraction is a positive number, because a negative number divided by a negative number produces a positive result. For example, $\dfrac{-1}{-2} = \dfrac{1}{2}$. The fraction $\dfrac{1}{2}$ is considered the simplified version of $\dfrac{-1}{-2}$. Therefore, it is not acceptable to leave a numerical result with a negative number in the numerator and in the denominator.

Problem Set B

1. Without using a calculator, evaluate each expression.
   a. 12 − 5
   b. ⁻10 + 4
   c. 16 − ⁻6
   d. ⁻14 − 3

2. Without using a calculator, evaluate each expression.
   a. ⁻30 − ⁻10
   b. 15 + ⁻8
   c. 42 − ⁻10
   d. ⁻100 + ⁻200

3. Rewrite each subtraction problem as an equivalent addition problem.
   a. 20 − 6
   b. ⁻14 − ⁻4
   c. ⁻10 − 25
   d. 14 − ⁻3

4. Rewrite each subtraction problem as an equivalent addition problem.
   a. 32 − 12
   b. 16 − ⁻7
   c. ⁻50 − ⁻15
   d. ⁻4 − 20

5. Without using a calculator, evaluate each expression.
   a. 2 ∗ 8
   b. ⁻3 ∗ 6
   c. 5 ∗ ⁻20
   d. ⁻4 ∗ ⁻30

6. Without using a calculator, evaluate each expression.
   a. ⁻4 ∗ ⁻30
   b. 10 ∗ ⁻9
   c. ⁻4 ∗ 10 ∗ ⁻5
   d. ⁻2 ∗ 5 ∗ 5

7. Without using a calculator, evaluate each expression.
   a. $\dfrac{-20}{4}$
   b. $\dfrac{42}{-7}$
   c. $\dfrac{-50}{-5}$
   d. $\dfrac{-28}{2 * 7}$

8. Without using a calculator, evaluate each expression.
   a. $\dfrac{-15}{-3}$
   b. $\dfrac{24}{-2 * 3}$
   c. $\dfrac{24}{-2 * -3}$
   d. $\dfrac{4 * -9}{-3 * -3}$

9. Without using a calculator, evaluate each expression.
   a. 18 − ⁻20
   b. 30 ∗ ⁻2
   c. ⁻5 ∗ ⁻8
   d. ⁻4 + ⁻23

10. Without using a calculator, evaluate each expression.
    a. $\dfrac{80}{-4}$
    b. ⁻30 − ⁻40
    c. ⁻2 ∗ 14
    d. ⁻15 − ⁻5

11. Without using your calculator, determine which of the following fractions are equivalent to $-\dfrac{1}{4}$.
    a. $\dfrac{-1}{-4}$
    b. $\dfrac{1}{-4}$
    c. $\dfrac{-1}{4}$
    d. $-\dfrac{-1}{-4}$

12. Without using your calculator, determine which of the following fractions are equivalent to $\dfrac{2}{5}$.
    a. $\dfrac{-2}{5}$
    b. $\dfrac{2}{-5}$
    c. $\dfrac{-2}{-5}$
    d. $-\dfrac{-2}{5}$

13. A team of students needs to solve the following problem.

    On February 7 in Tera's home town, the low temperature was ⁻10°F and the high was 17°F. What was the change in temperature?

    Each member of the team used a different expression to determine the answer to the question. Identify the methods that produce correct results.
    a. 17 − 10
    b. 17 − ⁻10
    c. ⁻10 − 17
    d. 10 + 17

14. Choose all of the following expressions that can be used to produce a correct response to the following problem.

    The low temperatures for the first week of January in Tera's home town were $^-12°F$, 5°F, $^-26°F$, 12°F, $^-7°F$, $^-15°F$, and 0°F. What was the range of low temperatures for that week?

    a. $12 - {^-7}$ b. $12 - 26$ c. $12 + 26$ d. $12 - 7$ e. $12 - {^-26}$

15. For each part, an expression is described in words. Decide whether the result of the expression is positive or negative.
    a. The expression is the sum of two negative numbers.
    b. The expression is the product of a negative number and a positive number.
    c. The expression is the quotient of two negative numbers.
    d. The expression is the sum of a negative number and a positive number, for which the magnitude of the negative number is larger than the magnitude of the positive number.
    e. The expression is the product of three numbers, for which two of the factors are negative and the third factor is positive.

16. Write a subtraction problem, that contains a negative number and a positive number, whose result is positive.

17. Write a subtraction problem, that contains a negative number and a positive number, whose result is negative.

Appendix C Reading and Measuring with Rulers

Discussion C

Centimeter Rulers

We use two types of rulers in this course: a centimeter ruler and an inch ruler. Let's start with the centimeter ruler.

On a centimeter ruler, each unit represents 1 centimeter. The following figure shows a centimeter ruler with only the centimeter marks.

Then each centimeter is divided into ten equal parts. This means each of these parts is $\frac{1}{10}$ of a centimeter, or 0.1 cm. The next figure shows the first 8 centimeters of a complete centimeter ruler.

Because each interval is divided into tenths, we record the measurements as decimals rather than fractions. Whenever we record measurements, we must include the unit name or abbreviation because there are many different units of measurement.

Example 1

Record the measurement indicated by each arrow.

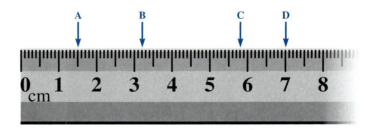

Solution

Arrow A points at the fifth mark past 1 cm. Therefore, the measurement indicated by mark A is 1.5 cm.

The measurement indicated by arrow B is 3.2 cm.

The measurement indicated by arrow C is 5.8 cm.

The measurement indicated by arrow D is 7.0 cm. Notice that we included the 0 in the tenths place. This indicates that we know the place value in the tenths place.

One-tenth of a centimeter is called a millimeter (mm). Therefore, we could record each of our results from the last example using millimeters. The measurement indicated by arrow A could be recorded as 15 mm, arrow B could be recorded as 32 mm, arrow C 58 mm, and arrow D 70 mm.

Inch Rulers

Next, let's look at an inch ruler. On this ruler each unit represents 1 inch. An inch is longer than a centimeter. The following figure shows an inch ruler with only the inch units marked.

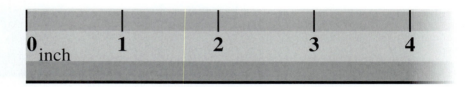

Then each interval is divided, but the way it is divided is quite different from the centimeter ruler. We start by dividing each interval into halves. To help make the lines for the halves distinct from the whole numbers, they are typically, but not always, drawn shorter. Here we draw the line to divide each interval in half about half as long as the inch marks.

Then we divide each subinterval into halves. Again, to make the ruler easier to read, we make these new lines shorter by drawing them about half as long as the half marks. This divides each inch into fourths. The following figure shows a ruler with each inch divided into fourths.

When reading an inch ruler, we must count the spaces that are of equal width. Consider the arrow that is pointing at the following ruler.

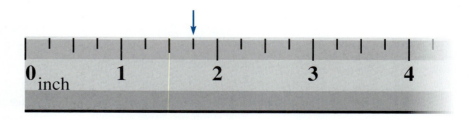

This arrow is pointing at a quarter mark, so we start by determining the size of the space to the first quarter mark of the interval between 1 and 2. Then we count equal spaces up to the arrow. This arrow indicates a measurement of $1\frac{3}{4}$ inches as shown in the following figure.

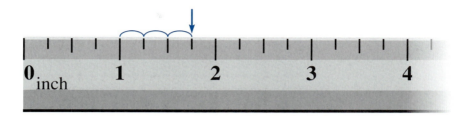

However, this ruler is not complete. We must further divide each subinterval in half by drawing lines that are about half as long as the quarter marks. This divides each inch into eighths.

We continue this process again to create a ruler for which each inch is divided into sixteenths. The following figure is the first 4 inches of a standard inch ruler. The longest lines represent inches, the second longest lines represent the halves, then the quarters, eighths, and sixteenths.

It is important to understand the relationship between the length of the lines and the type of fraction the lines represent. However, some rulers use the same length line for different fractions. This makes reading an inch ruler much more difficult. We suggest that you look for a ruler for which each type of fraction is marked with a different length of line.

Example 2

For each part, record the measurement indicated by the arrow.

a.

b.
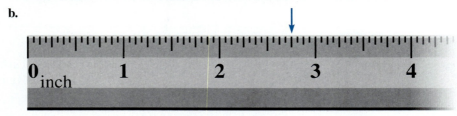

c.
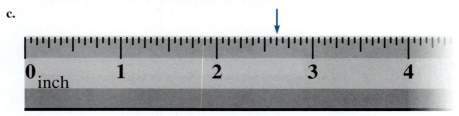

d.
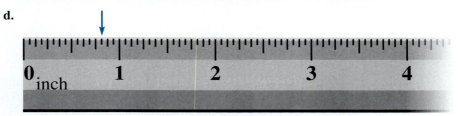

Solution   a. On this ruler, the arrow is pointing to a quarter mark. Because this is the first quarter mark of the interval, the arrow is indicating a measurement of $1\frac{1}{4}$ inches.

b. The arrow in this figure is also pointing to a quarter mark. Because it is not the first quarter mark between 2 and 3, we must count the spaces to determine the measurement.

We use the first quarter mark of the interval to determine the size of the space. Then count equal spaces up to the arrow. This arrow indicates a measurement of $2\frac{3}{4}$ inches.

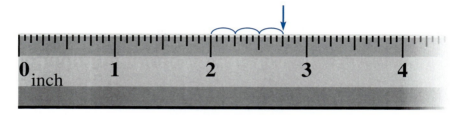

c. Because the length of the line that the arrow is pointing to is the second shortest, it is pointing to an eighth mark. Again, because it is not the first eighth mark of the interval, we must count the spaces to determine which eighth.

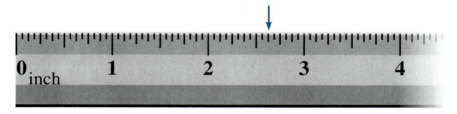

We use the first eighth mark of the interval to determine the size of the space. Then count equal spaces up to the arrow. There are five spaces. Therefore, this arrow indicates a measurement of $2\frac{5}{8}$ inches.

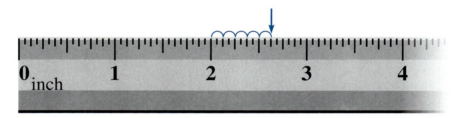

d. On this ruler, the arrow is pointing at the shortest mark, which is a sixteenth.

There are 13 spaces. Therefore, this arrow indicates a measurement of $\frac{13}{16}$ inches.

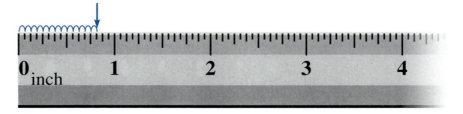

On part d of Example 2, some people only count the spaces after the halfway mark. They know that $\frac{8}{16}$ is equal to $\frac{1}{2}$, so they start counting at eight. There are 5 sixteenths after the halfway mark; this means the arrow indicates a measurement of $\frac{13}{16}$ inches.

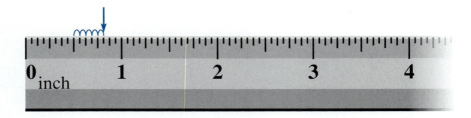

Similarly, $\frac{16}{16}$ is equal to 1, so we could start at 1 inch and count down to the arrow. There are 3 sixteenths if we count down from 1; this again takes us to $\frac{13}{16}$ inches.

1. Determine the measurement indicated by each arrow. Record your results using both centimeters and millimeters.

   a.

   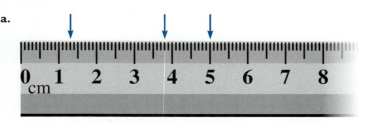

   b.

   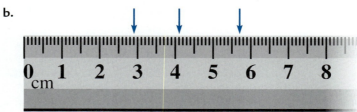

   c.

   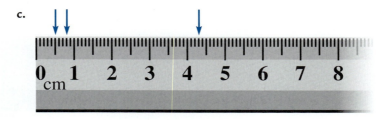

2. Measure the length of each line. Record your results in centimeters.
   a. _____
   b. _____

3. Record the measurement indicated by each arrow.

   a.

   b.

   c.

   d.

4. Record the measurement indicated by each arrow.

   a.

   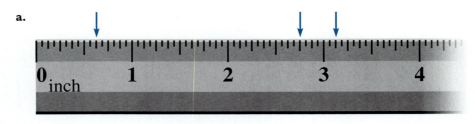

   b.

   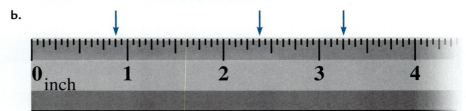

   c.

   d.

   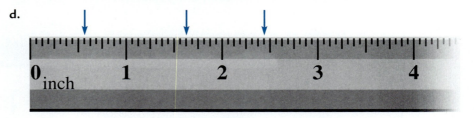

5. Measure the sides of the following figure. Record your results in inches.

6. Measure the sides of the following figure. Record your results in centimeters.

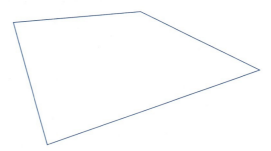

Appendix D Perimeter and Area Concepts

Many applications of mathematics make use of the concepts of perimeter and area. In this appendix, we will look at the meaning of these concepts and then apply them.

Perimeter

The perimeter of a figure is the distance around the outside of the figure. For example, if our unit of length is the length shown here

⊢—⊣   1 unit

then finding the perimeter of the following figure means finding the number of those units needed to trace around the outside of the figure.

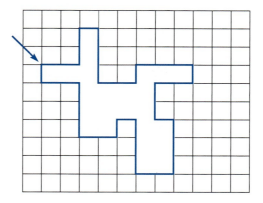

If we start at the arrow and go clockwise around the figure, we can add up the lengths as we go. Perimeter = 2 + 2 + 1 + 3 + 2 + 1 + 3 + 1 + 2 + 2 + 1 + 3 + 2 + 3 + 1 + 1 + 2 + 3 + 2 + 1 = 38 units. The perimeter of the figure is 38 units.

> **Definition**
>
> The **perimeter** of a two-dimensional figure is the length of the outer boundary of the figure.

To find the perimeter of a figure, we need the lengths of all of the pieces that make up the boundary of that figure. Sometimes these length measurements are given. At other times, we are expected to make the appropriate measurements. In some situations we can only estimate the needed lengths. Because perimeter is a length, it is measured in units such as inches, miles, centimeters, kilometers, and so on.

Example 1

Determine the perimeter of the following figure.

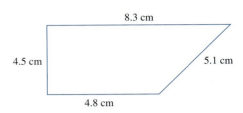

Solution We need to find the sum of the lengths of the four sides that form this figure.

Perimeter = 4.5 cm + 8.3 cm + 5.1 cm + 4.8 cm
= 22.7 cm

The perimeter of the figure is 22.7 centimeters.

The perimeter of a circle is called the circumference. The following formula can be used to find the circumference of a circle.

Circumference of a Circle

Circumference = $\pi * d$

or

Circumference = $2\pi * r$

where r is the radius of the circle and
d is the diameter of the circle

Example 2

Determine the perimeter of the following figure.

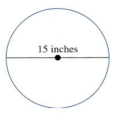

Solution Here, we are given the diameter of the circle. Using the formula, we get

circumference = $\pi *$ Diameter
= $\pi *$ 15 inches
$\approx$ 47 inches

The circumference of the circle is approximately 47 inches.

In Example 2 we rounded the answer to the nearest inch because that was the degree of accuracy of the diameter. We will talk more in the text about how to round our results. Until then, we just round to what seems reasonable.

Sometimes the boundary of a figure is made up of both straight edges and curves. If the curve is part of a circle, we use the circumference formula to determine the length of that part.

Appendix D   Perimeter and Area Concepts   A-25

Example 3

Determine the perimeter of the following figure. The curved portion of the figure is a semicircle.

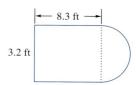

Solution Remember that the perimeter is just the outside boundary of the figure. This means we need to add the three sides of the rectangle and the circumference of the semicircle. The diameter of the circle is the same length as the left side of the rectangle, 3.2 ft.

Perimeter = [rectangle] + [semicircle]

$$= 8.3 \text{ ft} + 3.2 \text{ ft} + 8.3 \text{ ft} + \frac{\pi * 3.2 \text{ ft}}{2}$$

$$\approx 24.8 \text{ ft}$$

Notice that we used the circumference formula and then divided by 2 to get the perimeter of the semicircle.

We conclude that the perimeter of the figure is approximately 24.8 feet.

Area

The area of a figure is a measure of the surface of the figure and is measured in square units. To find the area of a figure, we must find the total number of some square unit that cover the figure. If our square unit is

☐   1 unit² or 1 square unit

then the area of the figure we looked at earlier is found by counting the number of square units that cover this figure.

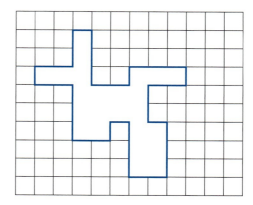

The area of the figure is 24 square units, or 24 units².

> **Definition**
> The **area** of a two-dimensional figure is the measure of the surface inside the boundary of the figure.

The units of area are square units such as square feet (ft^2, cm^2, and yd^2). There is no difference between saying square feet or ft^2. Both refer to a unit of area, a square that is 1 foot on each side.

You should be familiar with the formulas for finding the areas of many common geometric figures, such as rectangles, triangles, parallelograms, trapezoids, and circles. These formulas are in a reference section of this text.

Most of the areas you will be asked to find are either one of the common figures or combinations of these figures. Because area is a measure of the surface, we can find the area of simple figures and add them together—or subtract them—to find the area of more complex figures.

Example 4

Determine the area of the following figure. The curved portion of the figure is a semicircle.

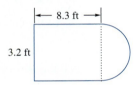

Solution To determine the area of this figure, find the sum of the area of the rectangle and the area of the semicircle.

$$\text{Area} = \text{length} * \text{width} + \pi * \frac{\text{radius}^2}{2}$$

$$= 8.3 \text{ ft} * 3.2 \text{ ft} + \frac{\pi (1.6 \text{ ft})^2}{2}$$

$$\approx 31 \text{ ft}^2$$

Notice that, for the circle, we used the area formula and then divided by 2 to get the area of the semicircle. When we multiply feet by feet we get square feet, an appropriate unit for area.

We conclude that the area of the figure is approximately 31 square feet.

Example 5

Determine the shaded area of the following figure. The circle has a diameter of 1.8 meters.

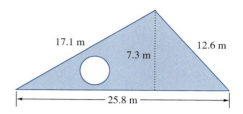

Solution

To determine the area of this figure, we find the area of the triangle and subtract the area of the circle. To determine the area of the triangle we need the base and the height. Let's use the bottom of the triangle as the base. Then the base is 25.8 meters and the height is 7.3 meters. For the circle, we need the radius, which is half of 1.8 meters, or 0.9 meters.

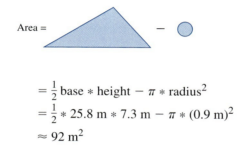

$$= \tfrac{1}{2} \text{ base} * \text{height} - \pi * \text{radius}^2$$
$$= \tfrac{1}{2} * 25.8 \text{ m} * 7.3 \text{ m} - \pi * (0.9 \text{ m})^2$$
$$\approx 92 \text{ m}^2$$

We conclude that the area of the figure is approximately 92 square meters.

In this appendix, we reviewed the concepts of perimeter and area. We found that the perimeter measures the boundary of a figure. It is a length measurement. To find the perimeter of a figure, we add up all the segments of the boundary. In general, it is easier to just add the segments than to use a formula. However, when part of the perimeter is curved, we may need to use a formula such as the one for the circumference of a circle or we may need to estimate the length of the curve.

Area measures the surface of a figure. It is measured in square units. We explored how to combine the formulas for common geometric shapes to find the areas of more complex shapes.

Problem Set D

1. Determine the area and perimeter of each of the following figures.

    a.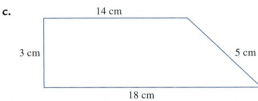

       24.6 m × 15.7 m rectangle

    b. Right triangle with legs 3 in. and 4 in., hypotenuse 5 in.

    c. Trapezoid: top 14 cm, left 3 cm, bottom 18 cm, right side 5 cm

    d. Circle with diameter 8 ft

2. Decide if perimeter or area is appropriate to use in each of the following situations.
    a. Carpeting a room
    b. Putting molding around a window
    c. Building a picture frame
    d. Buying fertilizer for your garden
    e. Fencing a yard
    f. Staining a deck

3. Which of the following are appropriate units for perimeter?
    a. meters
    b. ft^2
    c. kilometers
    d. in^3
    e. acres
    f. inches
    g. miles
    h. square centimeters

4. Which of the following are appropriate units of area?
    a. m^2
    b. meters
    c. yards
    d. acres
    e. square miles
    f. $in.^2$
    g. cubic inches
    h. liters

5. For each of the following situations, select an appropriate unit of measurement, such as feet, kilometers, or square inches.
    a. Ordering carpeting
    b. Framing a picture
    c. Fencing a garden
    d. Roofing a house
    e. Determining the distance from Seattle, Washington, to Portland, Oregon
    f. Measuring a person's height

6. Determine the area and perimeter of each of the following figures.

   a.
   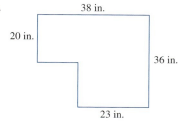

   b.
   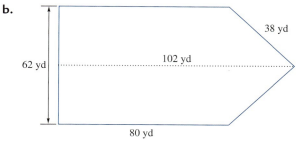

7. Determine the area and perimeter of each of the following figures.

   a.
   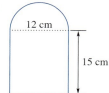

   b.
   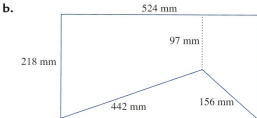

8. Determine the perimeter and area of the shaded part of the following figure.

   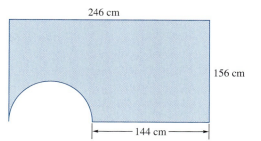

Geometry Reference

Definitions

angle   A figure formed by two line segments extending from a common point called the vertex

    central   An angle whose vertex is at the center of a regular polygon or a circle

    interior   An angle formed by two adjacent sides of a polygon and lying inside the polygon

circumference   The distance around the outside of a circle

congruent figures   Geometric figures having the same size and shape; orientation may be different

diagonal   A straight line joining two nonadjacent vertices of a polygon

diameter   A line joining the two sides of a circle or sphere and passing through the center

n-gon   A polygon with n sides, where n is a natural number

parallel lines   Two or more lines in a plane that never meet or intersect no matter how far they are extended

perimeter   The distance around the outside edge of a figure

polygon   A closed geometric figure made up of straight-line segments

    convex   A polygon in which none of the interior angles is greater than or equal to 180° (see illustration)

    concave   A polygon in which at least one of the interior angles is greater than or equal to 180° (see illustration)

    regular   A polygon in which all sides are the same length and all interior angles are the same size

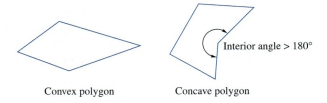

Convex polygon   Concave polygon

radius   A line from the center of a circle or sphere to the edge of the circle or sphere

similar figures   Geometric figures having the same shape but not necessarily the same size; corresponding angles are equal and corresponding sides are proportional

Formulas for Two-Dimensional Objects

A **rectangle** is a four-sided figure with four right angles.

area = LW

A **triangle** is a three-sided figure.

area = $\frac{1}{2}bh$

NOTE: The height of any figure must be perpendicular to the base.

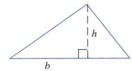 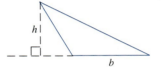

The Pythagorean Theorem
In a *right triangle*, the sum of the squares of the legs is equal to the square of the hypotenuse.

$\text{leg}_1^2 + \text{leg}_2^2 = \text{hypotenuse}^2$

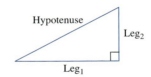

A **parallelogram** is a four-sided figure with opposite sides parallel.

area = bh

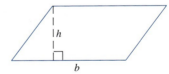

A **trapezoid** is a four-sided figure with two parallel sides, called bases.

area = $\dfrac{h(b_1 + b_2)}{2}$

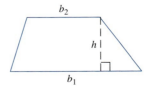

A **circle** is the set of all points in a plane that are equidistant from the center.

area = πr^2

circumference = $2\pi r$

or

circumference = πd, where d = diameter

Formulas for Three-Dimensional Objects

A **rectangular solid** is a three-dimensional figure in which all sides are rectangles.

volume = LWH

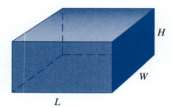

A **prism** is a polyhedron with two parallel congruent faces, called the bases of the prism. If the lateral sides are perpendicular to the bases, the prism is called a **right prism.**

*volume = area of the base * height*

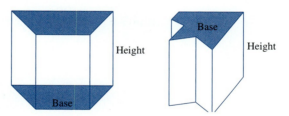

A **pyramid** is a polyhedron that has a base consisting of any polygon and triangular sides.

*volume = $\frac{1}{3}$ * area of base * height*

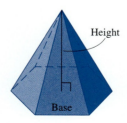

A **circular cylinder** is a three-dimensional object with two parallel faces that are circles. These are the bases of the cylinder. If the lateral sides are perpendicular to the bases, it is called a **right circular cylinder.** If not otherwise specified, a cylinder is assumed to be a right circular cylinder.

volume $= \pi r^2 h$

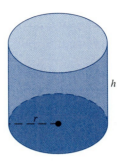

A **circular cone** has a circular base and the lateral surface forms a single point.

volume $= \dfrac{1}{3}\pi r^2 h$

surface area $= \pi r^2 + \pi r s$

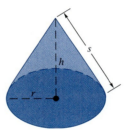

A **sphere** is the set of all points in space that are equidistant from the center.

volume $= \dfrac{4}{3}\pi r^3$

surface area $= 4\pi r^2$

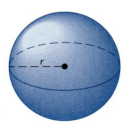

Conversion Tables

AMERICAN–AMERICAN CONVERSIONS

Distance	Volume
12 inches (in.) = 1 foot (ft)	3 teaspoons (tsp) = 1 tablespoon (tbsp)
3 feet = 1 yard (yd)	1 tablespoon = 0.5 fluid ounces
5280 feet = 1 mile	8 fluid ounces (fl oz) = 1 cup (c)
8 furlongs = 1 mile	2 cups = 1 pint (pt)
6 feet = 1 fathom	2 pints = 1 quart (qt)
100 links = 1 chain	4 quarts = 1 gallon (gal)
66 feet = 1 chain	1 ft^3 ≈ 7.481 gallons

Area	Weight
640 acres = 1 mile^2	480 grains = 1 ounce (oz)
	16 ounces = 1 pound (lb)
	2000 pounds = 1 ton

METRIC–METRIC CONVERSIONS

Distance	Volume
1 kilometer (km) = 1000 meters (m)	1000 milliliter (mL) = 1 liter
100 centimeters (cm) = 1 meter	1 cm^3 = 1 milliliter
1000 millimeters (mm) = 1 meter	
10 decimeters (dm) = 1 meter	

Area	Weight
1 hectare = 10,000 m^2	1 kilogram (kg) = 1000 grams
	1000 milligrams (mg) = 1 gram (g)
	1000 kilograms = 1 metric ton

AMERICAN–METRIC CONVERSIONS

Distance	Volume	Weight
2.54 cm = 1 inch	1.06 qt ≈ 1 liter	454 g ≈ 1 lb
39.37 in ≈ 1 meter		2.2 lb ≈ 1 kg
1.609 km ≈ 1 mile		

TIME CONVERSIONS

60 seconds (sec) = 1 minute (min)
60 minutes = 1 hour (h)
24 hours = 1 day
7 days = 1 week
365 days = 1 calendar year (nonleap year)
365.242199 days = 1 tropical year

METRIC PREFIXES

Prefix	Factor		Prefix	Factor	
tera	10^{12}	T	deci	10^{-1}	d
giga	10^9	G	centi	10^{-2}	c
mega	10^6	M	milli	10^{-3}	m
kilo	10^3	k	micro	10^{-6}	μ
hecto	10^2	h	nano	10^{-9}	n
deca	10^1	da	pico	10^{-12}	p

Selected Answers

Chapter 1

SECTION 1.1

1.
   a. $\underline{5 + \frac{1}{2} * 6} = 8$
   b. $\underline{16} - \underline{14 \div {}^-2} = 23$
   c. $\underline{{}^-14 + 2(2-9)} = {}^-28$
   d. $\underline{(5*2)^3} + \underline{6} = 1006$
   e. $\underline{{}^-3^2} - \underline{12 \div 3} = {}^-13$
   f. $\underline{500} - \underline{100 * 3^2} = {}^-400$
   g. $\underline{27} - \underline{3(4^2 - 20)} = 39$
   h. $\underline{23} - \underline{2\sqrt{25}} = 13$
   i. $\underline{10} + \underline{\frac{2}{3}\sqrt{36}} - \underline{2^4} = {}^-2$
   j. $\underline{4800 \div 40} - \underline{0.5\sqrt{6400}} = 80$
   k. $\underline{\frac{-12}{3}} * \underline{(2 + {}^-5)} = 12$
   l. $\underline{\left(\frac{4-12}{4}\right)^3} = {}^-8$
   m. $\underline{30 * 10^3} - \underline{\frac{4000 - 40{,}000}{3}} = 42{,}000$

3. The two terms are ${}^-73.56 * 13.08$ and $78.96 \div 2.45$.

4. The four subterms of the numerator are $3 * 57$, $7 * 88$, $4 * 62$, and $4 * 98$.

6.
   a. $15 + 4$
   b. $15 * 12$
   c. $\frac{20}{-4}$
   d. $35 - 14$
   e. $2 * 25$
   f. $\frac{1}{14}$
   g. ${}^-({}^-5)$

ANS-1

ANS-2 Selected Answers

8.
a. $20 - (5 + 4)$
b. $12(21 + 15)$
c. $25 + 9 * {}^-5$
d. $\sqrt{7 + 9}$
e. $(^-7 + 9)^2$
f. $\dfrac{36}{5 + 4}$
g. $\dfrac{24}{4 * 2}$
h. $5 + (^-3)^2$
i. $(5 * 7)^3$
j. $^-5 + \dfrac{1}{10}$
k. $\dfrac{1}{^-5 + 10}$
l. $(8 + 16)(11 - 3)$

SECTION 1.2

1. The two terms are 63.8 and $12.4\sqrt{22.3 + 28.1}$.

3. The two subterms of the numerator are 3.7 and $9.1 * 2$.

4. e. Actual ≈ 40,581.96   f. Actual ≈ 151.83   i. Actual ≈ 8.00

SECTION 1.3

2. a. 13.5 meters, 14.5 meters   c. 5.5 miles, 6.5 miles
   b. 3.015 cm, 3.025 cm   d. 11.95 mm, 12.05 mm

4. $3\frac{1}{4}$ in., $3\frac{3}{4}$ in.

6. a. 3   b. 2   c. 3   d. 4

9. b and d

10. a. $8.01 * 10^{-3}$ b. $7.8 * 10^4$ c. $5.03 * 10^8$ d. $5.600 * 10^{-5}$

14. e, c, a, d, b (from smallest to largest)
    ($9.90 * 10^{-4}$; $7.98 * 10^2$; $2.19 * 10^3$; $5.46 * 10^3$; $1.08 * 10^4$)

18. The smallest possible perimeter is 403.4 ft, and the smallest possible area is 9419.9625 ft².

22. a. 1031 ft, 14.24 cm, ≈ 0.0158 cm
    b. ≈ 17,000 ft², ≈ 12.7 cm², ≈ 0.0000200 cm²

SECTION 1.4

1. ≈ 15,000 feet

3. ≈ 9.02 feet

5. ≈ 5.28 grams

7. 31,536,000 seconds

9. ≈ 873,000 square feet

11. ≈ $196,000 per American ton

13. ≈ 180 cubic inches

21. a. About 10.3 photo inches   b. About 3.2 miles

25. The square is about 209 feet on a side.

Chapter 2

SECTION 2.1

1. The answer to part iii is given for each pattern.
   a. 2550 blocks in 50th structure
   b. 150 blocks in 50th structure
   c. 256 blocks in 50th structure

SECTION 2.2

1. $A = (2, 5), B = (-5, 6), C = (-4, 0), D = (-4, -8), E = (4, -10)$

3. a. 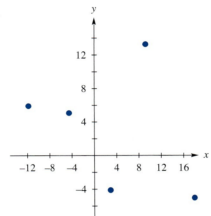   b.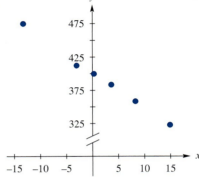

4. a. ≈ 1950, ≈ 1995
   b. ≈ 1967
      ≈ 4200 cigarettes
   c. ≈ 1984

7. a. December; 6.1 inches
   b. Jan. ≈ 5.3 inches; Feb. ≈ 3.9 in.; Mar. ≈ 3.6 in.; Apr. ≈ 2.3 in.; May ≈ 2.0 in.; June ≈ 1.5 in.; July ≈ 0.7 in.; Aug. ≈ 1.1 in.; Sept. ≈ 1.8 in.; Oct. ≈ 2.7 in.; Nov. ≈ 5.3 in.; Dec. ≈ 6.1 in.
   c. ≈ 3.0 inches
   d. 7 months

SECTION 2.3

2. a. $(0, -5), (2, -3), (10, 5)$ (other answers possible)
   b. $(0, 10), (1, 9), (-1, 9)$ (other answers possible)
   c. $(0, 0), (1, 2), (4, 4)$ (other answers possible)

5. a.

x	y
-7	24
-5	0
-3	-16
0	-25
3	-16
5	0
7	24

b.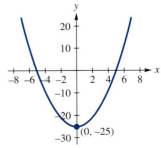

Selected Answers

SECTION 2.4

5. a. W is the independent variable; S is the dependent variable.
  b. y is the independent variable; P is the dependent variable.
  c. N is the independent variable; T is the dependent variable.

7. f. The actual nightly cost is $211.68. **g.** The total cost is $496.05.

SECTION 2.5

1. b. 600 in.^{2} **e.** ≈ 110 ft^2 **l.** ≈ 59 gal
   d. ≈ 79 in.^{2} **h.** ≈ 478 K   **n.** 5.2 ft/sec^2

3. 18.3 cm $*$ 8.9 cm $+$ 18.3 cm $* \sqrt{(8.9 \text{ cm})^2 + (14.3 \text{ cm})^2} + 0.5 * 8.9$ cm $* 14.3$ cm $+$ 14.3 cm $* 18.3$ cm $+ 0.5 * 8.9$ cm $* 14.3$ cm ≈ 860 cm^2

6. b, c, d, e are correct entries.

8. b and d are equivalent

9. c

PROBLEM SET 2.6

1. a. General   **c.** General   **e.** General
   b. Unit-specific   **d.** Unit-specific   **f.** General

4. a. 29 m/sec   **b.** -35 ft/sec

6. ≈ 32 miles/h

13. ≈ 181 in.^{2}

Chapter 3

PROBLEM SET 3.1

1. a. 3, m, n, and k **b.** $\frac{4}{5}$, x, y, and y **c.** 5, T, and $(T - 6)$ **d.** a, a, b, b, and b

3. a, b, c, and e

5. a. $\frac{2}{3}$ **b.** $\frac{1}{2}$ **c.** $-\frac{1}{5}$ **d.** $\frac{1}{4}$

6. a. $\frac{5}{3}$ **b.** $\frac{5xy^2}{3}$ **c.** $\frac{5y^2z}{3}$ **d.** $\frac{5xz}{3}$

8. a, c, and f are like terms; b and e are like terms

PROBLEM SET 3.2

2. a. $15x + 6y$ **e.** $10x^2 - 15xy$ **i.** $M - 6MN$
   c. $-8x - 6a - 2$ **g.** $4.8k + 9.0p$ **k.** $4 - 2Q^2 + 2Q$

4. c. $2m^3 + 9m^2p + 12mp$ **e.** $9.73 * 10^4 RT - 8.1 * 10^3 R^2$

PROBLEM SET 3.3

6. $13\frac{3}{4}$ miles

12. a. Contradiction **b.** Identity **c.** Identity **d.** Contradiction

16. ≈ 362 pieces

PROBLEM SET 3.4

1. a. -5 **b.** $-5WR$ **c.** $-5WT$

3. a. 1 **b.** $P(K + 5)$ **c.** $M(K + 5)$

4. a. $r = \dfrac{A - P}{Pt}$ **g.** $a = -\dfrac{b}{2}$

c. $S = \dfrac{F}{3}$ **i.** $t_4 = 4G - t_1 - t_2 - t_3$

e. $b_2 = \dfrac{2A - hb_1}{h}$ or $b_2 = \dfrac{2A}{h} - b_1$

10. $6,500,000 1998 budget; $975,000 increase

PROBLEM SET 3.5

1. b. $a(6b - 13)$ **d.** $K(1 - C)$ **f.** $x^2(5 + 3y)$ **i.** $G(4K - 9K^2 - 1)$

3. a. $I = \dfrac{E}{R_1 + R_2 + R_3}$ **g.** $r = \dfrac{P - 2L}{2\pi + 2}$

c. $P = \dfrac{A}{1 + rt}$ **i.** $b = \dfrac{D - ad}{-c}$ or $b = \dfrac{ad - D}{c}$

e. $P = \dfrac{-12}{Q + 5}$

8. $(0, 2)$, $(1, 3)$, and $(2, 4)$ (many other possibilities)

11. $2808

Chapter 4

PROBLEM SET 4.1

2. a. $\frac{1}{4}$ **b.** -3 **3. a.** $-\frac{5}{2}$ **b.** $\frac{1}{3}$

9. a.

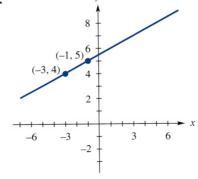

c.

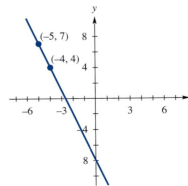

11. a. ≈ $6 per hour

b. The slope means that the cost of an item increases $6 for each additional hour it takes to produce the item.

PROBLEM SET 4.2

1. a, b, d, f, g, and h graph as straight lines.

2. c. Vertical intercept is (0, 20); slope is $\frac{-5}{2}$;

$$y = \frac{-5}{2}x + 20$$

4. b. Slope is $-\frac{2}{3}$; vertical intercept is (0, 4)

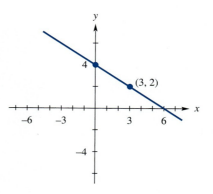

e. Slope is 55; vertical intercept is (0, −35)

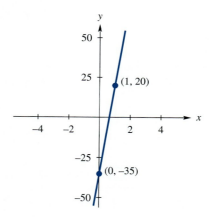

g. $y = {}^-3x + 5$; slope is $^-3$; vertical intercept is (0, 5)

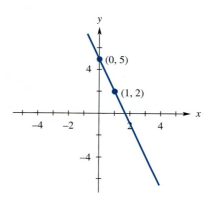

5. a. $(0, {}^-9)$ and $(15, 0)$

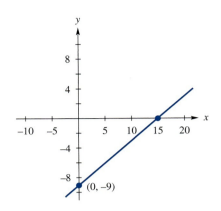

7. a.

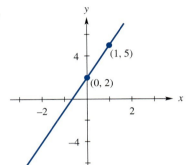

c.

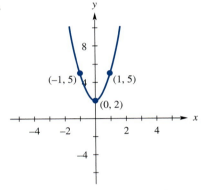

10. b. ≈ 2 hours, \$112 for Pat's and \$113 for Flow Right

Chapter 5

PROBLEM SET 5.1

1. a. x^6 **b.** $(^-x)^5$ **c.** $(^-4)^2$ **d.** $^-4^2$ **e.** $(^-x)^2$ **f.** $^-x^2$

2. a. $^-512$ **b.** $^-512$ **c.** $^-64$ **d.** 64 **e.** $^-1$

5.

x	x^2	x^3	x^4	x^5
$^-3$	$(^-3)^2 = 9$	$(^-3)^3 = {}^-27$	$(^-3)^4 = 81$	$(^-3)^5 = {}^-243$
0	$0^2 = 0$	$0^3 = 0$	$0^4 = 0$	$0^5 = 0$
3	$3^2 = 9$	$3^3 = 27$	$3^4 = 81$	$3^5 = 243$

14. a. $^-108$ **b.** ≈ 395.79 **c.** ≈ 564.58 **d.** $^-4050$ **e.** 64 **f.** 192

PROBLEM SET 5.2

2. a, d, and e are like terms.

4. a. Always true   **c.** Not always true   **e.** Not true
b. Not always true   **d.** Not always true   **f.** Not true

5. a. $9m^3$ **d.** a^2b^2 **g.** $\dfrac{2\pi L^4}{g^6}$

7. a. Not always true **c.** Not always true **e.** Always true
   b. Always true **d.** Always true **f.** Not always true

12. $\approx 1.27 * 10^{14}$ square meters

Chapter 6

PROBLEM SET 6.1

1. a. $3x^2 + 16x + 21$ **e.** $x^3 + 10x^2 + 22x - 15$
   c. $12c^2 - 11ac - 5a^2$ **g.** $2p^3 + 13p^2 + 24p + 9$

2. a. $6r^2 + 20r + 4$ **d.** $24x - 120$

3. b. $m = \dfrac{5k - 6}{1 - 3k}$ or $m = \dfrac{6 - 5k}{3k - 1}$

8. a. 2706 blocks

PROBLEM SET 6.2

1. b. $13t + 8dt^2 - \dfrac{4t}{d}$ **d.** $9x^4 + 3x$

2. a. $y = \dfrac{-3}{4}x + 3$;
slope $= \dfrac{-3}{4}$;
vertical intercept $(0, 3)$

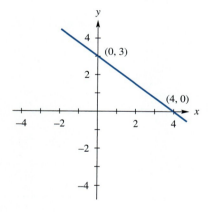

e. $y = \dfrac{5}{3}x$;
slope $= \dfrac{5}{3}$;
vertical intercept $(0, 0)$

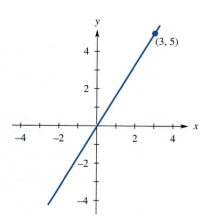

4.
   a. Always true
   b. Never true
   c. Always true
   d. Not always true
   e. Not always true
   f. Always true
   g. Never true
   h. Not always true

8. a. 7.4 inches

PROBLEM SET 6.3

1. a. $x = 10$ or -10 b. $R = 4$ or -4 c. $m \approx \pm 4.472$

2. a. $f = \pm\sqrt{\dfrac{J}{k}}$ b. $s = \pm\sqrt{\dfrac{1+K}{K}}$

5. a. ≈ 3400 kg-m²/sec²   b. ≈ 21.4 m/sec

6. a. ≈ 6.7 in.   b. ≈ 4.0 cm   c. ≈ 2.12 ft   d. ≈ 53.62 m

11. ≈ 12 inches

13. The radius is approximately 6.8 inches.

PROBLEM SET 6.4

1.
   a. $6(x - 4)$
   b. $2t$
   c. $10m$
   d. $10k$
   e. $3R^2$
   f. $p(p + 4)$
   g. $3(2x - 1)$

2.
   a. $x = 16$
   b. $t = 4.8$
   c. $m = -0.3$
   d. $k = 0.75$
   e. $R = 3.375$
   f. $p = -8$
   g. $x = 1.1$

3. i. $M \pm 1.414$ j. No real solution

6. b. $f = \dfrac{1}{2\pi CX}$

   d. $A = \dfrac{RB + RC}{B - R}$

   f. $R_1 = \dfrac{R_{eq}R_2}{R_2 - R_{eq}}$

   h. $M = \dfrac{rav^2}{2aG - Gr}$

   j. $M_1 = \dfrac{MM_3 + MM_2 - M_2M_3}{M_2 - M}$

7. $L = \dfrac{4n^2}{n^2R - 4R}$
   Red: $6.563 * 10^{-7}$ m
   Green: $4.862 * 10^{-7}$ m
   Blue: $4.341 * 10^{-7}$ m

Chapter 7

PROBLEM SET 7.1 **2. a.** (0, 8), (2, 10), (5, 13) (Many different answers are possible.)

PROBLEM SET 7.2 **1. a.** $y = \dfrac{4}{3}x - \dfrac{16}{3}$ **b.** $y = \dfrac{-2}{15}x + \dfrac{2}{3}$ **c.** $y = 3x - 2$ **d.** $y = -80x - 220$

2. a. $y = -2x + 2$ **c.** $y = -x + 1$

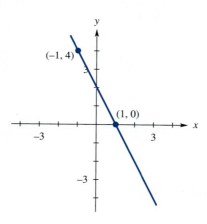
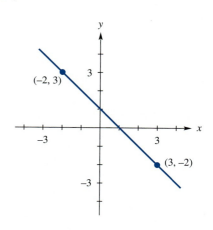

3. a. $y = 2x + 5$ **b.** $y = \dfrac{-1}{2}x$

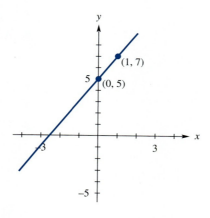
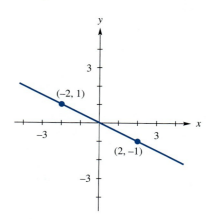

5. a. $y = 4x$, $y = 4x + 2$, $y = 4x - 3$ are three possible answers.

6. a. $G \approx 22.64n + 1447$, where n is the number of years after 1980 and G is the number of tons of grain in millions.

  c. The slope is 22.64 million tons per year. The slope gives the rate of growth of the production of grain per year.

8. a. The suburb 2 data can be modeled by a linear equation. The change in the population is constant for a constant change in the input.

  b. $P = 290.5x + 23{,}112$, where P is the population and x is the number of years after 1980.

  c. The model predicts the population will be 31,827 in the year 2010.

  d. The population will go above 30,000 in the 23rd year after 1980, that is, in 2003.

PROBLEM SET 7.3

1. a. (−2, 5) is a solution to the system.

3. a. $x = 4$ and $y = 3$ **b.** $x = -1.5$ and $y = -3$

11. a. $\begin{cases} 0.0264S + 0.089M = 750 \\ S + M = 10{,}000 \end{cases}$

where S is the amount of money put in savings and M is the amount of money put in the mutual fund.

b. $2236 should be put in savings and $7764 in the mutual fund.

c. She cannot earn $1000 because even if all of the $10,000 were in the mutual fund, the maximum earned would be $890.

14. She should mix 1.3 liters of the 10% solution with 0.7 liters of the 30% solution.

PROBLEM SET 7.4

1. a. $x > -2$ **b.** $k \geq 0$ **c.** $M \leq 8$

2. a.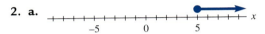

b.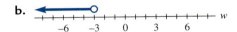

c.

d.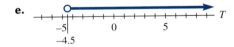

e.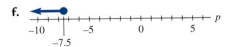

f.

6. a. They can afford to spend $16.50 total on the movie tickets.

b. They can afford the economy-hour tickets.

c. $3t + 5 \leq 21.5$, where t is the cost of a movie ticket. The solution is $t \leq 5.5$.

Chapter 8

PROBLEM SET 8.1

2. a. $(-4)^2$
b. -4^2
c. $(-x)^2$
d. $-x^2$
e. $2 * 3^2$
f. $(2 * 3)^2$
g. $(-3)^{-2}$
h. $(2n)^{-4}$

4. a. -4 **b.** -8 **c.** -16 **d.** -32

5. **a.** 4 **b.** −8 **c.** 16 **d.** −32

7. **a.** 125 **b.** $\frac{1}{125}$ **c.** 1

8. **a.** −8 **b.** $-\frac{1}{8}$ **c.** 64 **d.** $-\frac{1}{64}$ **e.** 1

9. **a.** −27 **b.** $-\frac{1}{27}$ **c.** −81 **d.** $\frac{1}{81}$

13. **a.** −116 **d.** ≈ 72.08 **g.** 0.125
    b. 463 **e.** ≈ 2660.02 **h.** 0.03125
    c. 136,070,766 **f.** 0.25

15. b, c, and e are in scientific notation.

PROBLEM SET 8.2

3. **b.** $18k^4$ **d.** $\frac{12a^2}{25}$, or $0.48\,a^2$ **f.** x^6 **h.** $x^6 + x^7$

5. **a.** Actual is ≈ $9.91 * 10^{18}$ **c.** Actual is ≈ $1.09 * 10^{29}$

6. **a.** $8036.96 **b.** $9600; $1563.04 **c.** $223.97

Chapter 9

PROBLEM SET 9.1

1. $\angle BDC = 65°$

3. $x = 43°; y = 137°; z = 43°$

5. **a.** Angles 4 and 8 are alternate interior angles; angles 2 and 7 are alternate interior angles.
   b. Angles 1 and 8 are corresponding angles; angles 5 and 9 are corresponding angles.
   c. Angles 7 and 8 are nonadjacent complementary angles; angles 2 and 4 are nonadjacent complementary angles. (Others are possible.)

8. $\angle DBC = 65°$

PROBLEM SET 9.2

1.

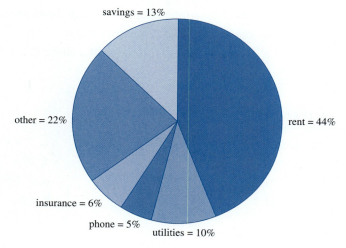

5. **a.** ∠DOC, ∠DOA, ∠AOB, ∠COA (Other answers are possible.)

   b. ∠EAD, ∠EAC, ∠EAB, ∠ADB (Other answers are possible.)

   c. ∠COD and ∠DOA, ∠DOA and ∠AOB, ∠AOB and ∠BOC, ∠BOC and ∠COD are pairs of supplementary angles (Other answers are possible.)

   d. ∠DAC and ∠CAB are complementary angles or ∠ADB and ∠ABD are complementary angles (Other answers are possible.)

   e. m ∠EAD = m ∠ADB (Other answers are possible.)

10. 135°

12. **a.** 90°   **b.** ≈ 9   **c.** 8

PROBLEM SET 9.3

1. $TU \approx 5.1$ cm; $UV \approx 8.6$ cm

5. $CD = 8.4$ ft

6. **a.** ∠1 and ∠5 (Other answers possible)   **d.** ∠6 and ∠7 (Other answers possible)
   b. ∠5 and ∠8 (Other answers possible)   **e.** ∠6 and ∠3 (Other answers possible)
   c. ∠2 and ∠9 (Other answers possible)

8. △ABE ~ △CBD

PROBLEM SET 9.5

1. **a.** 664 in.^{2} **b.** ≈ 79 cm^2 **c.** ≈ 1660 ft^2 **d.** ≈ 340 cm^2

2. **a.** III   **b.** IV   **c.** II   **d.** I

3. **b.** $20T^6$ **e.** $x + x^2$ **h.** $\dfrac{10}{p^2}$

6. 423 in.^{2}

11. ≈ $628

15. ≈ 37.1 cm

PROBLEM SET 9.6

1. **a.** ≈ 280 in.^{3} **c.** ≈ 380 ft^3 **e.** ≈ 72 in.^{3}
   b. ≈ 790 cm^3 **d.** ≈ 548 cm^3

2. **a.** ≈ 240 in.^{2} **b.** ≈ 690 cm^2 **c.** ≈ 250 ft^2 **d.** ≈ 447 cm^2 **e.** ≈ 130 in.^{2}

3. 2640 ft^3; 19,700 gallons

6. ≈ 230 cubic yards (Answer rounded up)

8. ≈ 2.8 cords

10. ≈ 480 ft^3

13. 5 cm

17. ≈ 31 ft

20. ≈ 29,700 cm^3

Chapter 10

PROBLEM SET 10.2

1. a. $90   b. $20   c. $20   d. $40

2. a. In Chicago two buildings have heights between 800 and 899 feet; they are about 850 and 860 feet tall. In New York, four buildings are in this range; they are about 810, 810, 810, and 850 feet tall.
   b. The tallest building is about 1450 feet tall, and it is in Chicago.
   c. New York has more buildings over 700 feet tall. If the building in the diagram listed as about 1000 feet tall is actually above that, New York has more buildings over 1000 feet. Otherwise New York and Chicago have the same number of buildings over 1000 feet.

5. a. Number of People Visiting the Cafeteria between 4 and 6 P.M.

30	0					
31	0					
32	2	4	4	4	9	
33	3	9				
34	0	2	2	5	5	9
35	0	2	4			
36	0					
Key: 31:0 means 310						

   b. 341
   c. We need more information.

6. a. 35   b. 7   c. ≈ 21.1   d. 22

9. a. ≈ 40   b. ≈ 42%

10. a. ≈ $265   b. ≈ $170   c. ≈ $228   d. 50%   e. 75%
    f. Because this amount is above the median, it is a good idea to shop around for a better rate.

14. c. 282,912
    d. 24.0 persons per square kilometer, or 62.1 persons per square mile
    e. 81.5 persons per square mile
    f. Yamhill County

PROBLEM SET 10.3

1. a. $\frac{1}{13} \approx 0.077$ d. $\frac{12}{51} \approx 0.235$
   b. $\frac{3}{13} \approx 0.231$ e. $\frac{1}{52} \approx 0.019$
   c. $\frac{1}{4} = 0.25$

2. a. $\frac{7}{10} = 70\%; \frac{3}{10} = 30\%$ b. $\frac{7}{10} = 70\%; \frac{3}{10} = 30\%$ c. $\frac{2}{3} \approx 66.7\%; \frac{1}{3} \approx 33.3\%$

3. a.

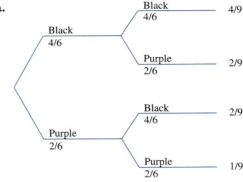

b. $\frac{4}{9} \approx 0.444$ **c.** $\frac{2}{9} \approx 0.222$ **d.** $\frac{8}{9} \approx 0.889$

4. a.

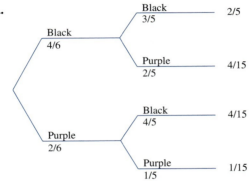

b. $\frac{2}{5} = 0.4$ **c.** $\frac{4}{15} \approx 0.267$ **d.** $\frac{14}{15} \approx 0.933$

5. a. $\frac{1}{16} = 6.25\%$ **b.** $\frac{1}{16} = 6.25\%$ **c.** $\frac{1}{2} = 50\%$

6. a. $\frac{7}{10} = 70\%$ **b.** $\frac{2}{7} \approx 28.6\%$ **c.** $\frac{5}{8} = 62.5\%$ **d.** 8:2 or 4:1

7. a. 1000   **b.** 650; 190   **c.** 19%; 46.5%   **d.** $\approx 20.8\%; \approx 15.7\%$ **e.** $\approx 47.7\%$

Appendix A

1. a. 15

  b. 0, 15

  c. 0, 15, $^-10$, $^-\sqrt{9}$

  d. $2\frac{3}{5}$, 0, 2.5, $-\frac{1}{4}$, 15, 0.667, $^-10$, 0.151515 . . . , $^-\sqrt{9}$

  e. $\sqrt{7}$, 0.010120123 . . .

  f. All of the given numbers

3. a. False   **d.** False   **g.** False
  b. True   **e.** False   **h.** True
  c. True   **f.** True   **i.** True

5. b

7. b

Appendix B

1. a. 7   b. −6   c. 22   d. −17
3. a. 20 + −6   b. −14 + 4   c. −10 + −25   d. 14 + 3
5. a. 16   b. −18   c. −100   d. 120
7. a. −5   b. −6   c. 10   d. −2
9. a. 38   b. −60   c. 40   d. −27
11. b, c, and d are equivalent to $-\frac{1}{4}$
13. b and d

Appendix C

1. a. 1.3 cm = 13 mm, 3.8 cm = 38 mm, 5.0 cm = 50 mm
   b. 2.9 cm = 29 mm, 4.1 cm = 41 mm, 5.7 cm = 57 mm
   c. 0.5 cm = 5 mm, 0.8 cm = 8 mm, 4.3 cm = 43 mm

3. a. $1\frac{1}{2}$ in., $2\frac{3}{4}$ in., 4 in.   c. $1\frac{1}{8}$ in., $2\frac{1}{4}$ in., $3\frac{1}{2}$ in.
   b. $\frac{1}{4}$ in., $1\frac{3}{4}$ in., $3\frac{1}{4}$ in.   d. $\frac{3}{8}$ in., $1\frac{5}{8}$ in., $2\frac{3}{4}$ in.

Appendix D

1. a. area ≈ 386 m²; perimeter = 80.6 m   c. area = 48 cm²; perimeter = 40 cm
   b. area = 6 in.²; perimeter = 12 in.   d. area ≈ 50 ft²; perimeter ≈ 25 ft

3. a, c, f, and g

6. a. area ≈ 1100 in.²; perimeter = 148 in.   b. area ≈ 5600 yd²; perimeter = 298 yd

Index

2×2 linear system 304, 324

A

acute angle 349
add. & subt. property of inequalities 318
adjacent angles 350
algebra pieces 128
algebraic model 95, 101
alternate interior angles 351, 414
angle 349, A-31
    acute 349
    adjacent 350
    alternate interior 351, 414
    bearing 383
    central 361, 362, 414, A-30
    complementary 350, 414
    corresponding (parallel lines) 351, 414
    corresponding (similar triangles) 369, 414
    exterior 351
    inscribed 361, 362, 414
    interior 351, A-30
    obtuse 349
    opposite 350, 414
    protractor 349
    right 349
    straight 349
    supplementary 350, 414
    vertex 349
    vertical 350, 414
approximate numbers 34, 61
    guidelines 43, 61
arc 360, 362
area A-26
axis
    break 81
    horizontal 71
    vertical 71

B

base of exponent 222
bearing angle 383
border point 319, 325
box-and-whisker plot 436
box plot 436
break 81

C

Cartesian coordinate system 80
centimeter ruler A-15
central angle 361, 362, 414, A-30
central tendency 429
circle
    area A-32
    central angle 361, 362, 414, A-30
    circumference A-24, A-30, A-32
    inscribed angle 361, 362, 414, A-30
    major arc 361, 362
    minor arc 361, 362
circumference A-24, A-30, A-32
coefficient 131, 172
    numerical 131, 172
coincidental 312
common denominator 267
common factor 165
complementary 350
conditional equation 152
conditional probability 458
cone 394
    surface area 394, A-33
    volume 406, A-33
congruent 350, A-30
constant 67, 126
contradiction 152
coordinate graph 71
conversion 52

conversion tables A-35
corresponding angles (parallel lines) 351, 414
corresponding angles (similar triangles) 369, 414
corresponding sides 369
cross multiplication 271
cylinder 404, A-33
    volume 404, A-33

D

data point 73
decimal
    repeating A-2
    terminating A-2
denominator
    least common 267
dependent variable 72, 95, 126
diagonal 367, A-30
diameter A-30
difference 14
dimensional analysis 50
distributive property 139
    division 253

E

edges 390
elimination method 307, 325
equality
    properties of 146
equation
    conditional 152
    contradiction 152
    identity 152
    linear 146, 187, 203
        general 204, 219
        slope–intercept 203, 219
    literal 158, 248
    proportion 151, 270, 277
    rational 266, 277
    simple quadratic 258, 277
    solution 87
    solve 87
error 38
estimate 24, 61
evaluate 15
evenly divisible 266
exact numbers 33, 61
experimental probability 451, 467
exponent 222
    negative integer 330
    positive integer 228, 266
    properties
        integer 336
        positive integers 229, 239
    zero power 329, 330

expression
    algebraic 132
    estimating 24
    numerical 14
    simplifying 132, 141
exterior angles 351
eyeball fit line 297

F

faces 390
factor 14, 61, 133
    common 165
false statement 311
five number summary 435
FOIL 244, 277
formula 158
    evaluate with exponents 235
    general 116, 117
    measurements in 116
    solve 158, 167
    unit specific 117, 119

G

general equation of a line 204, 219
general formula 116, 117
geometry formulas A-31, A-32, A-33
graph 70, 126
    break 81
    Cartesian coordinate system 80
    coordinate 71
    data point 73
    equation 126
    horizontal axis 71
    label 74
    linear 187
    number line
        solution of linear inequality 317
    ordered pair 71
    picture 421
    quadrant 79
    rectangular coordinate system 79
    scale 71, 73
    vertical axis 71
graphical model 95, 101
ground rules 6
grouping symbols 14

H

heading 383
horizontal axis 71
horizontal intercept 191, 205
hypotenuse 260

I

identity 152
inch ruler A-16
independent variable 72, 95, 125
inequality
    addition and subtraction property 318
    multiplication and divison property 318
    linear 318
        number line graph 317
inscribed angle 361, 362, 414
integers A-2
intercept
    horizontal 191, 205
    vertical 191, 203, 205
intercepted arc 361
interior angles 351
interquartile range 435
irrational numbers A-3

L

label 74
learning team 6
least common denominator 267
legs 260
like terms 131, 172
    combining 132
linear equation 187, 203
linear inequality 318
lines
    parallel 296
    perpendicular 296
literal equation 158, 248
literal symbol 67, 126
    constant 67, 126
    variable 67, 126

M

major arc 361
mean 429, 467
measurement
    centimeter ruler A-15
    inch ruler A-16
measures of central tendency 429
median 429, 467
metric prefixes A-35
minor arc 361
mode 429, 467
model
    algebraic 95, 101
    graphical 95, 101
    numerical 95, 101
mult. & div. property of inequalities 318

N

natural numbers A-2
negative integer exponent 330
number
    approximate 34, 61
    exact 33, 61
    integer A-2
    irrational A-3
    natural A-2
    rational A-2
    signed A-6
    whole A-2
number line 317
numerical coefficient 131
numerical expression 14
numerical model 95, 101

O

oblique prism 390
obtuse angle 349
operation
    addition 14
    division 14
    exponentiation 14
    multiplication 14
    root extraction 14
    subtraction 14
opposite angles 350
ordered pair 71
origin 71

P

parallel lines 296, A-30
    transversal 351
parallelogram A-31
    area A-31
perimeter A-23, A-30
perpendicular lines 296
picture graph 421
polygon
    concave A-30
    convex A-30
    diagonal 367, A-30
    regular A-30
polyhedron 390
    edges 390
    faces 390
    vertices 390
power 14, 222
precision 34, 37

prism 390
    oblique 390
    right 390, A-32
    volume 402, A-32
probability 451
    conditional 458
    experimental 451, 467
    sample space 453, 468
    theoretical 451, 468
    tree diagram 453
probability tree 455
product 14
production team 5
properties of integer exponents 336
properties of positive integer exponents 229, 239
properties of similar triangles 371
proportion 151, 270, 277
protractor 349
pyramid 390
    volume 406, A-32
Pythagorean theorem 261, 277, A-31

Q

quadrant 79, 80
quadratic
    simple equation 258, 277
quartiles 434, 467
quotient 14

R

radius A-30
range 433, 467
rational equation 266
rational numbers A-2
real numbers A-3
real number system A-3
rectangle
    area A-31
rectangular coordinate system 79
rectangular solid A-32
right angle 349
right prism 390
right triangle
    hypotenuse 260
    legs 260
rise 190, 219
root 14
run 190, 219

S

sample space 453
scale 71, 73
scale drawing 482

scientific notation 36
signed numbers A-6
    addition A-7
    division A-11
    multiplication A-10
    subtraction A-9
significant digits 34
similar figures A-31
similar triangles 369, 414
    corresponding angles 369, 414
    corresponding sides 369, 414
    properties 371
slope 189, 190, 219
    parallel lines 296
    perpendicular lines 296
slope–intercept equation of a line 203, 219
solve 87
sphere 394
    surface area 394, A-34
    volume 407, A-34
stem–and–leaf diagram 427, 467
straight angle 349
subscript 119
substitution method 308, 325
subterms 17
subtraction as addition of opposite A-9
sum 14
supplementary 350, 414
surface area
    cone 394
    sphere 394
system of equations 304, 324
    2×2 linear system 304, 324

T

tangent 364
team 5
    ground rules 6
    learning 6
    production 5
term 15, 61
    algebraic 130
    combining 132
    identify 15
    like 131, 172
    subterm 17
theoretical probability 451, 468
transversal 351
trapezoid
    area A-32
tree diagram 453
triangle
    area A-31
    obtuse 347
    right 260
    similar 369, 414

U

unit analysis 50
unit conversion 52
unit fractions 50, 52
unit-specific formula 117, 119

V

variable 67
    dependent 72, 95, 126
    independent 72, 95, 126
    subscript 119
verify
    literal equation 160
    simplification 134
vertex 349
vertical angles 350
vertical axis 71
vertical intercept 191, 205
vertices 390
volume
    cone 406, A-33
    cylinder 404, A-33
    prism 402, A-32
    pyramid 406, A-32
    rectangular solid A-32
    sphere 407, A-33

W

whole numbers A-2

Z

zero power 329, 330

INDEX OF STRATEGIES

Converting Between Units 52
Estimating a Numerical Expression 24
Evaluating an Expression Term by Term 16
Identifying the Terms of an Expression 15
Simplifying Algebraic Expressions 141
Solving a Linear Literal Equation 167, 173
Solving Linear Equations in One Variable
    149, 172
Using a Table to Create an Equation 100
Verifying the Results to a Literal Equation
    160, 173